ALGORITHMIC LOGIC

T0332648

ERRATA

Page, line	For	Read
18_1	$\alpha(x)$	$\sim \alpha(x)$
21_6	$u = y$	$u \neq y$
35^1	$\gamma_{\mathfrak{A}}(v) = 0$	$\gamma_{\mathfrak{A}}(v) = 1$
67^{18}	$M((x := y)\alpha \Rightarrow \beta)$	$\big(M((x := y)\alpha) \Rightarrow \beta\big)$
67_9	$v' = M_{\mathfrak{A}}(v)$	$v' = M'_{\mathfrak{A}}(v)$
$91_{10}, 91_8$	$M'\alpha$	$M'\beta$
$91_{10}, 91_8$	$M''\alpha$	$M''\beta$
$91_2, 91_1, 92^5$	α	β
$97^9, 97^{12}, 97_{10}$	$(\sim\gamma \wedge \alpha)$	$\sim\gamma$
111_{13}	$M_{\mathfrak{A}}(v)$	$M'_{\mathfrak{A}}(v)$
112^9	$M_{\mathfrak{A}}(K'_{\mathfrak{A}}(v))$	$M'_A(K'_A(v))$
142_4	$(s := \text{in}(e, s))\,(\text{mb}(e, s)$	$(s' := \text{in}(e, s))\,(\text{mb}(e, s')$
142_3	$(s := \text{del}(e, s))\,(\sim \text{mb}(e, s)$	$(s' := \text{del}(e, s))\,(\sim\text{mb}(e, s')$
148_{11}	$\text{em}(s)$	$\sim\text{em}(s)$
149^6	$\text{eq}\big(\text{in}(e, \text{del}(e, s), s\big)$	$\text{mb}(e, s) \Rightarrow \text{eq}\big(\text{in}(e, \text{del}(e, s)), s\big)$
150_5	$(\exists e)\text{mb}(e, s) \wedge \sim\text{mb}(e, s')$	$(\exists e)\,(\text{mb}(e, s) \wedge \sim\text{mb}(e, s'))$
263^1	$q \in \text{Car}(K)$	$q \notin \text{Car}(K)$
263^8	$= v(q_i)$	$= v'(q_i)$
277^3	$\mathfrak{A}, v \models (\gamma \vee \sim$	$\mathfrak{A}, \bar{\bar{v}} \models (\gamma \wedge \sim$
282_{16}	$\sim (\exists x)\alpha(x)$	$\sim (\exists x) \sim\alpha(x)$
329_2	$f = \textbf{none}$	$f \neq \textbf{none}$
330^7	$s = \text{allfree}$	$s \neq \text{allfree}$
330^8	$f_i = \text{newfr}(s')$	$f := \text{newfr}(s')$

In parts of the text concerning the Boolean algebra the signs $\vee$, $\wedge$, $\Rightarrow$, $\sim$, should be replaced by $\cup$, $\cap$, $\rightarrow$, $-$ respectively.

G. Mirkowska, A. Salwicki, *Algorithmic Logic*

G. MIRKOWSKA and **A. SALWICKI**
Institute of Mathematics *Institute of Informatics*
University of Warsaw *University of Warsaw*

ALGORITHMIC LOGIC

D. REIDEL PUBLISHING COMPANY

A MEMBER OF THE KLUWER ACADEMIC PUBLISHERS GROUP

DORDRECHT / BOSTON / LANCASTER / TOKYO

PWN-POLISH SCIENTIFIC PUBLISHERS

WARSZAWA

Library of Congress Cataloging-in-Publication Data

Mirkowska-Salwicka, Grażyna.
  Algorithmic logic.

  Bibliography: p.
  Includes index.
  1. Formal languages.   2. Algorithms.   3. Logic,
Symbolic and mathematical.   I. Salwicki, Andrzej.
II. Title.
QA267.3.M57   1986   511.3   85-2201
ISBN 90-277-1928-4

This edition published by PWN—Polish Scientific Publishers, Warszawa, Poland,
in co-publication with
D. Reidel Publishing Company, P.O. Box 17, 3300 AA Dordrecht, Holland

Distributors for Albania, Bulgaria, Cuba, Czechoslovakia, German Democratic
Republic, Hungary, Korean People's Democratic Republic, Mongolia, People's
Republic of China, Poland, Romania, the U.S.S.R., Vietnam and Yugoslavia

ARS POLONA
Krakowskie Przedmieście 7, 00-068 Warszawa 1, Poland

Sold and distributed in the U.S.A. and Canada
by Kluwer Academic Publishers,
101 Philip Drive, Assinippi Park, Norwell, MA 02061, U.S.A.

in all other countries, sold and distributed
by Kluwer Academic Publishers Group,
P.O. Box 322, 3300 AH Dordrecht, Holland

PRINTED IN POLAND

CONTENTS

VIII CONTENTS

PREFACE

The purpose of this book is manyfold. It is intended both to present techniques useful in software engineering and to expose results of research on properties of these techniques.

The major goal of the book is to help the reader in elaboration of his own views on foundations of computing. The present authors believe that semantics of programs will always be the necessary foundation for every student of computing. On this foundation one can construct subsequent layers of skill and knowledge in computer science. Later one discovers more questions of a different nature, e.g. on cost and optimality of algorithms. This book shall be mainly concerned with semantics.

Secondly, the book aims to supply a new set of logical axioms and inference rules appropriate for reasoning about the properties of algorithms. Such tools are useful for formalizing the verification and analysis of algorithms. The tools should be of quality—they should be consistent and complete. These and similar requirements lead us toward metamathematical questions concerning the structure of algorithmic logic.

Algorithmic properties are expressed by algorithmic formulas in a straigthforward way. Therefore the analysis of algorithms, i.e. their verification and evaluation of their effectiveness can be based on algorithmic logic. Our third aim is to expose the possible applications of algorithmic logic in the description of structures and systems, especially those appearing in computer science.

Finally, we wish to stress strong connections between the formal methods described in this book and the methodologies supported by modern programming languages. This phenomenon has two aspects commercial and scientific. Scientific—since modern tools of programming inspire many problems. Commercial—since formal methods of AL can be used in software engineering for creating industrial means of production of software.

We are aware that algorithmics, i.e. the creation of new more efficient algorithms and the discovery of new data structures, differs from study of rules of reasoning about algorithms. The book may be useful for those who wish to learn about formal, logical methods of computer science, but we cannot assure, however, that the reader will learn how to conduct a research in computer science. The topics presented in this book belong to the mathematical foundations of computer science. The main questions considered are: analysis of algorithms and the analysis of the process of analysing algorithms. The formal counterparts of these notions are the notions of proof of a semantical property of a program and metamathematical properties of the system of algorithmic logic. The formal tools developed by algorithmic logic have many applications in specification of abstract data types, in verification of algorithms and the implementation of data structures, and in defining the semantics of programming languages.

This book can serve as a textbook for a course on the theory of programs or logic of programs or as a textbook of logic for computer scientists. It does not assume any special mathematical background from the reader, but skill in programming and experience with mathematical reasoning are desirable.

The book can also serve as an auxiliary textbook for courses on programming languages and on methods of programming. Indicating the elements of the logic of programs may be helpful in courses for beginners.

This book arose from lectures that both authors have given on algorithmic logic at the University of Warsaw, Christian-Albrechts Universität in Kiel, Université Paris 6 and in IAC Roma.

During one semester course we skip Chapters V, VI and final parts of the Chapters III and IV. For a two semester course it is advisable to add material on the logic of recursive procedures. An introductory course of computer science or a course on methods of programming can use the material contained in Chapters II (methods of verification) and IV (specification of data structures and related topics). In these lectures we stress the relationship between ideas of hierarchical and modular programming and the ideas contained in the book.

The defects of this book are caused by the authors. One such defect is the omission of recursion and procedures. The authors presented elsewhere the results concerned to algorithmic logic of programs with block structures and recursive procedures and also their own approach toward semantics of functional procedures. We do not include these results here since, so far, they have found little application in the practice of verification. We hope that future research will bring answer to our doubts.

We are sure that new branches of algorithmic logic will appear in connection with new methods and tools of programming, especially a logic of concurrent programs. One can foresee a broader, commercial application of AL in specification of data types and their implementation leading toward production of software modules in programming languages which allow extension of modules by their concatenation.

We would like to express our gratitude to dr L. Banachowski and prof. Z. Pawlak for their critical remarks which helped us to improve several parts of the manuscript. We have also profited from the comments of many colleagues and students, we thank to all of them.

The book would never appear without the sympathetical help and patience of the Polish publishers. We thank Mrs K. Regulska and others for the help in preparation of the manuscript and Mr J. Roguski for the help in proof-reading.

CHAPTER I

INTRODUCTION

1. THE MOTIVATIONS

The design and applications of algorithms must be accompanied by analysis and verification. We shall try to answer a few questions which can arise in connection with this claim.

(i) Why is analysis needed? When should one start this analysis?

(ii) What does the word "algorithm" mean? How do we conceive the process of programming?

(iii) What kind of analysis should we ask for?

Let us begin with a few remarks. The last years have brought in an enormous increase not only in the number of algorithms designed, but also in the magnitude of computational processes determined by those algorithms, in the speed of application (the time which elapses between the construction of a new program and its applications is now very short compare this with nineteenth-century science and technology) and in mass production an algorithm can be copied and used many times in various circumstances. This means that, the cost of an error can be enormous; its practical consequences might be disastrous. Hence, analysis and verification ought to be included in the process of programming from the very beginning.

Algorithms have long been in use in mathematics and technology. However, for most of the time the meaning of the term has been imprecise. It has been assumed that the notion of "algorithm" and the notion of "function" (also not defined precisely) are identical. In the nineteenth century the difference between these two notions was recognized. In mathematical research, the way indicated by Frege, Cantor (cf. Fraenkel, 1958), and others led to many beautiful and important results and theories. Nevertheless, the notion of the algorithm, and of computability, were overlooked. They became the centre of attention in metamathematics around 1930, in connection with the works of Hilbert, Gödel, Church, Turing, Kleene, Markov, Herbrandt, Post, and others (cf. Machtey, 1978). It was necessary to have a definition of an effectively computable function in order to answer questions like "is there an algorithm for solving a given problem?". A negative answer needed a formal definition of the notion of an algorithm. As a result, many equivalent definitions of an algorithm appeared, e.g., Markov's normal algorithms (Markov, 1954), μ-recursive functions, and recursive functions. In connection with this, Church formulated an important conjecture, namely that all formalized definitions of the notion of an algorithm coincide. Mathematical logic has been oriented towards negative results, proving that there is no algorithm for solving a given problem. In computer science, however, we have a positive program of research, not only a negative one. In this book we present various definitions of the notion of an algorithm, and we shall study the consequences of the difference between them.

As a practical example, consider the following well-known procedure known as Euclid's algorithm.

EXAMPLE. Finding the greatest common divisor of two integers involves the following computational process:

1. Divide a_1 by a_2, find the remainder a_3 and check whether it is zero or not; if $a_3 = 0$ then the process terminates and a_2 is the greatest common divisor of a_1 and a_2, if $a_3 \neq 0$ then

2. divide a_2 by a_3, find the remainder a_4; if $a_4 = 0$ then the process terminates and a_3 is the result; if $a_4 \neq 0$ then

3. divide a_3 by a_4, etc.

The process will terminate after at most a_2 steps (why?).

The algorithm itself reads as follows:

while the remainder of the division of x by y is not equal to zero repeat

    let r be the remainder;

    put y as new x;

    put r as new y;

otherwise (i.e., if the remainder is equal to zero)

          y is the greatest common divisor.

Observe that the same algorithm can also be used to find the maximal common length of two segments, or the greatest common divisor of two polynomials. What is needed is only a new understanding of the words "divide", "find the remainder" and "compare with zero".

A study of this simple algorithm leads us to the following conclusions:

(i) The notion of an algorithm is of a syntactical nature.

(ii) An algorithm must be interpreted in order to determine a computing process.

(iii) Interpretation of an algorithm consists in assigning meanings to operators (the meaning of an operator is an operation in the corresponding set), and in assigning initial data.

(iv) Once we have fixed the meanings of operators, we can apply the algorithm to many initial data sets.

Let us compare these remarks with the abstract definition of an algorithm proposed by Kolmogorov, Uspienski and Malcev (cf. Malcev, 1965). An algorithm should have the following features:

(i) The algorithm and the initial state determine (or accept) a sequence of states. A state is a finite object. For every state of the algorithm a finite set of possible next states is determined.

(ii) The relation of direct successorship of states is verifiable in finite time.

(iii) If there is no next state, then the total result should be indicated.

(iv) The initial state can be chosen from a potentially infinite set.

Every algorithm should be verified before its eventual application. There is no doubt about this. But we must first clarify which properties of the algorithm are to be verified, and which methods assure the appropriateness of an eventual answer.

Let us observe that before an algorithm is constructed the following question must be considered: "Does an algorithmic solution of the problem in question exist?". The history of science, especially of mathematics, provides many cases where a negative answer has been found. Often, attempts to solve a problem have yielded many elegant results before the final answer was reached "no, there exists no algorithm for doubling a cube, for trisecting an angle, for squaring a circle, for solving the word problem in semigroups, for deciding whether a given formula is a tautology of the predicate calculus", etc.

Much time has been wasted in the attempt to construct systems for the verification of software, optimization of programs, and so on. Research of this kind will not be fruitless if one starts with an awareness of the unsolvability of the problems in question. The systems arrived

at, can be of only limited use, or, possibly, they might work in an inter-
active manner indicating trouble spots to those who operate them.

Hence the first type of semantic questions met in algorithmics (the
name sometimes used for the field of design and analysis of algorithms)
can be called computability problems. These include, for example,
questions like: "Is a given function or relation computable?" More
precisely, suppose we are given an algebraic system $\mathfrak{A}$, also called
a data structure. (The system consists of a set called the universe, com-
prising certain operations and relations. Does there exist an algorithm
to compute a function f in $\mathfrak{A}$?).

This and similar problems can be treated if one defines the meaning
of the notion of algorithm.

Suppose we are given an algorithm and a requirement, also called
a specification. The second group of semantic questions can be called
correctness problems. Here one can find questions such as: "Is an algorithm correct with respect to a specification?" "Does the algorithm
in question terminate?" "Is an algorithm a proper implementation
of the system required?"

The third important class of semantic questions, optimality problems, contains questions like: "Is a given algorithm the best solution
of a problem?" "Does an optimal algorithm exist?" (From the abstract
theory of computational complexity we have learned that there exist
problems such that every algorithm solving one of those problems
can be replaced by a better algorithm which has asymptotically lower
computational complexity.)

The necessity of solving the above-mentioned problems in practice
makes it clear that we need to find a general mathematical theory of programs. One possible way to approach this problem is to present a theory
of programs as a logical formalized system: algorithmic logic is one
of the first attempts in this direction.

The status of computer science as a deductive or an empirical science
is of secondary importance. In any case, it seems obvious to us that
research in computer science and the development of its applications
necessarily require a proper deductive system. To reason about algorithms we need appropriate inference rules which describe the semantics
of programming constructs. This need has been explained in many
publications (cf. Dijkstra, 1976; Scott, 1970). The research program
of algorithmic logic takes into consideration the demand for the con-

struction of a deductive system suitable for algorithmics. This program contains many questions already known from metamathematics. Are these questions important in computer science? Professor A. Mostowski wrote "many mathematicians do successful research in mathematics without knowledge of mathematical logic, mathematical logic is not necessary for them" (cf. Mostowski, 1948). It is true, however, that mathematics had developed its logical tools long before metamathematical studies were initiated. For computer science the situation is radically different. It has had no time to elaborate its tools. Theories concerning the semantics of programming languages, and various logics of programs, have been developed almost simultaneously with new algorithms under the pressure of quickly growing demands. These theories have found many applications in the practice of designing new programming languages. Nevertheless, we must warn the reader that algorithmic logic is not a magic wand for solving the problems of computer science. It can help, however, in understanding them.

2. AN INFORMAL INTRODUCTION TO FORMALIZED LANGUAGES

There is no such thing as "The programming language", the best and the unique one. This will be obvious to the reader, who must have encountered a few languages in practice and have heard about dozens of others. Can we even hope that there is one general pattern in this rich variety of programming languages? After a short examination we find that the answer is "no". But we should still like to find a classification and, later, some tools facilitating the work of programmers, or some methodological hints on how to develop software.

After a little thought, one can propose a classification of languages built around the programming constructs allowed in a language. At the bottom of this classification we find deterministic, iterative languages. In this class, programs are built from certain atomic instructions by means of program connectives of composition, branching, and iteration. Two languages of this class can differ in the sets of functional and/or relational signs appearing in their alphabets. Higher in our classification are those languages which admit procedures.

The process of enriching a language can be continued. At the top of our hierarchy we should place a language which allows most of the constructions known today. Hence, a language of the highest quality (remember that we are discussing only the richness of the programming constructs offered!) should contain co-routines and parallel processes, classes and methods for their extensions, ability to signal between modules, etc. We do not know a language which could be called functionally complete. The criterion for functional completeness of a programming language which we would like to propose is the following: a language should contain all the known essential tools for composing algorithms (from the program connectives to the concurrent processes) and all the tools for defining data structures. (The LOGLAN programming language developed at the University of Warsaw seems to be a good approximation, cf. Bartol *et al.*, 1983).

REMARK. It is believed that all possible ways of defining algorithms are known. The most recent discoveries are co-routines, exception handlers, and parallel processes. There is no consensus of opinion as to which are the basic tools for the definition of data types. Arrays and records are not satisfactory. The present authors believe that classes extendable by the prefixing mechanism form a complete set of tools for data types. Research in this direction is far from complete.   □

We must emphasize here that the number of existing programming languages exceeds thousand. Can one define general rules of computation, independent of the varying details of orthography?

There is some hope. First, we can remark that programming languages have a common feature. Their main goal is to make communication among programmers possible. But programs have also to be communicated to a computer (equipped with an appropriate translator), and hence must be written in a formal way. Accordingly, we can conceive every programming language as a formal language, defined by its alphabet and the set of well-formed expressions. Every programming language has an intersubjective, mechanical way of deciding whether an expression is in the language or not.

Let us analyse alphabets. An alphabet is simply a set of signs. One can distinguish various subsets in it:

(i) sings of program connectives and constructions, for example **while do...od** (sign of iteration), **procedure**... and **call**... (signs of procedure declaration and procedure instruction),

(ii) logical signs, e.g., $\sim$ for negation, $\wedge$ for conjunction, $\vee$ for disjunction,

(iii) functional and relational signs,

(iv) variables,

(v) auxiliary signs, e.g., brackets.

These symbols have different roles. Variables and functional symbols allow us to construct arithmetical expressions. For example, if x, y, z are variables and $+$, $\cdot$ are two-argument signs of operations, then

$$x \cdot y + z$$

is an arithmetical expression.

In a formal approach we treat such expressions as patterns or definitions of new functions, whose values can be computed whenever we known the values of variables x, y, z and the meanings of the functional symbols. Such expressions will be called *terms*.

In a similar way we can create Boolean expressions. They assume logical values true or false, and they usually play the role of tests in programming languages. If $x \cdot y$ and $x + y$ are terms and $<$ is a sign of a two-argument relation, then

$$(x \cdot y) < (x + y)$$

is a Boolean expression, which may or may not be valid depending on the values of the variables and on the meaning of the symbols $+$, $\cdot$, $<$. For example, if x, y are subsets of a set A, $+$, $\cdot$ are the set-theoretical sum and intersection respectively, and $<$ is interpreted as inclusion, then the value of the above Boolean expression is true. However, it is not so if $<$ is interpreted as equality. (The problem of interpretation will be discussed with greater precision in the next section.)

Thus a Boolean expression can be treated as the definition scheme of a relation, which becomes a relation when one fixes the interpretation of the functional and relational symbols, and the interpretation of the variables.

Using logical operators such as the signs of conjunction, negation and disjunction we can construct more complicated Boolean expressions. We shall call these formulas.

The formalization of a programming language still requires a precise description of the notion of a program. In the sequel we shall consider various classes of programs. We shall analyse and compare programming languages with respect to the repertoire of the programming constructs they offer.

A rough classification of programming concepts allows us to distinguish the following classes of programs:

(i) the class of deterministic iterative programs,

(ii) the class of non-deterministic iterative programs,

(iii) the class of programs with recursive, non-functional procedures and blocks,

(iv) the class of programs with recursive, functional procedures,

(v) the class of programs which permit declaration of new types,

(vi) the class of parallel programs,

(vii) the class of schemes of programs.

In this book we shall consider only some of these classes. Moreover, we shall not discuss recursively enumerable programs, Friedman's schemes, or random assignments, which in the authors' opinion are mathematical abstractions having little in common with the programming practice. The reader is advised to study these concepts in the literature (Tiuryn, 1981c; Harel, 1978c).

In most existing programming languages a program is considered to be a sequence of instructions. The set of instructions consists of atomic actions, and some tools for composing them.

We shall look more closely at the structure of deterministic iterative programs. We shall explain the constructions by means of graphs, usually called flow-diagrams. Each flow-diagram has one entry and one exit.

We shall start with the simplest instruction, the assignment statement.

If x is a variable and τ is a term, then the graph shown in Figure 2.1 is a flow-diagram of the assignment instruction.

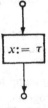

Fig. 2.1

If we are given the diagrams of two programs P_1 and P_2

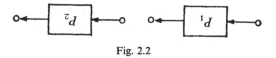

Fig. 2.2

(Figure 2.2), then we can compose them by putting one after the other. The flow-diagram of the composed program is described in Figure 2.3. It is obtained by identifying the exit of P_1 with the entry of P_2.

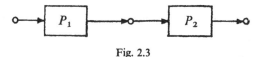

Fig. 2.3

Given two programs P_1 and P_2 and a formula γ, we can produce very useful constructions called *branching* (or *conditional instruction*) and iteration, as shown in Figure 2.4.

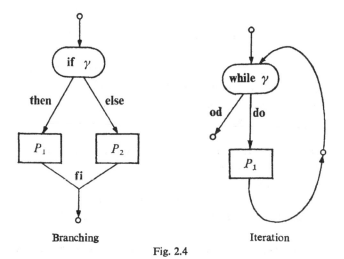

Branching   Iteration

Fig. 2.4

It is easy to see that the set of programs defined in this way forms an algebra, which is generated from assignments by means of the operations of composition, branching, and iteration. We shall call programs of this class *structural* or *modular* ones.

To recapitulate, we have seen that the set of well-formed expressions of a programming language can be split into three subsets: the set of terms (arithmetical expressions), the set of formulas (Boolean expressions), and the set of programs. These expressions have no meaning in themselves. They can be considered as patterns which allow us to compute different functions or relations, depending on the interpretation.

In order to illustrate the main assertion of this section, namely that programs by themselves have no meaning, we present a few examples. We use a Pascal-like orthography, in the hope that this will be understandable to the reader.

EXAMPLE 2.1. Consider the following program K (Kleene's algorithm) (cf. Aho, 1974):

K: **begin**

> **for** $i := 1$ to n **do** $C_{ii}^0 := \varepsilon \cup l(i, i)$ **od**;
> **for** $1 \leqslant i, j \leqslant n$ **and** $i \neq j$ **do** $C_{ij}^0 = l(i, j)$ **od**;
> **for** $k := 1$ to n **do**
> > **for** $1 \leqslant i, j \leqslant n$ **do**
> > $$C_{ij}^k := C_{ij}^{k-1} \cup C_{ik}^{k-1} \cdot (C_{kk}^{k-1})^* \cdot C_{kj}^{k-1} \text{ **od**}$$
>
> **od** :
> **for** $1 \leqslant i, j \leqslant m$ **do** $c(i, j) := C_{ij}^n$ **od**

end.

It is well known that there exist at least three interpretations of the above program, and each implies a different meaning.

(i) Let us interpret the program in the structure

$$\langle A, \cup, \cdot, *, \varepsilon \rangle$$

where

A the universe of the structure, is the family of all subsets of the set of finite words over an alphabet A_0,

$\cup$ is a set-theoretical union,

$\cdot$ is the operation of concatenation of languages,

$*$ is the star-operation on languages (i.e., for $X \in A$, $X^* = \varepsilon \cup X \cup$ $\cup X \cdot X \cup X \cdot X \cdot X ...$),

ε is a one-element set which contains the empty word over A_0.

Let $l(i, j)$ be a one-element set which consists of a symbol from the set A_0, produced while some automaton $\mathfrak{A}$ changes the state from i to j. Then the program K computes regular events. The meaning of the element $c(i, j)$ is the set of all words which lead from state i to state j in the automaton $\mathfrak{A}$.

(ii) Let us interpret program K in the two-element Boolean algebra

$$B_0 = \langle \{0, 1\}, \cup, \cap, *, 0 \rangle$$

where

$\cup$ the disjunction, is an interpretation of $\cup$,

$\cap$ the conjunction, is an interpretation of $\cdot$,

$*$ for every $x \in B$, $x^* = 1$,

0 is the interpretation of ε.

Let us assume that for a given graph G

$$l(i,j) = \begin{cases} 1 & \text{iff edge } (i,j) \text{ is in } G \\ 0 & \text{otherwise} \end{cases}$$

Then the results of program K are:

$C(i,j) = 1$ iff (i,j) belongs to the transitive closure of G, i.e., if there exists a path from i to j.

(iii) Consider the data structure

$$\mathfrak{C} = \langle R^+, \min, +, *, 0 \rangle$$

where

R^+ is the set of non-negative real numbers extended by the maximal element $+\infty$,

min the minimum operation, is the interpretation of $\cup$,

$+$ is the arithmetical sum and the interpretation of $\cdot$,

$*$ is a one-argument operation such that $n^* = 0$ for all possible $n \in R^+$,

0 is a constant zero, the interpretation of ε.

Let $l(i, j)$ be the cost of traversing the edge (i, j) in the given graph G and assume $l(i, j) = +\infty$ if there is no edge (i, j) in G. Then the results computed by program K can be interpreted as follows: $c(i, j)$ is the cost of the shortest path in G from i to j.

3. ASSIGNING MEANINGS TO PROGRAMS

We have seen in the previous section that one program may have many interpretations. The process of interpretation (i.e., semantics) is defined separately of the syntactical rules. Syntax decides which expressions are well-formed ones; it does not determine the meaning of an expression.

It is generally agreed that in order to define an interpretation of a programming language we have to fix the meaning of all symbols of the language. First, we ought to decide which elements will appear as the values of variables and, second, we ought to associate with every functional symbol the corresponding function (or partial function) and with every relational symbol the corresponding relation. In this way we can determine a relational system, also called a *data structure*.

A given data structure determines a mapping which with any expression of the language associates its meaning. This method of defining semantics can be attributed to the work of Tarski and of Mostowski (cf. Rasiowa, 1970).

For example, in the data structure of real numbers we can associate with the term $(x+y+z)/3$ the three-argument function of the arithmetical mean, where $+$ is interpreted as addition and $/$ as division.

Similarly, in the same data structure the formula $(x^2 > y^2 \Rightarrow x > y)$ can be conceived as a two-argument function which associates with every pair (x, y) of real numbers the logical value **true** when $|x| \leqslant |y|$ or $x > y$ and **false** otherwise.

In this book we shall assume that every formula has a defined value which is **true** or **false**; in other words, we shall work with a two-element Boolean algebra.

At this point let us observe that there are other possible concepts; for example, a multivalued logic can also be accepted as the semantic base of the logical part of a language. There are also systems which admit a third logical value (cf. MacCarthy, 1963), and systems which regard a Post algebra as an algebra of logical values (cf. Rasiowa, 1975c; Perkowska, 1972). Such systems will not be discussed in this book.

To complete our description of the interpretation of a programming language, it remains to assign meaning to programs. There is no unanimous opinion on how to understand particular constructions. Users and researchers are free to make their own choice.

However, there is almost common agreement in associating with every program a binary relation. Every program can be regarded as a mapping which transforms an initial memory state, i.e., data, into a final memory state, i.e., results. The connection between the input and the output states is called the *input-output relation* determined by the program and by the assumed data structure.

Now we must tackle the problem of how to define the input-output relation. The first approach is based on the modular structure of programs. We can define the meaning of a program step by step, putting together interpretations of simple instructions. For example the input--output relation associated with the program

begin K_1 ; K_2 **end**

is a composition of the input-output relation associated with K_1 and the input-output relation associated with K_2.

This method of assigning meaning to programs is called *operational semantics*.

A deeper insight into the method allows us to observe the process by which the initial state of memory is transformed into the result. This process is called *computation*. Usually we define the computation of a program in a given data structure as a sequence of configurations, each of which describes a valuation of variables, i.e., a memory state, and a list of instructions to be executed. Two consecutive configurations in this sequence ought to be in the relation of direct successorship.

This notion of computation is not the only possible one. Another definition is related to the notion of proof. One can ask whether there exists a proof that the results of a program K applied to data v are equal to w. This idea, originating in the papers of Herbrand and Gödel (cf. Hermes, 1965), continues to be used in the notion of formal computation and in the PROLOG programming language.

Consider the following example. The language admits two functors: a zero argument constant 0 and a one-argument functor s. The interpretation of the functors will be standard in the set of natural numbers. We shall introduce two new functors by means of the equations

$$f(x, 0) = x, \quad f(x, s(y)) = s(f(x, y)),$$
$$g(x, 0) = 0, \quad g(x, s(y)) = f(g(x, y), x).$$

Figure 3.1 shows a diagram which can be interpreted as a proof that $g(s(0), s(0))$ is equal to $s(0)$.

REMARK. This may seem an odd way to find that $1 \cdot 1 = 1$. The literature concerning PROLOG and other non-imperative languages, and also the discussion about the 'fifth generation' of computers, show that there are many computer scientists who are convinced of the future applicability of such a style of programming.   □

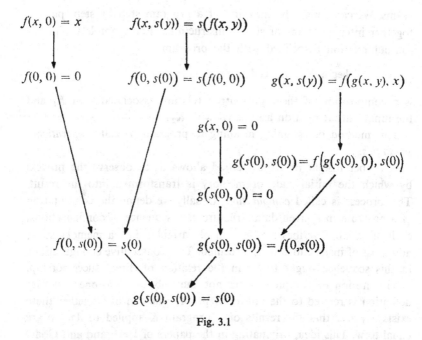

Fig. 3.1

Another look at the example can lead to the following observations. A proof-like computation is composed of subcomputations by the rules of computing, which resemble the rules of inference. In our example the rules used were simply

$$\frac{\tau(x) = \eta(x)}{\tau(x/w) = \eta(x/w)} \quad \text{rule of substitution,}$$

$$\frac{\tau(\tau_1) = \eta, \quad \tau_1 = \tau_2}{\tau(\tau_1/\tau_2) = \eta} \quad \text{rule of replacement,}$$

where τ, η, τ_1, τ_2 are terms, $\tau(\tau_1)$ means that τ_1 is a subexpression of τ, and $\tau(\tau_1/\tau_2)$ is the result of replacing one or more occurrences of τ_1 in τ by τ_2.

The method of defining the meaning of a program by means of the notion of computation is very useful for the class of deterministic and the class of non-deterministic iterative programs. It is not obvious whether this method can be used to define the semantics of more developed programs, e.g. programs with recursion or objects of types declared in a program.

For this reason another method has been suggested by Scott and Strachey (1971). Their proposal is to treat a program as an implicit definition of an input-output mapping between states. The mapping (i.e., semantics) is the least solution of a system of functional equations which can be associated with every program. The elegance and simplicity, of this method, which is called *denotational semantics*, have attracted many researchers. The programmers can comment that, when this method of identification of a mathematical object is used, its application in verifying properties of programs is not always possible.

The third method of defining semantics, the *axiomatic semantics* is similar to denotational semantics. A semantics is axiomatically defined whenever a set of axioms and inference rules is given such that every true semantic property of a program can be proved in the system. Obviously, we require that the system should be consistent. Denotational semantics can be placed half way towards axiomatic semantics. One can regard implicit equations as axioms. There are no syntactic rules of inference; instead, the method offers a powerful semantic tool—the least fixed point of the system of equations is proposed as a solution.

For us, operational semantics based on the notion of computation seems the most natural. Axiomatic semantics or mathematical identification of meaning are secondary for a programmer who deals with computations in his everyday practice. The programmers intuitions are formed by computations. We realize that for the designer or implementer of a programming language, denotational and axiomatic semantics may be very attractive. However, even the designer of a language should not overlook questions of effectiveness of implementation connected with the operational semantics.

Practice allows us to make experiments and to develop our intuitions about a computational process. However, this is not enough. What we need is the possibility of formulating a specification before the software is designed, and verifying the correctness of software with respect to this specification. Let us quote here the well-known assertion that computational experiments can help us to find a bug in our program, but no experiment can prove correctness of the program with respect to a potentially infinite set of initial data. The verification should be made before applying the program to the data. This is the proper place for axiomatic semantics. It offers axioms and inference rules which can be used in the process of verification of the properties of a program. It is written in a language of logical formulas, and the same language can be used for specifications. The language of axiomatization differs from the language of programs. It is unlikely that the first-order predicate calculus could serve as a logical basis for the axiomatization of semantics. We shall explain this in the next section.

4. SEMANTIC PROPERTIES OF PROGRAMS

Having chosen a definition of the notion of computation, one can observe various semantic phenomena. Their nature differs, according to the definition of computation. In the case of formal computations, the crucial problem is whether a computation exists. In the case of computations which are sequences of states, the most important question is whether a computation is finite or infinite. If a computation is understood as an algebraic process of composing the meanings of subexpressions to obtain the meaning of the whole expression, the question would be: "Does the process give a result?"

Let us survey the properties of programs which will be discussed in this book. For the case of computations defined as sequences of configurations the relevant properties of deterministic iterative programs are termination, correctness and equivalence.

Termination. Does the program in question have finite computations? Are all computations of the program finite? If they are not, then what is the suffcient and necessary condition for the finiteness of the computations?

EXAMPLE 4.1. In some interpretations Euclid's algorithm always terminates, e.g., in the structure of rational numbers. For the ancient Greeks the discovery that the algorithm does not necessarily terminate if interpreted in the structure of segments of planar geometry was a shock. □

The problem of termination can be stated in various circumstances. The question whether a given program M terminates in any interpretation and for any data differs from the question whether the same program M will terminate in a data structure $\mathfrak{A}$.

Correctness. Does the program compute the results which were expected from it? Our requirements (specification) can be given as a pair of conditions, an input condition (precondition) for the data and an output condition for the results (postcondition).

EXAMPLE 4.2. Suppose that the specification is:
(precondition) a and b are two positive integers,
(postcondition) the result is the greatest common divisor of a and b.
Suppose the program considered is Euclid's algorithm. One should be convinced *a priori*—before possible computation—that the final value of the computation is the greatest common divisor of a and b.   □

In order to prove correctness one uses:
(i) the structure of the specification and of the program,
(ii) certain properties of the data structure, i.e., of interpretation.

Equivalence. Do two programs M and K compute the same results? This question is connected with the classification of programs as 'better' and 'worse'. Two programs are equivalent if for equal data either both do not terminate (= diverge) or both terminate and give results satisfying the same postcondition. Hence, if one program is correct with respect of the specification (α, β) the other is also correct with respect to (α, β). In this case one can begin an analysis of costs of the two algorithms in order to find the better program.

When the programming language considered is more developed and admits classes and deallocation of objects (cf. LOGLAN), one should ensure the property that no computation will lead to a situation in which reference is made to a non-existent object.

This survey of various semantic properties can be continued. In the sequel we shall study several of the already mentioned properties, and introduce many others.

As regards prooflike computations, the main question is not the termination of a computation. By definition all formal computations are finite. The main problem is whether a computation exists. Another kind of problem is the reasons for the non-existence of a prooflike computation. It may be caused by an inconsistency in the system of procedures (axioms), e.g.,

$$f(x) = f(x)+1,$$

or, by circular definitions, e.g.,

$$g(x) = g(x).$$

It is worthwhile to distinguish the two cases; in the second case the functional equation can be solved by an arbitrary choice of the function g, whereas the first case is a hopeless one—there is no function f which will satisfy such an inconsistent system of axioms.

We should like to end this section leaving the reader with the conviction that the variety of interesting and important semantic phenomena is great and worth studying.

5. EXPRESSIVITY. AN INTRODUCTION TO THE LANGUAGE OF ALGORITHMIC LOGIC

Semantic properties of programs should be an object of study. We should like to prove or disprove them, just as in mathematics we prove or disprove various theorems.

Before we try to construct a system for reasoning about the semantic properties of programs, we should find a way to express them as formulas (logical or mathematical ones, according to the reader's preference). The natural candidate is a language of first-order logic. Can we express properties like termination, correctness, etc., as formulas of the first-order predicate calculus? After a closer examination we find that we cannot. The termination property allows us to express many properties known as non-expressible in the language of first-order logic.

As one of many possible examples, let us mention the property of a number being a natural number.

There is no formula of first-order logic defining natural numbers; on the other hand, the property

> *the program*
>
> **begin** $y := 0$; **while** $x \neq y$ **do** $y := y+1$ **od** **end**
>
> *terminates*

holds iff the number x is a non-negative integer.

Consider the loop

> **while** $\alpha(x)$ **do** $x := f(x)$ **od**.

The termination property is equivalent to an infinite disjunction

$$\alpha(x) \quad \text{or} \quad \alpha(x/f(x)) \quad \text{or} \quad \alpha(x/f(f(x))) \quad \text{or} \dots$$

This observation was first made by Engeler, 1967, who proposed the use of $L_{\omega_1\omega}$ logic. The language of $L_{\omega_1\omega}$ allows any infinite disjunctions and conjuctions. It can be observed that this language is too rich. For example, there is an algorithm to construct the i-th component of the infinite disjunction above.

Another possibility is the use of weak second-order logic, WSL. The termination property of a program can be expressed as follows: "there is a finite sequence of states such that...". This expression is typical of weak second-order logic. Again, WSL seems much richer than is necessary for an analysis of programs.

One can certainly study the properties of programs in $L_{\omega_1\omega}$ or in WSL, but we suggest considering a minimal extension of first-order logic which will allow us to investigate the properties of programs, i.e., algorithmic logic.

The language of algorithmic logic will be the least extension of the language of first-order logic such that expressions of the form

$$\langle \text{program} \rangle \ \langle \text{formula} \rangle$$

are also regarded as formulas.

The proposed meaning of the formula $K\alpha$, where K is a program and α is a formula, would read "the formula $K\alpha$ is satisfied in a data structure $\mathfrak{A}$ at a valuation v iff the computation of the program K which starts from the initial data v in the structure $\mathfrak{A}$ is finite and the results satisfy the formula α".

Let us look at a few of the semantic properties:

(i) a program K terminates iff the formula (**K**true) is valid,

(ii) a program K is correct with respect to a precondition α and a postcondition β iff the formula $(\alpha \Rightarrow K\beta)$ is valid,

(iii) two programs K and M are equivalent with respect to a postcondition α iff the formula $(K\alpha \equiv M\alpha)$ is valid.

The cases of non-deterministic or concurrent programs require slight modifications. A non-deterministic program can possess more than one computation. It is then natural to split the question about results into two different problems: "Do all results satisfy the required property?" and "Is there a result which satisfies the property?". Accordingly, we assume in the algorithmic language the following two modal constructions:

$\Diamond K\alpha$ with the meaning "it is possible that after a finite computation of program K the property α holds",

$\Box K\alpha$ with the meaning "it is necessary that all computations of the program K should be finite and all results should have the property α".

The property of strong termination of a non-deterministic program K can be expressed in such a language by the formula $\Box$ K**true**. Various notions of program correctness can be expressed by formulae like $(\alpha \Rightarrow \Diamond K\beta)$, $(\alpha \wedge \sim\Diamond K\sim\beta)$, $(\alpha \wedge \Diamond K$**true** $\Rightarrow K\beta)$. Now, the goal of the verification of the properties of programs has a formal counterpart. In order to verify that a non-deterministic program K meets the conditions of a specification α, β, it is enough to prove, or disprove, a corresponding formula, e.g. $(\alpha \Rightarrow \Diamond K\beta)$.

Communications like "the deterministic program K applied to the data d gives the result r" can be verified by a repetition of the computing experiment. However, one can also make the more general statements "for every data satisfying a precondition α the program K will terminate", etc. The validity of such statements cannot be checked by experiment. In order to prove or disprove such statements it is necessary to use more general tools, such as inference rules or axioms.

Let us remark that, as in mathematics, it is not necessary to present a complete formal proof with all the details. It is often more convincing simply to present arguments that a proof exists. In this way we can exchange communications about software and its properties, much as chemists exchange communications about experiments and deductions based upon them. The development of software can be treated as a social scientific skill, with intersubjective methods of verifying the communications about the properties of software.

6. ON APPLICATIONS

In this section we shall discuss the practical consequences of research on algorithmic logic.

Algorithmic logic (AL) can be applied in the analysis of semantic properties of programs. The completeness property of AL makes the objective program sound. AL offers methods of verification of partial and total correctness, and moreover it permits the analysis of on-going processes. Even the estimation of the complexity of algorithms can be formalized in AL. Let us observe that the formulas

$$\textbf{(if } \beta \textbf{ then } K \textbf{ fi)}^n \beta \quad \text{and} \quad \textbf{(if } \beta \textbf{ then } K \textbf{ fi)}^{n+1} \sim \beta$$

assert that the number of iterations of the loop-statement

$$\textbf{while } \beta \textbf{ do } K \textbf{ od}$$

will not exceed the number $n+1$.

The whole system of Floyd–Hoare logic is included in AL, and thus all examples of the proofs in this systems are in AL. Floyd–Hoare logic (cf. Hoare, 1969) is not complete: not every valid semantic property has a proof. Algorithmic logic supplements the missing parts of axiomatization. There is an ω-rule in AL, i.e., a rule of inference with infinitely many premises which is necessary for the completeness of the system. However, we do not intend to present formal proofs in all cases. In most examples it is enough to have reasonable arguments for the validity of the assertion, i.e., it is enough to prove that the proof exists.

There are numerous arguments showing that data structures can be specified with the help of AL (cf. Chapter IV). What is the importance of this? Not only do algorithmic formulas allow us to define data structures which are not axiomatizable in first-order logic, not only is the axiomatization of these structures compact, but also—and this is much more important—algorithmic axioms facilitate the task of proving the correctness of many algorithms.

EXAMPLE. Algorithmic specification (axiomatization) of the data structure of natural numbers consists of three formulas:

$$s(x) = 0,$$
$$s(x) = s(y) \Rightarrow x = y,$$
$$(y := 0; \textbf{ while } x \neq y \textbf{ do } y := s(y) \textbf{ od}) \textbf{ true}.$$

One can prove the termination of a program, e.g.,

$$M\colon u := 0; z := x; \textbf{ while } u = y \textbf{ do } u := s(u); z := s(z) \textbf{ od}$$

by a natural transformation of the program appearing in the axiom. Since we have assumed that this program terminates, and since the program M can be obtained by transformations which do not spoil the termination property, the program M also has the termination property. In this way we can hide induction in algorithmic reasoning. □

The implementation of data structures can also be considered on the basis of AL. It finds a formal counterpart in the notion of an interpretation of one algorithmic theory within another. Chapter IV contains more details and examples illustrating the method of development of type declarations (in LOGLAN), together with the proof of their correctness, which is based on this idea.

AL can be treated as an axiomatic method of defining semantics (we deal with this problem in Chapter III). Axiomatization of AL can be used by implementors as a test in an assessment procedure for an implementation of a programming language.

For more developed languages one can propose a method of defining language semantics by constructing a collection of algorithmic theories. Various theories can define different aspects of LOGLAN's semantics. When put together, they will form a system completely describing the semantics of a rich programming language (cf. Chapter VII).

Another application of AL is in the definition of semantics based on formal proofs. The notion of formal computation can lead to a new, non-imperative programming language (Salwicki, 1975).

Algorithmic logic and other logics of programs can be used in the process of teaching programmers and even mathematicians. It may be that in the long run AL will help us to a better understanding not only of programming, but also of mathematics. It is AL which provides us with another viewpoint on data structures. Algorithmic properties have equal rights with first-order properties: they may simplify reasoning about data structures.

CHAPTER II

LOGIC OF DETERMINISTIC ITERATIVE PROGRAMS

The main questions dealt with in this chapter are: "What are the semantic properties of programs?" and "How can they be expressed in a formalized language?"

We start with the definition of a *class of algorithmic languages which admit iterative programs*. Iterative programs are built from assignment statements by means of program connectives like *composition, branching* and *iteration*. Each program is interpreted as a binary input-output relation in the set of all computer memory states. We then define the notion of *computation*. This allows us to discuss semantic properties of programs, like *termination, correctness*, etc. The importance of these notions for the analysis of programs is obvious.

To express semantic properties of programs we shall use *algorithmic formulas*, i.e., the constructions of a form $K\beta$ where K is a program and β is a 'formula. The intuitive meaning of this formula is "after execution of program K the property β holds". Such constructions allow us to express properties of programs and data structures which are not expressible in the first-order language.

The next step is to formulate laws and rules concerning computational processes. They provide us with formal tools for reasoning about programs. We aim to construct a formal system in which all valid sentences are provable and all provable sentences are valid. However, a more strict investigation of the semantics of the algorithmic language leads to the conclusion that the compactness property does not hold. This means that there exists a sentence which is a semantic consequence of an infinite set of formulas and which is not a consequence of any finite subset of this set. The most important consequence of this fact is that the logical system we are going to construct cannot be a finitistic one.

In this chapter we shall present a formal system of algorithmic logic in the Hilbert style which uses infinitistic rules of inference of the ω-type. We shall prove that all provable formulas of this system are valid.

We conclude the chapter with some examples of formal proofs in the formalized system of algorithmic logic.

1. LANGUAGE

We shall now consider the algorithmic language L of deterministic **while**-programs. There are three kinds of well-formed expressions in L: *terms, formulas* and *programs*. In this section we shall introduce these three notions formally.

Let us assume that the *alphabet* of the language L contains enumerable sets of signs of relations P (predicates for short), signs of functions Φ (functors for short) and variables V. There are two kinds of variables, *propositional* and *individual*. Hence the set V is a set-theoretical union of two disjoint sets of propositional variables V_0 and of individual variables V_i.

DEFINITION 1.1. *By the type of language L we shall understand the system $\langle \{n_\varphi\}_{\varphi \in \Phi}, \{m_\varrho\}_{\varrho \in P}\rangle$ of two families of natural numbers such that for every $\varphi \in \Phi$, n_φ is an arity of the functor φ and for every $\varrho \in P$, m_ϱ is an arity of the predicate ϱ.* □

The notion of *term* is just the same as in classical logic. We shall recall the definition below.

DEFINITION 1.2. *The set of terms T is the smallest set which contains the set of individual variables V_i and is closed with respect to the rule that if φ is an n-argument functor, $\varphi \in \Phi$, and $\tau_1, ..., \tau_n$ are terms, then the expression $\varphi(\tau_1, ..., \tau_n)$ is a term.* □

REMARK. In most examples throughout this book we shall consider two-argument functors and two-argument predicates. In keeping with tradition we shall then write $x < y$, $x+y$ instead of $< (x, y)$, $+(x, y)$ as in the definition above. □

EXAMPLE 1.1. Assuming x, y, z, i are individual variables and $\cdot$, $+$ are two-argument functors, then $(i \cdot y)+z$, $(x \cdot y)+(x \cdot z)$ are terms. □

LEMMA 1.1. *The system $\mathfrak{A} = \langle T, \{\varphi_\mathfrak{A}\}_{\varphi \in \Phi}\rangle$ is an abstract algebra with the set V_i being the set of free generators in T, such that for every*

n-argument functor $\varphi \in \Phi$, $\varphi_{\mathfrak{A}}$ *is an operation in T and for arbitrary terms*
$\tau_1, \ldots, \tau_n$ *we have*

$$\varphi_{\mathfrak{A}}(\tau_1, \ldots, \tau_n) = \varphi(\tau_1, \ldots, \tau_n). \qquad \square$$

The set of all formulas F will be described later after the definition
of programs. We now recall the notion of an *open formula*.

DEFINITION 1.3. *The set of open formulas* F_0 *is the least set that contains the set of propositional variables* V_0 *and such that*
  (i) *if* α, β *belong to* F_0 *then the expressions* $(\alpha \vee \beta)$, $(\alpha \wedge \beta)$, $(\alpha \Rightarrow \beta)$,
$\sim \alpha$ *also belong to* F_0,
  (ii) *if* $\tau_1, \ldots, \tau_n$ *are terms and* ϱ *is an n-argument predicate, then*
$\varrho(\tau_1, \ldots, \tau_n)$ *belongs to* F_0. $\qquad \square$

The formulas defined in (ii) above are called *elementary formulas*.
In other words every propositional variable is an open formula; every
elementary formula is an open formula, the conjunction, the alternative, the implication of open formulas is an open formula, and the
negation of an open formula is an open formula.

EXAMPLE 1.2. Assume p, q are propositional variables and $<$, $\leqslant$, $=$ are
two-argument predicates. Let x, y, z and $+$, $\cdot$ be as in Example 1.1.
The expressions

$$(1) \qquad \begin{aligned} & \big(x = ((i \cdot y) + z) \wedge (z < y \wedge 0 \leqslant i)\big), \\ & \big((\sim q \wedge p) \Rightarrow (x + y) \cdot z \leqslant x + (y \cdot z)\big) \end{aligned}$$

are then open formulas. $\qquad \square$

LEMMA 1.2. *The system* $\langle F_0, \cup, \cap, \rightarrow, - \rangle$ *is a free abstract algebra
in the class of all algebras* $\langle A, o_1, o_2, o_3, o_4 \rangle$ *with three binary operations* o_1, o_2, o_3 *and one unary operation* o_4, *and such that for arbitrary*
α, $\beta \in F_0$ *we have*

$$\begin{aligned} \alpha \cup \beta &= (\alpha \vee \beta), \\ \alpha \cap \beta &= (\alpha \wedge \beta), \\ \alpha \rightarrow \beta &= (\alpha \Rightarrow \beta), \\ -\alpha &= \sim \alpha. \end{aligned}$$

*The set of all propositional variables and elementary formulas is a set
of free generators of the algebra* $\langle F_0, \cup, \cap, \rightarrow, - \rangle$.

The proof is analogus to that in classical logic, and also to the proof of Lemma 1.3 below.   □

DEFINITION 1.3. *The set of all programs Π is the least set such that*:
(i) *Every expression of the form $(x := \tau)$ or $(q := \gamma)$ is a program, where x is an individual variable, τ is a term, q is a propositional variable, γ is an open formula.*
(ii) *If γ is an open formula and M and M' are programs then the expressions* **if γ then M else M' fi, while γ do M od, begin M; M' end** *are programs.*   □

The set of all expressions defined in (i) shall be called the *set of assignment instructions* and will be denoted by S. Note that the pairs of words **then—else, else—fi, do—od, begin—end**, play the role of parentheses similar to (,). To avoid superfluous parentheses we shall write for example
1° **begin M_1 ; ... ; M_n end** instead of **begin M_1 ; begin M_2 ; ... begin M_{n-1} ; M_n end ... end end**;
2° **while γ do M_1 ; M_2 od** instead of **while γ do begin M_1 ; M_2 end od**;
3° **if γ then M_1 ; M_2 else M_1' ;M_2' fi** instead of **if γ then begin M_1 ; M_2 end else begin M_1' ; M_2' end fi**. According to the definition the expression $(x := x)$ is a program for every variable x. We shall denote such a program by Id.
For the sake of simplicity we shall write **if γ then M fi** instead of **if γ then M else Id fi**.
If M is a program and i— a natural number, then M^i is a shortened form of the program **begin $\underbrace{M; ...; M}_{i \text{ times}}$ end**; $M^0 \overset{\text{df}}{=}$ Id. The program **begin M; M' end** is called the *composition* of programs M, M'; the program **if γ then M else M' fi** is called the *branching* between the two programs M and M', the program **while γ do M od** is called the *iteration of* the program M.

EXAMPLE 1.3. Let x, y, z, i be individual variables, $+$, $-$ two-argument functors, $\geqslant$ a two-argument predicate and 0, 1 zero-argument functors (i.e. constants). The following expression is then a program

> **begin**
> > $z := x$;
> > $i := 0$;
> > **while** $z \geqslant y$

do
$$z := z-y;$$
$$i := i+1$$
od;
end;   □

LEMMA 1.3. *The system $\Pi = \langle \Pi, \circ, \{*_\gamma\}_{\gamma \in F_0}, \{if_\gamma\}_{\gamma \in F_0}\rangle$ is an abstract algebra such that for every $\gamma \in F_0$, $\circ$ and if_γ are two-argument operations in Π, and $*_\gamma$, is a one-argument operation in Π and for every M, $M' \in \Pi$ we have*

$$M \circ M' = \textbf{begin } M, M' \textbf{ end}$$
$$*_\gamma(M) = \textbf{while } \gamma \textbf{ do } M \textbf{ od}$$
$$if_\gamma(M, M') = \textbf{if } \gamma \textbf{ then } M \textbf{ else } M' \textbf{ fi.}$$

Moreover, Π is a free algebra in the class of similar algebras with the set S of assignment instructions being the set of free generators.

PROOF. Let $\mathfrak{A} = \langle A, \circ, \{*_\gamma\}_{\gamma \in F_0}, \{if_\gamma\}_{\gamma \in F_0}\rangle$ be an algebra similar to Π, and let $f: S \to A$ be any mapping from the set of generators S into A. The mapping f can be extended in a unique way to the function $h: \Pi \to A$, defined by induction on the length of programs:

$$h(s) = f(s) \quad \text{for every assignment instruction } s \in S.$$
$$h(\textbf{begin } M; M' \textbf{ end}) = h(M) \circ h(M'),$$
$$h(\textbf{if } \gamma \textbf{ then } M \textbf{ else } M' \textbf{ fi}) = if_\gamma(h(M), h(M')),$$
$$h(\textbf{while } \gamma \textbf{ do } M \textbf{ od}) = *_\gamma(h(M))$$

for each programs M, M' and every formula $\gamma \in F_0$.

By the definition, h is an extension of f and h is a homomorphism. The uniqueness of h follows from the property that every program in Π is of exactly one of the following forms: it is an assignment instruction, a composition of two programs, an iteration of a program or it is a branching between two programs. Moreover, the representation of a program is unique.   □

Now we can define the set of all formulas of the language L.

DEFINITION 1.4. *The set of all formulas F is the least extension of the set F_0 such that:*
  (i) *If M is a program and α is a formula, then $M\alpha$ is a formula.*

(ii) *If M is a program and α is a formula, then $\bigcup M\alpha$, $\bigcap M\alpha$ are formulas.*

(iii) *If x is an individual variable and α(x) is a formula, then $(\exists x)\alpha(x)$ and $(\forall x)\alpha(x)$ are formulas.*

(iv) *If α and β are formulas, then $(\alpha \vee \beta)$, $(\alpha \wedge \beta)$, $(\alpha \Rightarrow \beta)$ and $\sim\alpha$ are formulas.* □

We call the signs $\bigcap$, $\bigcup$ *universal* and *existential iteration quantifiers*, and the signs $\forall$, $\exists$ *universal* and *existential classical quantifiers*.

Any formula in which neither iteration quantifiers nor classical quantifiers appear is called a *quantifier-free algorithmic formula*.

EXAMPLE 1.4. Let M be the program defined in Example 1.3. The expression

$$(2) \qquad M\left(x = \left((i \cdot y) + z\right) \wedge (z < y \wedge 0 \leqslant i)\right)$$

is then a quantifier-free formula, and

$$(3) \qquad (\exists y)\left((x := y) \bigcup (x := x+1)\ z < x\right)$$

is a formula, where x, y, z, i are individual variables; $+$, $\cdot$ are two-argument functors; $<$, $=$, $\leqslant$ are two-argument predicates; 0, 1 are zero-argument functors. □

As in the case of the previous definitions we can formulate a theorem about the algebraic structure of the set F. Indeed, every program $M \in \Pi$ can be treated as a one-argument operation in F such that for a given formula α it gives as a result the formula $M\alpha$. However the problem with quantifiers is much more difficult since they in fact define the generalized operations with infinitely many arguments.

At the end of this section we shall introduce some auxiliary notions.

Let w be any well-formed expression of the language L. By $V(w)$ we shall denote the set of all variables that appear in w.

Let s be an assignment instruction of the form $(u := w')$. By $\overline{s}w$ we shall then denote the expression which is obtained from w by the simultaneous replacement of all occurrences of the variable u, in the expression w, by the expression w'.

EXAMPLE 1.5. Let s be an assignment instruction $(y := x+y)$.

1. As a first simple example let us consider the case where w is a term $((i \cdot y) + z)$ from Example 1.1. Then $\overline{sw}$ is a term of the form

$$((i \cdot (x+y)) + z).$$

2. As a second example let us take the formula (1) to be w. The expression $\overline{sw}$ is then a formula of the form

$$(x = ((i \cdot (x+y)) + z) \wedge (z < (x+y) \wedge 0 \leqslant i)).$$

Obviously if x does not appear in the expression w then $\overline{sw}$ is identical to w.

3. Note also these negative examples. When w is the formula (2) or (3) then the expression $\overline{sw}$ is not a well-formed expression. □

The observations from the above examples can be summed up in the following lemma.

LEMMA 1.4. *For every assignment instruction s:*
1° *If w is an open formula, then $\overline{sw}$ is an open formula.*
2° *If w is a term, then $\overline{sw}$ is a term.*

The easy proof is left to the reader.   □

Now we give the strict definitions of the free and bounded occurrence of an individual variable in a formula.

DEFINITION 1.5. *The occurrence of an individual variable x in a formula α is bounded by a classical quantifier iff x occurs in a part of α of the form $(\exists x)\beta$ or $(\forall x)\beta$ for some formula β. In the opposite case an occurrence of x is called free.*   □

EXAMPLE 1.6. The occurrence of z in formula (3) is free; the occurrence of y in this formula is bounded by the existential quantifiers $(\exists y)$.

In the formula $((\exists y)x < y \vee x = y)$ both occurrences of x are free; the first occurrence of y is bounded and the second, free.   □

We write $\alpha(x)$ indicating that the variable x is free in α.

Let us denote by **true** the formula $(p \vee \sim p)$ and by **false** the formula $(p \wedge \sim p)$, for a fixed propositional variable p; let $\alpha \equiv \beta$ be a shortened form of $((\alpha \Rightarrow \beta) \wedge (\beta \Rightarrow \alpha))$.

2. SEMANTICS

In this section we shall define precisely the algebraic semantics of algorithmic language. We shall start from the interpretation of the language signs in the corresponding relational system and two-element Boolean algebra. Then we shall extend this interpretation to all well-formed expressions of language, terms, formulas and programs.

Let L be an algorithmic language of the type $\langle \{n_\varphi\}_{\varphi \in \Phi}, \{m_\varrho\}_{\varrho \in P} \rangle$ where Φ is the set of functors in L and P is the set of predicates in L.

DEFINITION 2.1. *By a data structure for L we shall understand a relational system $\mathfrak{A}$ which consists of the universe A and operations and relations such that*:

1° *For every n_q-argument functor φ, there exists n_φ-argument operation $\varphi_{\mathfrak{A}}$ in A.*

2° *For every m_ϱ-argument predicate ϱ, there exists m_ϱ-argument relation $\varrho_{\mathfrak{A}}$ in A.*

Hence a data structure for L is a relational system

$$\mathfrak{A} = \langle A, \{\varrho_{\mathfrak{A}}\}_{q \in \Phi}, \{\varrho_{\mathfrak{A}}\}_{\varrho \in P} \rangle$$

of the type $\langle \{n_q\}_{q \in \Phi}, \{m_\varrho\}_{\varrho \in P} \rangle$. □

The given data structure for L determines the interpretation of function and relation signs in the language. We shall call $\varphi_{\mathfrak{A}}$ the *interpretation of a functor* φ, and $\varrho_{\mathfrak{A}}$ the *interpretation of a predicate* ϱ. Let $\mathfrak{A}$ be a data structure for L, and let both $\mathfrak{A}$ and L be fixed for the rest of this section.

Individual variables of L will be interpreted as elements of A and propositional variables of L will be interpreted as elements of two-element Boolean algebra:

$$B_0 = \langle \{\mathbf{1}, \mathbf{0}\}, \vee, \wedge, \Rightarrow, \sim \rangle.$$

DEFINITION 2.2. *By a valuation in the given data structure $\mathfrak{A}$ we shall mean a mapping*

$$v: V_0 \cup V_i \to A \cup \{\mathbf{0}, \mathbf{1}\}$$

such that

$$v(p) \in \{\mathbf{0}, \mathbf{1}\} \quad for\ p \in V_0,$$
$$v(x) \in A \quad for\ x \in V_i. \qquad \square$$

The set of all possible valuations will always be denoted by W. The given data structure $\mathfrak{A}$ for L determines in a unique way the interpretation of a term as a mapping in A.

For every term τ we have a corresponding function

$$\tau_{\mathfrak{A}} : W \to A$$

which is defined recursively as follows:

$$x_{\mathfrak{A}}(v) = v(x) \quad \text{for } x \in V_i,$$

$$\varphi(\tau_1, \ldots, \tau_n)_{\mathfrak{A}}(v) = \varphi_{\mathfrak{A}}\big(\tau_{1\mathfrak{A}}(v), \ldots, \tau_{n\mathfrak{A}}(v)\big).$$

Here we have used the fact that every term τ is either a variable or is in the form $\varphi(\tau_1, \ldots, \tau_n)$ and the representation is unique.

Note that we have in fact defined a homomorphism h between the algebra of terms and the algebra $\langle A, \{\varphi_{\mathfrak{A}}\}_{\varphi \in \Phi} \rangle$ such that

$$h(\tau) = \tau_{\mathfrak{A}}(v).$$

By Theorem 1.1 the homomorphism is uniquely determined by the given valuation v.

EXAMPLE 2.1. Let R be the data structure of real numbers and let addition $(+)$ and multiplication $(\cdot)$ be an interpretation of functors $+$, $\cdot$ of the language L.

The term $((i \cdot y) + z)$ then determines the three-argument function $f(i, y, z)$ in R such that for every valuation v in R

$$f\big(v(i), v(y), v(z)\big) = \tau_{\mathfrak{A}}(v).$$

In particular, $f(1, 2, 3) = 5$.   □

The element $\tau_{\mathfrak{A}}(v)$ of A is called the *value of the term τ in the structure $\mathfrak{A}$ at the valuation v*.

Analogously, every formula α of the language L determines a mapping $\alpha_{\mathfrak{A}}$ from the set of all valuations W into the Boolean algebra B_0,

$$\alpha_{\mathfrak{A}} : W \to B_0.$$

Every program M of the language L determines a partial function $M_{\mathfrak{A}}$ from set W into itself, called *interpretation of program M*,

$$M_{\mathfrak{A}} : W \to W.$$

Both mappings will be defined by simultaneous induction with respect to the length of expressions:

$$p_{\mathfrak{A}}(v) = v(p) \quad \text{for } p \in V_0$$
$$\varrho(\tau_1, \ldots, \tau_n)_{\mathfrak{A}}(v) = \mathbf{1} \quad \text{iff} \quad (\tau_{1\mathfrak{A}}(v), \ldots, \tau_{n\mathfrak{A}}(v)) \in \varrho_{\mathfrak{A}}$$

for n-argument predicate ϱ and arbitrary terms $\tau_1, \ldots, \tau_n$.

If $\alpha_{\mathfrak{A}}(v)$ and $\beta_{\mathfrak{A}}(v)$ are defined, then

$$(\alpha \vee \beta)_{\mathfrak{A}}(v) = \alpha_{\mathfrak{A}}(v) \cup \beta_{\mathfrak{A}}(v),$$
$$(\alpha \wedge \beta)_{\mathfrak{A}}(v) = \alpha_{\mathfrak{A}}(v) \cap \beta_{\mathfrak{A}}(v),$$
$$(\alpha \Rightarrow \beta)_{\mathfrak{A}}(v) = \alpha_{\mathfrak{A}}(v) \rightarrow \beta_{\mathfrak{A}}(v),$$
$$(\sim \alpha)_{\mathfrak{A}}(v) = -\alpha_{\mathfrak{A}}(v).$$

Let s be an assignment instruction of the form $(u := w)$, $s_{\mathfrak{A}}(v)$ is then a valuation v' such that

$$v'(u) = w_{\mathfrak{A}}(v) \quad \text{and} \quad v'(z) = v(z) \quad \text{for } u \neq z.$$

Assume that the mappings $\gamma_{\mathfrak{A}}$, $M_{\mathfrak{A}}$ and $M_{\mathfrak{A}}'$ have been defined

$$\textbf{if } \gamma \textbf{ then } M \textbf{ else } M' \textbf{ fi}_{\mathfrak{A}}(v) = \begin{cases} M_{\mathfrak{A}}(v) & \text{if } \gamma_{\mathfrak{A}}(v) = \mathbf{1} \text{ and} \\ & M_{\mathfrak{A}}(v) \text{ is defined,} \\ M_{\mathfrak{A}}'(v) & \text{if } \gamma_{\mathfrak{A}}(v) = \mathbf{0} \text{ and} \\ & M_{\mathfrak{A}}'(v) \text{ is defined,} \\ \text{undefined} & \text{otherwise,} \end{cases}$$

$$\textbf{begin } M; \ M' \textbf{ end}_{\mathfrak{A}}(v) = \begin{cases} M_{\mathfrak{A}}'(M_{\mathfrak{A}}(v)) & \text{if } M_{\mathfrak{A}}(v) \text{ is defined} \\ & \text{and } M_{\mathfrak{A}}'(M_{\mathfrak{A}}(v)) \text{ is} \\ & \text{defined,} \\ \text{undefined} & \text{otherwise,} \end{cases}$$

$$\textbf{while } \gamma \textbf{ do } M \textbf{ od}_{\mathfrak{A}}(v) = \begin{cases} M_{\mathfrak{A}}^i(v) & \text{if } M_{\mathfrak{A}}^j(v) \text{ is defined for} \\ & \text{all } j \leqslant i \text{ and } \gamma_{\mathfrak{A}}(M_{\mathfrak{A}}^j(v)) \\ & = \mathbf{1} \text{ for } j < i, \gamma_{\mathfrak{A}}(M_{\mathfrak{A}}^j(v)) \\ & = \mathbf{0} \text{ for } j = i, \\ \text{undefined} & \text{otherwise,} \end{cases}$$

We continue the definition of the interpretation of formulas:

$$(M\alpha)_{\mathfrak{A}}(v) = \begin{cases} \alpha_{\mathfrak{A}}(v') & \text{if } M_{\mathfrak{A}}(v) \text{ is defined and } v' = M_{\mathfrak{A}}(v), \\ \mathbf{0} & \text{otherwise,} \end{cases}$$

$$(\bigcup M\alpha)_{\mathfrak{A}}(v) = \underset{i \in N}{\text{l.u.b.}} (M^i \alpha)_{\mathfrak{A}}(v) \quad \text{(cf. Appendix A),}$$

$$(\bigcap M\alpha)_{\mathfrak{A}}(v) = \underset{i \in N}{\text{g.l.b.}} (M^i \alpha)_{\mathfrak{A}}(v) \quad \text{(cf. Appendix A),}$$

$$((\exists x)\alpha(x))_{\mathfrak{A}}(v) = \underset{a \in A}{\text{l.u.b.}} \ \alpha_{\mathfrak{A}}(v_a^x),$$

$$((\forall x)\alpha(x))_{\mathfrak{A}}(v) = \underset{a \in A}{\text{g.l.b.}} \ \alpha_{\mathfrak{A}}(v_a^x),$$

where v_a^x is a valuation such that

$$v_a^x(x) = a \quad \text{and} \quad v_a^x(z) = v(z) \text{ for all } z \neq x.$$

REMARK 2.1. According to the definitions given above, the mappings $\alpha_{\mathfrak{A}}$, $\tau_{\mathfrak{A}}$, $M_{\mathfrak{A}}$ depend on the finite set of variables that occur in the formula α, the term τ or the program M. Hence only a finite part of the arguments described by the valuation is used in order to establish the values $\alpha_{\mathfrak{A}}(v)$, $\tau_{\mathfrak{A}}(v)$, $M_{\mathfrak{A}}(v)$. In order to simplify our definitions we shall treat these mappings as defined on the set W.   □

For a given data structure $\mathfrak{A}$ and valuation v, $\alpha_{\mathfrak{A}}(v)$ will be called the *value of the formula* α *in the structure* $\mathfrak{A}$ *at the valuation* v. Analogously whenever $M_{\mathfrak{A}}(v)$ is defined, we shall call it the *result of a program* M *in the structure* $\mathfrak{A}$ *at the initial data (valuation)* v.

REMARK 2.2. If a program M does not contain **while** then for every data structure $\mathfrak{A}$, the mapping $M_{\mathfrak{A}}$ is total.   □

EXAMPLE 2.2. We shall consider the program M described in Example 1.3. Let the set of natural numbers be the universe of a data structure, and let the interpretation of functors $-$, $+$ and predicate $\geqslant$ be the obvious one.

Below we shall describe the process of evaluating the result of the program M at an initial valuation v in the data structure $\mathfrak{A}$.

$$M_{\mathfrak{A}}(v) = (\textbf{while } z \geqslant y \textbf{ do } z := z-y; \ i := i+1 \textbf{ od})_{\mathfrak{A}}(v_0)$$

where

$$v_0 = (z := x)_{\mathfrak{A}} \big((i := 0)_{\mathfrak{A}}(v) \big),$$

i.e.

$$v_0(z) = v(x), \quad v_0(u) = v(u) \quad \text{for all } u \neq z \text{ and } u \neq i$$
$$v_0(i) = 0.$$

Let n be the quotient obtained on dividing $v(x)$ by $v(y)$. This gives $v(z) - j \cdot v(y) \geqslant 0$ for all $j \leqslant n$.

Let

$$v_j = (\textbf{begin } z := z-y; \ i := i+1 \textbf{ end})_{\mathfrak{A}}^j(v_0).$$

Thus for all $j \leqslant n$

$$v_j(i) = j, \quad v_j(z) = v(z) - j \cdot v(y)$$

and for $j < n$,

$$(z \geqslant y)_{\mathfrak{A}}(v_j) = 1, \quad (z \geqslant y)_{\mathfrak{A}}(v_n) = 0.$$

Hence,

$$M_{\mathfrak{A}}(v) = (\textbf{begin } z := z - y; \; i := i + 1 \textbf{ end})_{\mathfrak{A}}^a(v_0) = v',$$

where

$$v'(i) = n, \quad v'(z) = v(x) - n \cdot v(y),$$
$$v'(u) = v(u) \quad \text{for } u \neq z, \; u \neq i. \qquad \qquad \square$$

The strict analysis of the example allows us to observe that the process of evaluating a result of a program consists of consecutive steps in accordance with the structure of the program.

The notion of computation defined below captures the intuition of the evaluation process.

DEFINITION 2.3. *By a* configuration *we shall mean any ordered pair* $\langle v; \sigma \rangle$ *such that v is a valuation and σ is a finite sequence of programs.* $\quad \square$

DEFINITION 2.4. *By a* computation *of a program M in a data structure $\mathfrak{A}$ and an initial valuation v, we shall understand a sequence of configurations such that the initial configuration is of the form $\langle v; M \rangle$ and any two consecutive configurations satisfy the successorship relation defined in $1°$–$5°$ below:*
Assume that $\langle v'; M_1, ..., M_n \rangle$ is a configuration of the computation.
$1°$ *If M_1 is an assignment instruction s, then the next configuration is*

$$\langle s_{\mathfrak{A}}(v'); \; M_2, ..., M_n \rangle.$$

$2°$ *If M_1 is in the form* **begin** $M_{11}; \; M_{12}$ **end**, *then the next configuration is*

$$\langle v'; \; M_{11}, M_{12}, M_2, ..., M_n \rangle.$$

$3°$ *If M_1 is in the form* **if** γ **then** M_{11} **else** M_{12} **fi**, *then the next configuration is*

$$\langle v'; \; M_{11}, M_2, ..., M_n \rangle \quad \text{if} \quad \gamma_{\mathfrak{A}}(v) = 1,$$
$$\langle v'; \; M_{12}, M_2, ..., M_n \rangle \quad \text{if} \quad \gamma_{\mathfrak{A}}(v) = 0.$$

$4°$ *If M_1 is in the form* **while** γ **do** M **od**, *then the next configuration is either*

$$\langle v'; \; M_2, M_3, ..., M_n \rangle, \quad \text{when } \gamma_{\mathfrak{A}}(v) = 0$$

or

$$\langle v'; \; M, \; M_1, M_2, M_3, \ldots, M_n \rangle, \quad \text{when } \gamma_\mathfrak{N}(v) = \mathbf{0}.$$

5° *If a configuration of a computation is in the form* $\langle \bar{v}; \; \rangle$ *i.e., if it has an empty list of programs, then it is the last configuration of the computation and the computation is called finite. The valuation* $\bar{v}$ *is called the result of the computation.* □

EXAMPLE 2.3 (Evaluation of a formula value). Let α be the formula $(x := 0)(\bigcup (x := x+1)y < x)$ where 0 is a zero-argument functor, $+1$ is a one-argument functor and $<$ is a two-argument predicate. Let $\mathfrak{N}$ be a data structure such that the set of natural numbers is its universe and 0 is the number zero, $+1_\mathfrak{N}$ is the successor, $<_\mathfrak{N}$ is the ordering relation in the set of natural numbers. If v is any valuation in $\mathfrak{N}$, then

$$\begin{aligned}\alpha_\mathfrak{N}(v) &= \left(\bigcup (x := x+1)y < x\right)_\mathfrak{N}(v_0^x) \\ &= \underset{i \in N}{\text{l.u.b.}} \left((x := x+1)^i y < x\right)_\mathfrak{N}(v_0^x).\end{aligned}$$

Assume that

$$v_i = (x := x+1)^i_\mathfrak{N}(v_0^x),$$

i.e.

$$v_i(x) = i \text{ and } v_i(z) = v(z) \quad \text{for } z \neq x.$$

Then

$$\begin{aligned}\alpha_\mathfrak{N}(v) &= \underset{i \in N}{\text{l.u.b.}} (y < x)_\mathfrak{N}(v_i) = \underset{i \in N}{\text{l.u.b.}} \big(v_i(y) < v_i(x)\big) \\ &= \underset{i \in N}{\text{l.u.b.}} \big(v_i(y) < i\big) = \mathbf{1}.\end{aligned}$$

Hence, for every valuation v in the structure $\mathfrak{N}$, the formula α has the value $\mathbf{1}$. □

We shall now state some simple properties of a semantic character, which will be useful in the sequel.

LEMMA 2.1. *For every term* τ, *open formula* γ, *assignment instruction* s *and program* M, *for every data structure* $\mathfrak{A}$ *and valuation* v, *we have the following:*

(1) $\tau_\mathfrak{A}\big(s_\mathfrak{A}(v)\big) = \bar{s}\tau_\mathfrak{A}(v),$

(2) $(s\gamma)_\mathfrak{A}(v) = \overline{s\gamma}_\mathfrak{A}(v),$

(3) *If $V(M) \cap V(\alpha) = \emptyset$ and $M_\mathfrak{A}$ is defined at v, then*

$$\alpha_\mathfrak{A}(v) = (M\alpha)_\mathfrak{A}(v).$$

(4) *For every formula α in which the signs of quantifiers and* **while** *do not appear, there exists an open formula α' such that*

$$\alpha_\mathfrak{A}(v) = \alpha'_\mathfrak{A}(v),$$

for every data structure $\mathfrak{A}$ and every valuation v in $\mathfrak{A}$.

PROOF. Let $\mathfrak{A}$ be any data structure and v any valuation.

(1) The proof is by induction on the length of the expression. Assume that s is of the form $(u := w)$ and $s_\mathfrak{A}(v) = \bar{v}$.

Let x be an individual variable. By Definition 2.2 of valuation and by the definition of the mapping $s_\mathfrak{A}$ we then have

$$x_\mathfrak{A}\big(s_\mathfrak{A}(v)\big) = \begin{cases} x_\mathfrak{A}(v) & \text{iff } u \neq x \\ w_\mathfrak{A}(v) & \text{iff } u = x \end{cases} = \overline{sx}_\mathfrak{A}(v).$$

Let φ be an n-argument functor, and let us assume property (1) for the terms $\tau_1, \ldots, \tau_n$, i.e.

$$\tau_{i\mathfrak{A}}\big(s_\mathfrak{A}(v)\big) = (\overline{s\tau_i})_\mathfrak{A}(v) \quad \text{for } i = 1, 2, \ldots, n.$$

Thus

$$\varphi(\tau_1, \ldots, \tau_n)_\mathfrak{A}\big(s_\mathfrak{A}(v)\big) = \varphi_\mathfrak{A}\big(\tau_{1\mathfrak{A}}(\bar{v}), \ldots, \tau_{n\mathfrak{A}}(\bar{v})\big)$$

$$= \varphi_\mathfrak{A}\big(\overline{s\tau_1}_\mathfrak{A}(v), \ldots, \overline{s\tau_n}_\mathfrak{A}(v)\big) = \overline{s\varphi(\tau_1, \ldots, \tau_n)}_\mathfrak{A}(v).$$

Hence for every term τ and every assignment instruction s (1) holds.

The proofs of (2) and (3) although a little longer, but are based on the same idea and are therefore omitted.

(4) It is sufficient to prove property (4) for formulas of the form $M\beta$, where β is an open formula and M is a while-free program.

The proof is by induction on the length of M.

(a) Suppose M is an assignment instruction s. According to property (2) for every data structure $\mathfrak{A}$ and valuation v, $(s\beta)_\mathfrak{A}(v) = \overline{s\beta}_\mathfrak{A}(v)$. By Lemma 1.4, $\overline{s\beta}$ is an open formula. Thus $\overline{s\beta}$ is the formula we need. The inductive assumption is: suppose that for the programs M_1, M_2 and every formula β there exist open formulas β_1, β_2 such that

$$(M_1 \beta)_{\mathfrak{A}}(v) = \beta_{1\mathfrak{A}}(v) \quad \text{and} \quad (M_2 \beta)_{\mathfrak{A}}(v) = \beta_{2\mathfrak{A}}(v),$$

for every data structure $\mathfrak{A}$ and every valuation v.

(b) Let M be the program **if** γ **then** M_1 **else** M_2 **fi**. By the definition of the mapping $M_{\mathfrak{A}}$ we have

$$(M\beta)_{\mathfrak{A}}(v) = (\gamma \wedge M_1 \beta)_{\mathfrak{A}}(v) \vee (\sim \gamma \wedge M_2 \beta)_{\mathfrak{A}}(v).$$

Hence by the inductive assumption

$$(M\beta)_{\mathfrak{A}}(v) = (\gamma \wedge \beta_1)_{\mathfrak{A}}(v) \vee (\sim \gamma \wedge \beta_2)_{\mathfrak{A}}(v),$$

for every data structure $\mathfrak{A}$ and valuation v. Thus the open formula we need is in this case of the form $((\gamma \wedge \beta_1) \vee (\sim \gamma \wedge \beta_2))$.

(c) Let M be of the form **begin** M_1 ; M_2 **end**.

Let β_2 be an open formula such that

$$(M_2 \beta)_{\mathfrak{A}}(v) = \beta_{2\mathfrak{A}}(v) \quad \text{for all } \mathfrak{A} \text{ and } v,$$

and let β_1 be an open formula such that

$$(M_1 \beta_2)_{\mathfrak{A}}(v) = \beta_{1\mathfrak{A}}(v) \quad \text{for all } \mathfrak{A} \text{ and } v.$$

Hence,

$$(\textbf{begin } M_1 ; \ M_2 \textbf{ end } \beta)_{\mathfrak{A}}(v) = \big(M_1(M_2\beta)\big)_{\mathfrak{A}}(v) =$$
$$= (M_1 \beta_2)_{\mathfrak{A}}(v) = \beta_{1\mathfrak{A}}(v).$$

Thus β_1 is the formula we need.

This concludes the proof of (4).   □

DEFINITION 2.5. *We shall say that the valuation v in a data structure $\mathfrak{A}$ satisfies the formula α, $\mathfrak{A}, v \models \alpha$ iff $\alpha_{\mathfrak{A}}(v) = 1$.*

The formula α can be satisfied iff there exists a data structure $\mathfrak{A}$ and a valuation v such that $\mathfrak{A}, v \models \alpha$.

The formula α is valid in the structure $\mathfrak{A}$, for short $\mathfrak{A} \models \alpha$, iff every valuation in $\mathfrak{A}$ satisfies the formula α.

The formula α is a tautology, $\models \alpha$, iff α is valid in every data structure $\mathfrak{A}$ for algorithmic language L.   □

REMARK. If α can not be satisfied, then $\sim \alpha$ is a tautology.   □

EXAMPLE 2.4. Let δ be a simple formula of the form

while $\sim (x = 0)$ **do** $x := x - 2$ **od** true.

If $\mathfrak{A}$ is a data structure with the set of real numbers as universe and the obvious interpretation of $-, 2, =, 0$, then every valuation v such that $v(x)$ is an even non-negative number satisfies the formula δ and any other valuation does not satisfy δ.

Hence δ is satisfiable, but is not valid in $\mathfrak{A}$ and is not a tautology.

Consider another simple example, in this case letting δ be the formula

(5) $M(\alpha \lor \beta) \equiv (M\alpha \lor M\beta)$

where M is a program and α, β are formulas.

Let $\mathfrak{A}$ be a data structure and v be a valuation. Then by the definition of semantics $\mathfrak{A}, v \models M(\alpha \lor \beta)$ iff $M_{\mathfrak{A}}(v)$ is defined and $\mathfrak{A}, v' \models (\alpha \lor \beta)$ for $v' = M_{\mathfrak{A}}(v)$. Hence $\mathfrak{A}, v \models M(\alpha \lor \beta)$ iff $\mathfrak{A}, v \models M\alpha$ or $\mathfrak{A}, v \models M\beta$ iff $\mathfrak{A}, v \models (M\alpha \lor M\beta)$. Since $\mathfrak{A}, v$ are arbitrarily chosen, then δ is valid in every data structure, i.e., δ is a tautology. $\qquad\square$

3. EXPRESSIVENESS

We should like to show how useful algorithmic language is and how strong it is in expressing the properties of programs, computations and data structures. Intuitively, we shall say that a property of semantic character is expressible in algorithmic language if there exists an algorithic formula α such that for every data structure and every valuation, the formula α is true if and only if the property holds.

Termination property

The most important property, and one of the easiest to describe, is the termination property expressed as "the program has a finite computation", (see also Chapter I, § 4). According to the definition of semantics (cf. § 2), $\mathfrak{A}, v \models M$ **true** means that the program M has a finite computation which starts from the initial valuation v in the data structure $\mathfrak{A}$.

Thus the termination property can be expressed by the formula M **true**.

This formula gives us no information about how the terminating property of a program depends on its structure, but it can be useful to verifying the termination property. The appropriate facts are summated up in the following lemma. We shall use fin(M) as a denotation of the formula M **true**, hoping that the wording of the lemma will thereby be more suggestive.

LEMMA 3.1. *For every data structure* $\mathfrak{A}$, *every open formula* γ, *every assignment instruction* s, *and arbitrary programs* M, M', *the following properties hold*:

(1) $\mathfrak{A} \models \text{fin}(s) \equiv \textbf{true}$,

(2) $\mathfrak{A} \models \text{fin}(\textbf{begin } M; \ M' \textbf{ end}) \equiv \text{fin}(M) \wedge M \ \text{fin}(M')$,

(3) $\mathfrak{A} \models \text{fin}(\textbf{if } \gamma \textbf{ then } M \textbf{ else } M' \textbf{ fi})$

$\equiv (\gamma \wedge \text{fin}(M)) \vee (\sim\gamma \wedge \text{fin}(M'))$,

(4) $\mathfrak{A} \models \text{fin}(\textbf{while } \gamma \textbf{ do } M \textbf{ od}) \equiv \bigcup M \sim\gamma$.

PROOF. The first three properties are very simple and easy to verify, so we shall not prove them here. We would like to call the reader's attention to property (4). Its character is a little different from that of the others.

By the definition of semantics (cf. § 2) for an arbitrary valuation v, we have

$$\mathfrak{A}, v \models \text{fin}(\textbf{while } \gamma \textbf{ do } M \textbf{ od})$$

iff there exists such a natural number i, that M^i is defined at v and $\gamma_\mathfrak{A}(M_\mathfrak{A}^j(v)) = 1$ for $j < i$, $\gamma_\mathfrak{A}(M_\mathfrak{A}^i(v)) = 0$ (i.e., after the i-th iteration of the program M the formula γ does not hold at the resulting valuation) iff there exists i, such that $\mathfrak{A}, v \models M^i \sim\gamma$ iff $\mathfrak{A}, v \models \bigcup M \sim\gamma$.   □

Observe that property (4) of Lemma 3.1 can be reformulated as follows: For every valuation v

$$\mathfrak{A}, v \models \text{fin}(\textbf{while } \gamma \textbf{ do } M \textbf{ od})   \text{iff}$$

there exists a natural number i such that

$$\mathfrak{A}, v \models \text{fin}(M^i)   \text{and}   \mathfrak{A}, v' \models \sim\gamma,   v' = M_\mathfrak{A}^i(v).$$

Sometimes it is convenient to have information as to whether the program diverges. Let loop (M) denote the formula $\sim M$ **true**. Obviously, for every data structure $\mathfrak{A}$ and valuation v

$$\mathfrak{A}, v \models \text{loop}(M)   \text{iff}$$

M has an infinite computation in the structure $\mathfrak{A}$ and the valuation v.

Under the assumptions of the previous lemma we have the following:

LEMMA 3.2.

$$\mathfrak{A} \models \text{loop}(s) \equiv \textbf{false},$$

$$\mathfrak{A} \models \text{loop}(\textbf{if } \gamma \textbf{ then } M \textbf{ else } M' \textbf{ fi}) \equiv (\text{loop}(M) \wedge \gamma) \vee$$
$$\vee (\sim\gamma \wedge \text{loop}(M')),$$
$$\mathfrak{A} \models \text{loop}(\textbf{begin } M; \ M' \textbf{ end}) \equiv \text{loop}(M) \vee M \text{ loop}(M'),$$
$$\mathfrak{A} \models \text{loop}(\textbf{while } \gamma \textbf{ do } M \textbf{ od}) \equiv \bigcap M\gamma \vee \bigcup \textbf{if } \gamma \textbf{ then } M \textbf{ fi}$$
$$(\gamma \wedge \text{loop}(M)).$$

PROOF. We omit the exact proof of the lemma. Let us note only that the program **while** γ **do** M **od** has an infinite computation either if the formula γ is true after each iteration of M, or if after some iteration of the program M the resulting valuation satisfies the formula γ and starting from that valuation, M has an infinite computation.   □

Observe now that the expression $\mathfrak{A}, \ v \models M$ **true** means that v is proper data for the program M in the structure $\mathfrak{A}$, i.e., there exists a valuation v' such that $M_{\mathfrak{A}}(v) = v'$. Hence, the formula M **true** describes the domain of the program M, i.e., the domain of the mapping $M_{\mathfrak{A}}$.

The strongest postcondition

The question naturally arises as to whether it is possible to describe the counter domain of M. The answer is positive, but an additional assumption on the algorithmic language is required.

Let us assume that the algorithmic language contains the predicate $=$ interpreted in the data structure $\mathfrak{A}$ as identity. Throughout this section it will be convenient to accept the following abbreviations: Let α_i, for $i = 1, ..., n$, be a formula; $\bigwedge_{1 \leq i \leq n} \alpha_i$ is then a shortened form of the formula

$$(\alpha_1 \wedge \alpha_2 \wedge \ ... \ \wedge \alpha_n),$$

and $\bigvee_{1 \leq i \leq n} \alpha_i$ is a shortened form of the formula

$$(\alpha_1 \vee \alpha_2 \vee \ ... \ \vee \alpha_n).$$

Let $\vec{u} = (u_1, ..., u_n)$ and $\vec{t} = (t_1, ..., t_n)$ be two vectors of different variables such that for every $i \leq m \leq n$, u_i and t_i are individual variables, and for every $i, m < i \leq n$, u_i and t_i are propositional variables, and $\{u_1, ..., u_n\} \cap \{t_1, ..., t_n\} = \varnothing$; $\vec{u} = \vec{t}$ is then a shortened form of the formula

$$\bigwedge_{1 \leq i \leq m} (u_i = t_i) \wedge \bigwedge_{m < i \leq n} (u_i \equiv t_i).$$

Moreover $(\exists \vec{u})\alpha$ is a shortened form of the formula

$$(\exists u_1) \ldots (\exists u_m) \bigvee_{\substack{e_j \in \{\text{true, false}\} \\ m < j \leqslant n}} \textbf{begin } u_{m+1} := e_{m+1} ; \ldots ; u_n := e_n$$

$$\textbf{end } \alpha.$$

Let M be a program and let $\vec{t} = (t_1, \ldots, t_n)$ be the vector of all variables that occur in M. We shall consider the formula $(\exists \vec{u}) \, M(\vec{t}/\vec{u})(\vec{t} = \vec{u})$.

$\mathfrak{A}, v \models (\exists \vec{u}) \, M(\vec{t}/\vec{u}) \, (\vec{t} = \vec{u})$ iff

there exists a corresponding vector $\vec{a}$ of values of $\vec{u}$ such that for $v' = M(\vec{t}/\vec{u})_{\mathfrak{A}}(v\frac{\vec{u}}{\vec{a}})$, $\mathfrak{A}, v' \models (\vec{t} = \vec{u})$ iff

there exists $\vec{a}$ such that

$v' = M(\vec{t}/\vec{u})_{\mathfrak{A}}(v\frac{\vec{u}}{\vec{a}})$ and $v'(u_i) = v(t_i)$ for all $i \leqslant n$ iff

there exists an initial valuation v' such that the valuation v is the result of a finite computation of M starting from the valuation v'.

Let $\vec{t}$ be the vector of all variables that occur in M and α. Denote by αM the formula $(\exists \vec{u})(\alpha(\vec{t}/\vec{u}) \wedge M(\vec{t}/\vec{u}) \, (\vec{t} = \vec{u}))$. By virtue of the above we have for every valuation v,

$\mathfrak{A}, v \models \textbf{true } M$ iff

v is a result of a computation of M in the data structure $\mathfrak{A}$.

Analogously,

$\mathfrak{A}, v \models \alpha M$ iff

there exists a valuation v' such that $\mathfrak{A}, v' \models \alpha$ and v is a result of a computation of the program M in the structure $\mathfrak{A}$ from the valuation v'.

The formula αM describes in a data structure $\mathfrak{A}$ the set of all valuations which are the results of computations of the program M from the initial valuations satisfying the formula α.

DEFINITION 3.1. *The formula δ is called the strongest postcondition of a formula α with respect to the program M iff the following conditions hold in every data structure $\mathfrak{A}$:*

(i) $\mathfrak{A} \models ((\alpha \wedge M \textbf{ true}) \Rightarrow M\delta)$,   *i.e. δ is a postcondition.*

(ii)   *For every formula β, if $\mathfrak{A} \models ((\alpha \wedge M \text{ true}) \Rightarrow M\beta)$ then*
        $\mathfrak{A} \models (\delta \Rightarrow \beta)$ *(δ is the strongest postcondition, cf. Chapter* I).   □

REMARK. The formula αM is the strongest postcondition of a formula α with respect to the program M.

For a valuation v let $\mathfrak{A}, v \models (\alpha \wedge M \text{ true})$. It then follows that $\mathfrak{A}, v \models \alpha$ and there exists $\bar{v}$ such that $v = M_{\mathfrak{A}}(v)$. By the definition of construction αM we have that there exists a valuation $\bar{v}$ such that $\bar{v} = M_{\mathfrak{A}}(v)$ and $\mathfrak{A}, \bar{v} \models \alpha M$. Hence $\mathfrak{A}, v \models M(\alpha M)$.

Suppose that β is an arbitrary formula and

$$\mathfrak{A} \models ((\alpha \wedge M \text{ true}) \Rightarrow M\beta).$$

For a given valuation v, let

$$\mathfrak{A}, v \models \alpha M \quad \text{and} \quad \text{non } \mathfrak{A}, v \models \beta.$$

Thus there exists a valuation v' such that

$$\mathfrak{A}, v' \models \alpha \quad \text{and} \quad v = M_{\mathfrak{A}}(v') \quad \text{and} \quad \text{non } \mathfrak{A}, v \models \beta.$$

Consequently, there exists a valuation v' such that

$$\text{non } \mathfrak{A}, v' \models M\beta \quad \text{and} \quad \mathfrak{A}, v' \models (\alpha \wedge M \text{ true})$$

which contradicts the assumption.   □

EXAMPLE 3.2. Let M be a program in the algorithmic language L such that

```
M: begin
      while (z−y) > 0
      do
         z := z−y;
         y := y+2
      od;
      if z = y then y := 0 else y := z fi
   end.
```

Let the data structure $\mathfrak{R}$ for the language L be the set of real numbers with the obvious interpretation of the signs $=, >, +, -, 2, 0$.

The formula $y = x - [\sqrt{x}]^2$ is the strongest postcondition of the formula $(y = 1 \wedge z = x \wedge x > 0)$ with respect to the program M. In fact, for every valuation v,

$$\mathfrak{R}, v \models (y = 1 \wedge z = x \wedge x > 0)M \quad \text{iff}$$

there exists a valuation v' such that

$$\mathfrak{R}, v' \models (y = 1 \wedge z = x \wedge x > 0) \text{ and } v = M_\mathfrak{R}(v').$$

However, $v = M_\mathfrak{R}(v')$ if and only if there exists v'' such that

$$v'' = (\mathbf{begin}\ z := z - y;\ y := y + 2\ \mathbf{end})^n_\mathfrak{R}(v'),$$

where

$$n = \max_{i \in N}\ (v'(x) - (1 + 3 + \dots + 2i - 1) > 0)$$

and

$$v(y) = \begin{cases} v''(z) & \text{iff } v''(z) < v''(y), \\ 0 & \text{iff } v''(z) = v''(y). \end{cases}$$

Hence

$$v(y) = \begin{cases} v(x) - (1 + 3 + 5 \dots + 2n - 1) & \text{iff } v(x) > n^2, \\ 0 & \text{iff } v(x) = n^2. \end{cases}$$

Thus

$$\mathfrak{R}, v \models (y = 1 \wedge z = x \wedge x > 0)M \quad \text{iff}$$
$$\mathfrak{R}, v \models (y = x - [\sqrt{x}]^2). \qquad \square$$

The following lemma shows some simple properties of the strongest postcondition.

LEMMA 3.3. *Let $\mathfrak{A}$ be a data structure such that the predicate $=$ is interpreted as identity.*

(a) *The following formulas are valid in $\mathfrak{A}$:*

(1) $(\alpha \vee \beta)M \equiv (\alpha M \vee \beta M),$

(2) $(\alpha \wedge \beta)M \Rightarrow (\alpha M \wedge \beta M),$

(3) $\alpha\ \mathbf{begin}\ M;\ M'\ \mathbf{end} \equiv ((\alpha M)M'),$

(4) $\alpha\ \mathbf{if}\ \gamma\ \mathbf{then}\ M\ \mathbf{else}\ M'\ \mathbf{fi} \equiv ((\alpha \wedge \gamma)M \vee (\alpha \wedge \sim\gamma)M').$

(b) *If $\mathfrak{A} \models (\alpha \Rightarrow \beta)$, then $\mathfrak{A} \models (\alpha M \Rightarrow \beta M).$*

PROOF. Let v be an arbitrary valuation in $\mathfrak{A}$.

(1) $\mathfrak{A}, v \models (\alpha \vee \beta)M$ iff
       there exists v' such that $\mathfrak{A}, v' \models (\alpha \vee \beta)$ and $M_\mathfrak{A}(v') = v$ iff
       there exists v' such that $\mathfrak{A}, v' \models \alpha$ and $M_\mathfrak{A}(v') = v$,   or

there exists a valuation v'' such that $\mathfrak{A}, v'' \models \beta$ and $M_{\mathfrak{A}}(v'') = v$ iff

$\mathfrak{A}, v \models \alpha M$ or

$\mathfrak{A}, v \models \beta M$ iff

$\mathfrak{A}, v \models (\alpha M \vee \beta M)$.

The validity of formulas (2) and (4) can be proved analogously.

(3) $\mathfrak{A}, v \models \alpha$ **begin** M; M' **end**   iff

there exists a valuation v' such that $\mathfrak{A}, v' \models \alpha$ and $v = (\textbf{begin } M;\ M'\ \textbf{end})_{\mathfrak{A}}(v')$ iff

there exists a valuation v' and a valuation v'' such that $\mathfrak{A}, v' \models \alpha$ and $M_{\mathfrak{A}}(v) = v''$, $M'_{\mathfrak{A}}(v'') = v$ iff

there exists a valuation v'' such that $\mathfrak{A}, v'' \models \alpha M$ and $M'_{\mathfrak{A}}(v'') = v$ iff

$\mathfrak{A}, v \models (\alpha M)M'$.

(b) Suppose for every valuation v,

$\mathfrak{A}, v \models (\alpha \Rightarrow \beta)$.

If $\mathfrak{A}, v \models \alpha M$, then there exists a valuation v' such that $\mathfrak{A}, v' \models \alpha$ and $v = M_{\mathfrak{A}}(v')$. According to the assumption, if $\mathfrak{A}, v' \models \alpha$, then $\mathfrak{A}, v' \models \beta$. Hence there exists a valuation v', such that $\mathfrak{A}, v' \models \beta$ and $v = M_{\mathfrak{A}}(v')$. Thus $\mathfrak{A}, v \models \beta M$. This proves that for every v in $\mathfrak{A}$,

$$\mathfrak{A}, v \models (\alpha M \Rightarrow \beta M), \quad \text{i.e.} \quad \mathfrak{A} \models (\alpha M \Rightarrow \beta M). \qquad \square$$

The weakest precondition

DEFINITION 3.2. *The weakest precondition (cf. Chapter I) of a formula α with respect to the program M is a formula δ such that for every data structure $\mathfrak{A}$*

(i) $\mathfrak{A} \models (\delta \Rightarrow M\alpha)$ *(i.e. δ is precondition),*

(ii) *for every formula β,*

if $\mathfrak{A} \models (\beta \Rightarrow M\alpha)$,   then $\mathfrak{A} \models (\beta \Rightarrow \delta)$

(i.e. δ is the weakest precondition). $\square$

Obviously, the formula $M\alpha$ satisfies both conditions (i) and (ii) and therefore $M\alpha$ is the weakest precondition.

The notion of weakest precondition is dual to the notion of the strongest postcondition, since the formula $M\alpha$ describes the maximal set of (data) valuations for which the program M has a finite computation with result satisfying the formula α.

Below, we shall mention some of the properties of the weakest precondition.

LEMMA 3.4.

(a) *In every data structure $\mathfrak{A}$ the following formulas are valid*

(1)   **begin** M; M' **end** $\alpha \equiv M(M'\alpha)$,

(2)   **if** γ **then** M **else** M' **fi** $\alpha \equiv ((\gamma \wedge M\alpha) \vee (\sim\gamma \wedge M'\alpha))$,

(3) $M(\alpha \vee \beta) \equiv (M\alpha \vee M\beta)$,

(4) $M(\alpha \wedge \beta) \equiv (M\alpha \wedge M\beta)$.

(b) *If the formula $(\alpha \Rightarrow \beta)$ is valid in a data structure $\mathfrak{A}$, then the formula $(M\alpha \Rightarrow M\beta)$ is valid in $\mathfrak{A}$.*

PROOF.

(a)

(1) Let v be a valuation in a data structure $\mathfrak{A}$.

$\mathfrak{A}, v \models$ **begin** M; M' **end** α iff

there exists a valuation v' such that (**begin** M; M' **end**$)_{\mathfrak{A}}(v)$ $= v'$ and $\mathfrak{A}, v' \models \alpha$ iff

there exist valuations v', v'' such that $M'_{\mathfrak{A}}(v'') = v'$, $M_{\mathfrak{A}}(v)$ $= v''$, $\mathfrak{A}, v' \models \alpha$ iff

there exists valuation v'' such that $v'' = M_{\mathfrak{A}}(v)$ and $\mathfrak{A}, v'' \models M'\alpha$ iff

$\mathfrak{A}, v \models M(M'\alpha)$.

The analogous proofs of (2), (3) and (4) are omitted (see also Example 2.4).

(b) Let us assume that

$$\mathfrak{A} \models (\alpha \Rightarrow \beta).$$

If $\mathfrak{A}, v \models M\alpha$ for some valuation v, then by the definition of semantics, there exists a valuation v' such that

$$M_{\mathfrak{A}}(v) = v' \quad \text{and} \quad \mathfrak{A}, v' \models \alpha.$$

Hence there exists a valuation v' such that $M_{\mathfrak{A}}(v) = v'$ and $\mathfrak{A}, v' \models \beta$, i.e. $\mathfrak{A}, v \models M\beta$.

As a consequence

$$\mathfrak{A} \models (M\alpha \Rightarrow M\beta). \qquad\qquad \square$$

Correctness

DEFINITION 3.3. *Program M is correct with respect to an input for-mula α and an output formula β in a data structure $\mathfrak{A}$ iff the formula $(\alpha \Rightarrow M\beta)$ is valid in $\mathfrak{A}$.* $\qquad\qquad \square$

DEFINITION 3.4. *Program M is partially correct with respect to an input formula α and an output formula β in a data structure $\mathfrak{A}$ iff $\mathfrak{A} \models ((\alpha \wedge M \text{ true}) \Rightarrow M\beta)$.* $\qquad\qquad \square$

EXAMPLE 3.3. The following program is partially correct with respect to the input formula $(z = x \wedge y = u)$ and the output formula $z = (x+u)$ and is not correct in the data structure $\mathfrak{R}$ (cf. Example 3.2)

while $y \neq 0$ **do** $z := z+1$; $y := y-1$ **od.**

For every valuation v in the data structure $\mathfrak{R}$, if $v(z) = v(x)$ and the program under consideration terminates, then $v(y)$ is a natural number and obviously the result of the computation satisfies the formula $z = (x+u)$. $\qquad\qquad \square$

LEMMA 3.5. *Let us denote by $L_=$ an algorithmic language with the binary relation $=$, and let $\mathfrak{A}$ be a data structure for $L_=$ such that $=$ is interpreted as an identity relation. A program M in the language $L_=$ is partially correct with respect to an input formula α and an output for-mula β iff $\mathfrak{A} \models (\alpha M \Rightarrow \beta)$.*

PROOF. By Definition 3.4 it is sufficient to prove that the following condition holds:

$$\mathfrak{A} \models (\alpha M \Rightarrow \beta) \quad \text{iff} \quad \mathfrak{A} \models ((M \text{ true} \wedge \alpha) \Rightarrow M\beta).$$

Let $\mathfrak{A} \models (\alpha M \Rightarrow \beta)$ and let v be an arbitrary valuation. If $\mathfrak{A}, v \models \models (M \text{ true} \wedge \alpha)$, then $\mathfrak{A}, v \models \alpha$ and there exists a valuation v' such that

$$M_{\mathfrak{A}}(v) = v'.$$

Hence, there exists a valuation v' such that $\mathfrak{A}, v' \models \alpha M$ and $v' = M_{\mathfrak{A}}(v)$. Since $\mathfrak{A} \models (\alpha M \Rightarrow \beta)$, then $\mathfrak{A}, v \models M\beta$.

Conversely, assume that

$$\mathfrak{A} \models ((M \text{ true} \wedge \alpha) \Rightarrow M\beta).$$

If for a valuation v, $\mathfrak{A}, v \models \alpha M$ then there exists a valuation v' such that $\mathfrak{A}, v' \models \alpha$ and $M_{\mathfrak{A}}(v') = v$. Hence there exists a valuation v', such that

$$\mathfrak{A}, v' \models (M \text{ true} \wedge \alpha) \quad \text{and} \quad v = M_{\mathfrak{A}}(v').$$

By assumption, there exists a valuation v' such that

$$\mathfrak{A}, v' \models M\beta \quad \text{and} \quad M_{\mathfrak{A}}(v') = v,$$

i.e. $\mathfrak{A}, v \models \beta$. As a result, $\mathfrak{A} \models (\alpha M \Rightarrow \beta)$. $\square$

Verification condition

DEFINITION 3.5. *By an annotated version of a program we shall understand an expression defined by induction with respect to the length of program as follows*:

(i) *For all formulas α, β, the expression $\{\alpha\}s\{\beta\}$ is an annotated version of an assignment instruction s.*

Let $\hat{M}_1$ *and* $\hat{M}_2$ *be annotated versions of the programs M_1 and M_2, respectively.*

For all formulas α, β and every open formula γ;

(ii) *The expression $\{\alpha\}$ **if** γ **then** $\hat{M}_1$ **else** $\hat{M}_2$ **fi** $\{\beta\}$ is an annotated version of the program* **if** γ **then** M_1 **else** M_2 **fi**.

(iii) *The expression $\{\alpha\}$ **while** γ **do** $\hat{M}_1$ **od** $\{\beta\}$ is an annotated version of the program* **while** γ **do** M_1 **od**.

(iv) *The expression $\{\alpha\}$ **begin** $\hat{M}_1$; $\hat{M}_2$ **end** $\{\beta\}$ is an annotated version of the program* **begin** M_1 ; M_2 **end**. $\square$

We shall write $\hat{M}$ to denote an annotated version of the program M. For short, we shall say that $\hat{M}$ is an annotated program.

Informally, by an annotated program we shall mean a modification of a program such that every instruction is provided with two comment-conditions. The intuition is that they describe the properties of states before and after execution of an instruction. We shall call them the *precondition* and *postcondition*.

EXAMPLE 3.4. The following expression $\hat{M}$ is an annotated version of the program M described in Example 3.2:

(α_1) $\quad \{y = 1 \wedge z = x \wedge x > 0 \wedge i = 0\}$
$\quad\quad\quad$ **begin**

(α_2) $\{y = 2i+1 \wedge z = x-i^2 \wedge x > 0 \wedge i \geqslant 0\}$
       while $z-y > 0$ **do**

(α_3) $\{z > y \wedge z = x-i^2 \wedge y = 2i+1 \wedge x > 0 \wedge i \geqslant 0\}$
       $i := i+1;$

(α_4) $\{z > y \wedge z = x-(i-1)^2 \wedge y = 2i-1 \wedge x > 0 \wedge i \geqslant 0\}$
       $z := z-y;$

(α_5) $\{z > 0 \wedge z = x-(i-1)^2 - (2i-1) \wedge y = 2i-1 \wedge x > 0 \wedge$
       $\wedge\, i \geqslant 0\}$
       $y := y+2;$

(α_6) $\{x > 0 \wedge i \geqslant 0 \wedge z > 0 \wedge y = 2i+1 \wedge z = x - \sum\limits_{j=1}^{i} (2j-1)\}$
       od;

(α_7) $\{z \leqslant y \wedge y = 2i+1 \wedge z = x-i^2 \wedge x > 0 \wedge i \geqslant 0\}$
(α_8) $\{x-i^2 \leqslant 2i+1 \wedge z = x-i^2 \wedge y = 2i+1\}$
       if $z = y$ **then**

(α_9) $\{x-i^2 = 2i+1 \wedge z = x-i^2\}$
       $y := 0$

(α_{10}) $\{y = 0 \wedge x = (i+1)^2\}$
       else

(α_{11}) $\{i^2 < x < (i+1)^2 \wedge z = x-i^2\}$
       $y := z$

(α_{12}) $\{y = x-i^2 \wedge i^2 < x < (i+1)^2\}$
       fi

(α_{13}) $\{y = x-[\sqrt{x}\,]^2\}$
       end

(α_{14}) $\{y = x-[\sqrt{x}\,]^2\}$

In this example, formulas (α_2)–(α_6) can be repeated in order to obtain a version of the annotated program formally corresponding to Definition 3.5. Observe that whenever a computation passes from one instruction to the other instruction then the following property holds: if a formula α written before the instruction M is satisfied by a state preceding the execution of the statement M, then the formula appearing after the instruction M is satisfied by the state resulting from the previous one after execution of the instruction M, cf. the formulas

$$(\alpha_3 \Rightarrow (i := i+1)\alpha_4),$$
$$((\alpha_2 \wedge z-y \leqslant 0) \Rightarrow \alpha_7).$$

$\square$

DEFINITION 3.6. *By the verification condition of an annotated program $\hat{M}$ we shall understand the formula* $\mathrm{VC}(\hat{M})$ *defined by induction as follows*:

(i) *If $\hat{M}$ is of the form $\{\alpha\}s\{\beta\}$ where s is an assignment instruction and α, β are arbitrary formulas, then* $\mathrm{VC}(\hat{M}) = (\alpha \Rightarrow s\beta)$. *For $i = 1, 2$, let $\hat{M}_i$ be an annotated program with the precondition α_i and the postcondition β_i.*

(ii) *If $\hat{M}$ is of the form $\{\alpha\}$ if γ then $\hat{M}_1$ else $\hat{M}_2$ fi $\{\beta\}$ then*

$$\mathrm{VC}(\hat{M}) = \mathrm{VC}(\hat{M}_1) \wedge \mathrm{VC}(\hat{M}_2) \wedge ((\alpha \wedge \gamma) \Rightarrow \alpha_1) \wedge$$
$$\wedge ((\alpha \wedge \sim\gamma) \Rightarrow \alpha_2) \wedge ((\beta_1 \vee \beta_2) \Rightarrow \beta).$$

(iii) *If $\hat{M}$ is of the form $\{\alpha\}$ begin $\hat{M}_1$; $\hat{M}_2$ end $\{\beta\}$ then*

$$\mathrm{VC}(\hat{M}) = \mathrm{VC}(\hat{M}_1) \wedge \mathrm{VC}(\hat{M}_2) \wedge (\alpha \Rightarrow \alpha_1) \wedge (\beta_1 \Rightarrow \alpha_2) \wedge$$
$$\wedge (\beta_2 \Rightarrow \beta).$$

(iv) *If $\hat{M}$ is of the form $\{\alpha\}$ while γ do $\hat{M}_1$ od $\{\beta\}$, then*

$$\mathrm{VC}(\hat{M}) = \mathrm{VC}(\hat{M}_1) \wedge (((\alpha \vee \beta_1) \wedge \gamma) \Rightarrow \alpha_1) \wedge$$
$$\wedge (((\alpha \vee \beta_1) \wedge \sim\gamma) \Rightarrow \beta). \qquad \square$$

DEFINITION 3.7. *The verification condition* $\mathrm{VC}(\hat{M})$ *of an annotated program $\hat{M}$ is proper in a data structure $\mathfrak{A}$ if and only if* $\mathrm{VC}(\hat{M})$ *is valid in $\mathfrak{A}$.* $\qquad \square$

EXAMPLE 3.5. A. Let us consider the following annotated program:

$$\{n < 0\}$$
$$n := n \cdot n$$
$$\{n > 0\}.$$

Its verification condition is the formula

$$(n < 0 \Rightarrow (n := n \cdot n)\, n > 0).$$

This verification condition is proper in the structure of integers with the usual interpretation of the predicates $<$, $>$ and functors $\cdot$, 0. Note that it is not proper in the structure of integers if the functor $\cdot$ is interpreted as addition.

B. The verification condition of the annotated program $\hat{M}$ of Example 3.4 is as follows:

$$VC(\hat{M}) = (\alpha_1 \Rightarrow \alpha_2) \wedge (\alpha_7 \Rightarrow \alpha_8) \wedge (\alpha_{13} \Rightarrow \alpha_{14}) \wedge$$
$$\wedge (\alpha_5 \Rightarrow (y := y+2)\alpha_6) \wedge (\alpha_3 \Rightarrow (i := i+1)\alpha_4) \wedge$$
$$\wedge (\alpha_4 \Rightarrow (z := z-y)\alpha_5) \wedge (((\alpha_2 \vee \alpha_6) \wedge z-y > 0)$$
$$\Rightarrow \alpha_3) \wedge (((\alpha_2 \vee \alpha_6) \wedge z-y \leqslant 0) \Rightarrow \alpha_7) \wedge$$
$$\wedge (\alpha_9 \Rightarrow (y := 0)\alpha_{10}) \wedge (\alpha_{11} \Rightarrow (y := z)\alpha_{12}) \wedge$$
$$\wedge ((\alpha_8 \wedge z = y) \Rightarrow \alpha_9) \wedge ((\alpha_8 \wedge z \neq y) \Rightarrow \alpha_{11}) \wedge$$
$$\wedge ((\alpha_{10} \vee \alpha_{12}) \Rightarrow \alpha_{13}). \qquad \square$$

LEMMA 3.6. *Let $\hat{M}$ be an annotated version of a program M with the precondition α and the postcondition β, and let $\mathfrak{A}$ be an arbitrary data structure.*

If the verification condition $VC(\hat{M})$ is proper in the structure $\mathfrak{A}$, then the program M is partially correct with respect to the input formula α and the output formula β, i.e.

(1) $\mathfrak{A} \models VC(\hat{M})$ *implies* $\mathfrak{A} \models ((\alpha \wedge M \textbf{ true}) \Rightarrow M\beta).$

PROOF. We shall proceed by induction with respect to the length of the program M.

Implication (1) is obvious when M is an assignment instruction (cf. Definition 3.6).

Let us assume that (1) has been proved for the annotated program $\hat{M}_i$ with the precondition α_i and the postcondition β_i, where $i = 1, 2$.

Let us consider the annotated program $\hat{M}$ of the form $\{\alpha\}$ **if** γ **then** $\hat{M}_1$ **else** $\hat{M}_2$ **fi** $\{\beta\}$.

Suppose that for a data structure $\mathfrak{A}$,

(2) $\mathfrak{A} \models VC(\hat{M})$

and non $\mathfrak{A} \models ((\alpha \wedge M \textbf{ true} \Rightarrow M\beta)$. Hence there exists a valuation v in $\mathfrak{A}$ such that $\mathfrak{A}, v \models (\alpha \wedge M \textbf{ true})$ and $\mathfrak{A}, v \models \sim M\beta$. This means that there exists a finite computation of the program M from the valuation v with the result v' such that

(3) $\mathfrak{A}, v \models \alpha$

and

(4)   non $\mathfrak{A}, v' \models \beta.$

By the inductive assumption and (2)

(5) $\mathfrak{A} \models ((\alpha_1 \wedge M_1 \textbf{ true}) \Rightarrow M_1\beta_1),$

$$\mathfrak{A} \models ((\alpha_2 \wedge M_2 \text{ true}) \Rightarrow M_2 \beta_2),$$
(6) $$\mathfrak{A} \models ((\alpha \wedge \gamma) \Rightarrow \alpha_1) \wedge ((\alpha \wedge \sim \gamma) \Rightarrow \alpha_2) \wedge ((\beta_1 \vee \beta_2) \Rightarrow \beta).$$

By (3) and (6), $\mathfrak{A}, v \models (\alpha_1 \vee \alpha_2)$, and since $\mathfrak{A}, v \models M$ **true**, then

$$\mathfrak{A}, v \models M_1 \text{ true} \quad \text{and} \quad \mathfrak{A}, v \models \gamma \quad \text{and} \quad v' = M_{1\mathfrak{A}}(v)$$

or

$$\mathfrak{A}, v \models M_2 \text{ true} \quad \text{and} \quad \mathfrak{A}, v \models \sim \gamma \quad \text{and} \quad v' = M_{2\mathfrak{A}}(v).$$

Thus by (5)

$$\mathfrak{A}, v' \models \beta_1 \quad \text{or} \quad \mathfrak{A}, v' \models \beta_2.$$

As a consequence of (6), $\mathfrak{A}, v' \models \beta$ which contradicts assumption (4). Hence

$$\mathfrak{A} \models ((\alpha \wedge M \text{ true}) \Rightarrow \beta).$$

The remaining cases can be discussed analogously.   □

4. PROPERTIES OF THE SEMANTIC CONSEQUENCE OPERATION

DEFINITION 4.1. *We shall say that a data structure* $\mathfrak{A}$ *is a model for the set of formulas Z, for short* $\mathfrak{A} \models Z$, *iff for every formula* $\alpha \in Z$, α *is valid in the structure* $\mathfrak{A}$, $\mathfrak{A} \models \alpha$.

EXAMPLE 4.1. Let Z be a set which consists of all formulas of the form

(1) $(M \sim \alpha \Rightarrow \sim M\alpha),$

where M is a program and α is a formula.

Let $\mathfrak{A}$ be an arbitrary data structure and v a valuation in $\mathfrak{A}$.

Suppose $\mathfrak{A}, v \models M \sim \alpha$ and non $\mathfrak{A}, v \models \sim M\alpha$. Hence $\mathfrak{A}, v \models M \sim \alpha$ and $\mathfrak{A}, v \models M\alpha$. Then there exists a finite computation of the program M such that its result satisfies the formula α and the formula $\sim \alpha$, which is a contradiction. Hence for every valuation v

$$\mathfrak{A}, v \models M \sim \alpha \quad \text{implies} \quad \mathfrak{A}, v \models \sim M\alpha.$$

Thus, every data structure $\mathfrak{A}$ is a model for the set of formulas Z.

For our next example let us take as Z the set

(2) $\{\text{while } \gamma \text{ do } M \text{ od true}, \bigcap M\gamma\},$

where γ is an open formula, and M is a program. We shall prove that there is no model for the set Z.

Let $\mathfrak{A}$ be a data structure and let v be a fixed valuation. If $\mathfrak{A}, v \models \bigcap M\gamma$, then according to the definition of semantics (cf. § 2), every time we execute the program M, the obtained valuation $M_{\mathfrak{A}}^i(v)$, $i \in N$, satisfies the formula γ. Hence the program **while** γ **do** M **od** has an infinite computation in the data structure $\mathfrak{A}$ starting from the valuation v. Thus (**while** γ **do** M **od**)$_{\mathfrak{A}}(v)$ is not defined. In consequence, v does not satisfy the formula

$$\textbf{while } \gamma \textbf{ do } M \textbf{ od true}$$

and therefore Z has no model.

For our third example let AR be the set which consists of the three formulas

$$\sim \mathrm{succ}(x) = 0,$$
(3) $$(\mathrm{succ}(x) = \mathrm{succ}(y) \Rightarrow x = y),$$
$$(x := 0)\big(\textbf{while } x \neq y \textbf{ do } x := \mathrm{succ}(x) \textbf{ od } x = y\big)$$

where succ is a one-argument functor, 0 is a constant, $=$ is a binary predicate and x, y are individual variables.

Let $\mathfrak{N}$ be a data structure such that its universe is the set of natural numbers N and

$$\mathrm{succ}_{\mathfrak{N}}(n) = n+1 \quad \text{for } n \in N,$$
$$=_{\mathfrak{N}} \text{ is the identity relation in } N,$$
$$0_{\mathfrak{N}} = 0.$$

Obviously $\mathfrak{N}$ is a model of the set AR. The first formula states that 0 is not the successor of any natural number; the second formula ensures that successors of different natural numbers are different natural numbers, and the third formula states that every natural number is obtained from 0 by applying the successor operation a finite number of times. $\square$

DEFINITION 4.2. *We shall say that a formula α is a semantic consequence of the set of formulas Z, for short $Z \models \alpha$, iff α is valid in every model of Z. In other words, for every data structure $\mathfrak{A}$, $\mathfrak{A} \models Z$ implies $\mathfrak{A} \models \alpha$.* $\square$

EXAMPLE 4.2. Let us consider the set of formulas Z,

$$Z = \{(M'(M^i\alpha) \Rightarrow \beta)\}_{i \in N}.$$

We shall show that the formula δ,

$$\delta = (M' \cup M\alpha \Rightarrow \beta)$$

is a semantic consequence of Z.

Let $\mathfrak{A}$ be a model of Z and suppose that for some valuation v we have

$$\text{non } \mathfrak{A}, v \models (M' \cup M\alpha \Rightarrow \beta).$$

Hence

(4) $\mathfrak{A}, v \models M' \cup M\alpha$

and

(5) $\mathfrak{A}, v \models \sim\beta.$

By the definition of semantics (cf. § 2) and by (4)

$$\underset{i \in N}{\text{l.u.b.}}(M^i\alpha)_{\mathfrak{A}}(v') = 1 \quad \text{for } v' = M'_{\mathfrak{A}}(v),$$

i.e., there exists a natural number i_0 such that

$$\mathfrak{A}, v' \models M^{i_0}\alpha.$$

Thus by (5),

$$\mathfrak{A}, v \models \sim\big(M'(M^{i_0}\alpha) \Rightarrow \beta\big).$$

This contradicts the assumption that $\mathfrak{A}$ is a model of the set Z.   □

DEFINITION 4.3. *By the semantic consequence operation we shall understand an operation* Cn *which assigns to every set of formulas Z the set* $\text{Cn}(Z)$ *of all formulas α such that $Z \models \alpha$.*   □

The following lemma shows some of the properties of the semantic consequence operation.

LEMMA 4.1. *For arbitrary sets of formulas Z and Z' the following properties hold*:

(i) $Z \subset \text{Cn}(Z),$

(ii)   *if* $Z \subset Z'$, *then* $\text{Cn}(Z) \subset \text{Cn}(Z'),$

(iii) $\text{Cn}\big(\text{Cn}(Z)\big) = \text{Cn}(Z).$

PROOF.

(i) This property is an immediate consequence of Definition 4.3,

(ii) Suppose $\alpha \in \text{Cn}(Z)$ and $Z \subset Z'$. Then every model of Z is a model of $\{\alpha\}$ and every model of Z' is a model of Z. Hence $Z' \models \alpha$ and therefore $\alpha \in \text{Cn}(Z'),$

(iii) By the first two properties

$$\text{Cn}(Z) \subset \text{Cn}\big(\text{Cn}(Z)\big).$$

To prove the converse, let $\alpha \in \text{Cn}\big(\text{Cn}(Z)\big).$

Let $\mathfrak{A}$ be a model for the set Z. The structure $\mathfrak{A}$ is then a model of the set $\mathrm{Cn}(Z)$. Since $\alpha \in \mathrm{Cn}(\mathrm{Cn}(Z))$, the formula α is valid in $\mathfrak{A}$. Hence $Z \models \alpha$, i.e. $\alpha \in \mathrm{Cn}(Z)$. $\qquad\qquad\qquad\qquad\qquad\qquad$ □

Lemma 4.1 juxtaposes the properties of the semantic consequence operation which are analogous to those of the classical consequence operation. We now indicate some of the differences.

One of the basic results of classical logic is the Compactness Theorem, which states that if Z is a set of formulas such that each of its finite subsets has a model, then the set Z also has a model.

The following considerations show that this result fails for the semantic consequence operation defined here.

EXAMPLE 4.3. Let Z be the set of formulas

$$Z = \{(x := 0)((x := \mathrm{succ}(x))^i \, 0 \leqslant x\}_{i \in N}$$

and let α be the formula $(x := 0) \bigcap (x := \mathrm{succ}(x)) \, 0 \leqslant x$, where 0 is a constant, succ is one-argument functor and $0 \leqslant$ is a one-argument predicate.

We shall prove that $Z \models \alpha$, but that there is no finite subset $Z_0 \subset Z$ such that $Z_0 \models \alpha$.

Let $\mathfrak{A}$ be a model for the set Z, it then follows that for every valuation v and every natural number i,

$$\mathfrak{A}, v' \models (x := \mathrm{succ}(x))^i \, 0 \leqslant x, \quad \text{where} \quad v' = v_0^x.$$

Thus

$$\operatorname*{g.l.b.}_{i \in N}\big(((x := \mathrm{succ}(x))^i \, 0 \leqslant x\big)_{\mathfrak{A}}(v') = \mathbf{1}.$$

By the definition of semantics we have

$$\mathfrak{A}, v \models \alpha.$$

Hence the formula α is valid in $\mathfrak{A}$ and therefore $Z \models \alpha$.

Let Z_I be a finite subset of the set Z,

$$Z_I = \{(x := 0)((x := \mathrm{succ}(x))^i \, 0 \leqslant x\}_{i \in I},$$

where I is a finite subset of the set of natural numbers N.

We shall define a data structure $\mathfrak{A}$ such that

1° the universe of $\mathfrak{A}$ is the set of natural numbers,

2° $0_{\mathfrak{A}} = 0$, $\mathrm{succ}_{\mathfrak{A}}$ is the successor operation in N and $(0 \leqslant)_{\mathfrak{A}}(n) = \mathbf{1}$ iff $n \in I$.

For all natural numbers i, let v_i be a valuation such that

$$v_i(x) = i \quad \text{and} \quad v_i(z) = v(z) \quad \text{for all } z \neq x.$$

Thus for every $i \in N$,

$$((x := 0)\,((x := \mathrm{succ}(x)))^i\, 0 \leqslant x)_{\mathfrak{A}}(v)$$
$$= \big(((x := \mathrm{succ}(x))^i 0 \leqslant x)_{\mathfrak{A}}(v') = (0 \leqslant x)_{\mathfrak{A}}(v_i).$$

It follows from condition 2° that for every natural number i,

$$((x := 0)\,((x := \mathrm{succ}(x))^i\, 0 \leqslant x)_{\mathfrak{A}}(v) = \mathbf{1} \quad \text{iff} \quad i \in I.$$

Thus $\mathfrak{A}$ is a model of Z_I and the formula α is not valid in $\mathfrak{A}$.

Since every finite subset of Z can be characterized by a corresponding subset I of the set of natural numbers, there is no finite subset Z_0 such that

$$Z_0 \models \alpha. \qquad\qquad\qquad\qquad \square$$

THEOREM 4.1. *It is not the case that whenever each finite subset of a given set of formulas has a model then the set has a model, i.e. the semantic consequence operation has no compactness property.*

PROOF. To prove the theorem it is sufficient to consider the set $Z \cup \{\sim \alpha\}$ from the above example. $\qquad\qquad \square$

Another difference between semantic consequence operation defined here and the classical one is the upward Löwenheim–Skolem Theorem (cf. Rasiowa and Sikorski, 1968). This states that if a set of statements has an infinite model, then it has models of any infinite cardinality. The following theorem shows that the last sentence fails in the algorithmic case.

Let AR be the set of formulas denoted by (3).

THEOREM 4.2 (on categoricity). *The set AR has one enumerable model up to isomorphism.*

PROOF. Example 4.1 shows that AR has an enumerable model $\mathfrak{N}$ in the set of natural numbers.

Let $\mathfrak{A}$ be any model of the set AR. We shall prove that $\mathfrak{A}$ is isomorphic to $\mathfrak{N}$, i.e., there exists a one-to-one mapping h from the set of natural numbers N onto the universe A of the structure $\mathfrak{A}$ such that

$$h(0) = 0_{\mathfrak{A}},$$
$$h(n+1) = \mathrm{succ}_{\mathfrak{A}}(h(n)), \quad \text{for all } n \in N.$$

Observe that by the third formula of (3) for every element $a \in \mathfrak{A}$ there exists a natural number i such that

$$\text{succ}_\mathfrak{A}^i(0_\mathfrak{A}) = a.$$

Moreover, if $\text{succ}_\mathfrak{A}^i(0_\mathfrak{A}) = \text{succ}_\mathfrak{A}^j(0_\mathfrak{A})$, then $i = j$, by the second formula of (3). Hence for every $a \in A$ there exists exactly one natural number i such that

$$\text{succ}_\mathfrak{A}^i(0_\mathfrak{A}) = a.$$

Conversely, for every natural number n, there exists an element $a \in N$ such that

$$\text{succ}_\mathfrak{A}^n(0_\mathfrak{A}) = a,$$

since $\text{succ}_\mathfrak{A}$ is an operation in A.

Let us take as h the mapping

$$h(n) = \text{succ}_\mathfrak{A}^n(0_\mathfrak{A}) \quad \text{for all } n \in N.$$

It follows from the above that h is a one-to-one mapping from the set N onto the set A.

By the definition we have for every $n \in N$,

$$h(n+1) = \text{succ}_\mathfrak{A}^{n+1}(0_\mathfrak{A}) = \text{succ}_\mathfrak{A}(\text{succ}_\mathfrak{A}^n(0_\mathfrak{A}))$$
$$= \text{succ}_\mathfrak{A}(h(n)).$$

Hence h is an isomorphism between $\mathfrak{N}$ and $\mathfrak{A}$.   □

COROLLARY. *The set AR has an infinite model and does not have a model of cardinality greater than $\aleph_0$.*   □

5. AXIOMATIZATION

In this section we shall discuss the problem of the syntactic characterization of the semantic consequence operation. For this purpose we shall introduce *axioms* and *rules of inference* which allow us to deduce syntactically valid formulas from the valid assumptions. Our aim is to construct a system in which the syntactical process of deduction will be equivalent to the semantic process of validation of formulas.

Let us assume that α, β, δ are arbitrary formulas, γ is an open formula, s is an assignment instruction and M, M' are arbitrary programs. We admit the following schemata of axioms;

Ax1. $((\alpha \Rightarrow \beta) \Rightarrow ((\beta \Rightarrow \delta) \Rightarrow (\alpha \Rightarrow \delta)))$,

Ax2. $(\alpha \Rightarrow (\alpha \vee \beta))$,

Ax3. $(\beta \Rightarrow (\alpha \vee \beta))$,

Ax4. $((\alpha \Rightarrow \delta) \Rightarrow ((\beta \Rightarrow \delta) \Rightarrow ((\alpha \vee \beta) \Rightarrow \delta)))$,

Ax5. $((\alpha \wedge \beta) \Rightarrow \alpha)$,

Ax6. $((\alpha \wedge \beta) \Rightarrow \beta)$,

Ax7. $((\delta \Rightarrow \alpha) \Rightarrow ((\delta \Rightarrow \beta) \Rightarrow (\delta \Rightarrow (\alpha \wedge \beta))))$,

Ax8. $(\alpha \Rightarrow (\beta \Rightarrow \delta)) \equiv ((\alpha \wedge \beta) \Rightarrow \delta)$,

Ax9. $((\alpha \wedge \sim \alpha) \Rightarrow \beta)$,

Ax10. $((\alpha \Rightarrow (\alpha \wedge \sim \alpha)) \Rightarrow \sim \alpha)$,

Ax11. $(\alpha \vee \sim \alpha)$,

Ax12. $s\gamma \equiv \overline{s\gamma}$,

Ax13. $s \sim \alpha \equiv \sim s\alpha$,

Ax14. $M(\alpha \wedge \beta) \equiv (M\alpha \wedge M\beta)$,

Ax15. $M(\alpha \vee \beta) \equiv (M\alpha \vee M\beta)$,

Ax16. $\bigcup M\alpha \equiv (\alpha \vee \bigcup M(M\alpha))$,

Ax17. $\bigcap M\alpha \equiv (\alpha \wedge \bigcap M(M\alpha))$,

Ax18. $s\big((\exists x)\alpha(x)\big) = (\exists y)\big(s\big((x :- y)\alpha(x)\big)\big)$,　where y is an individual variable not occurring in s,

Ax19. $(((x := \tau)\alpha(x)) \Rightarrow (\exists x)\alpha(x))$,　where τ is a term,

Ax20. $(\forall x)\alpha(x) \equiv \sim (\exists x) \sim \alpha(x)$,

Ax21. **begin** M; M' **end** $\alpha \equiv M(M'\alpha)$,

Ax22. **if** γ **then** M **else** M' **fi** $\alpha \equiv ((\gamma \wedge M\alpha) \vee (\sim \gamma \wedge M'\alpha))$,

Ax23. **while** γ **do** M **od** $\alpha \equiv ((\sim \gamma \wedge \alpha) \vee (\gamma \wedge M(\textbf{while } \gamma \textbf{ do } M$ **od** $\alpha)))$.

We shall denote the set of all axioms by Ax.

The inference rules are as follows:

r1. $\dfrac{\alpha, \ (\alpha \Rightarrow \beta)}{\beta}$,　　r2. $\dfrac{(\alpha \Rightarrow \beta)}{(M\alpha \Rightarrow M\beta)}$,

r3. $\dfrac{(M((x := y)\alpha(x)) \Rightarrow \beta)}{(M(\exists x)\alpha(x) \Rightarrow \beta)}$　　where y is an individual variable, occurring neither in α nor in β,

r4. $\dfrac{\{(M'(M^i\alpha) \Rightarrow \beta)\}_{i \in N}}{(M'\bigcup M\alpha \Rightarrow \beta)}$,　　r5. $\dfrac{\{(\beta \Rightarrow M'(M^i\alpha))\}_{i \in N}}{(\beta \Rightarrow M'\bigcap M\alpha)}$,

r6. $\dfrac{\{(M'(\textbf{if } \gamma \textbf{ then } M \textbf{ fi})^i(\alpha \wedge \sim \gamma) \Rightarrow \beta)\}_{i \in N}}{(M'(\textbf{while } \gamma \textbf{ do } M \textbf{ od } \alpha) \Rightarrow \beta)}$.

In a rule of inference of the form Z/β, where Z is a set of formulas and β is a formula, Z is called the *set of premises* and β the *conclusion*.

Note that some of the rules of inference have infinitely many premises; we shall call them ω-*rules*.

The set of axioms and rules of inference determines the notion of formal proof. In the presence of ω-rules this differs from the classical definition of proof. Intuitively, by a *formal proof* we understand a tree with all paths of finite length such that the leaves of the tree are labelled by axioms, and other vertices of the tree labelled in accordance with the inference rules.

DEFINITION 5.1. *By a tree we shall mean a set D of finite sequences of natural numbers called vertices, such that the empty sequence $\emptyset$ is an element of D and if a sequence $c = (i_1, \ldots, i_n) \in D$, then for every $k \leqslant n$, the sequence $c^k = (i_1, \ldots, i_k)$ is an element of D.*

The empty sequence $\emptyset$ is called the root of the tree D.

If $c = (i_1, \ldots, i_n) \in D$ and $c' = (i_1, \ldots, i_n, j) \in D$ for some $j \in N$, then the number n is called the level of a vertex c and the vertex c' is called a son of the vertex c (c' is the j-th son of c, to be exact).

By a path in the tree D we shall understand a finite or infinite sequence of vertices $c_1, c_2, \ldots, c_k, \ldots$ such that for every k, c_{k+1} is a son of c_k. The last element of a finite path is called a leaf of the tree.   □

DEFINITION 5.2. *By a proof of a formula from the set of formulas Z we shall understand the ordered pair $\langle D, d \rangle$ where D is a tree with all paths finite and where d is a mapping which assigns a formula $d(c)$ to every vertex c of D such that*

1° *for every leaf c of the tree D, $d(c) \in Z$ or $d(c) \in Ax$;*

2° *for every vertex $c = (i_1, \ldots, i_n)$, which is not a leaf, $d(c)$ is a conclusion in a rule of inference from all formulas $d(i_1, \ldots, i_n, j)$ such that $(i_1, \ldots, i_n, j)$ is a vertex in D;*

3° $d(\emptyset) = \alpha$.   □

DEFINITION 5.3. *We shall say that a formula α is a syntactic consequence of a set of formulas Z, $Z \vdash \alpha$ for short, iff there exists a proof of the formula α from the set Z.*   □

EXAMPLE 5.1. Let $Z = \{\alpha, M \text{ true}\}$, where α is an arbitrary fixed formula and M is a program. Figure 5.1 is a proof of the formula $M\alpha$ from the set Z.

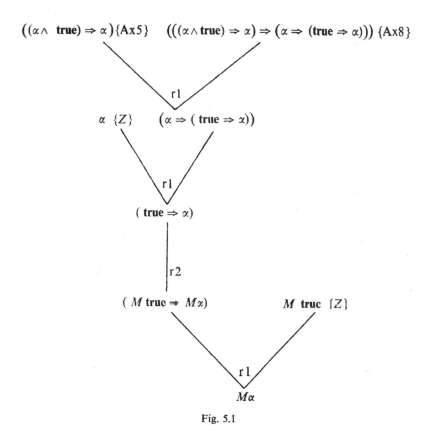

Fig. 5.1

Observe that in fact the relation ⊢ determines an operation in the set of all formulas, which to any set of formulas assigns the set of all its syntactic consequences. □

DEFINITION 5.4. *By the syntactic consequence operation we shall understand a mapping C which to every set of formulas Z assigns the least set of formulas C(Z) such that:*

(i) $Ax \cup Z \subset C(Z)$.

(ii)   *C(Z) is closed with respect to the rules of inference* r1–r6. □

REMARK. For every fromula α and every set of formulas Z, $Z \vdash \alpha$ iff $\alpha \in C(Z)$. □

As a simple corollary we can prove that the syntactic consequence operation has properties similar to the semantic consequence operation. We shall mention these below.

LEMMA 5.1. *For arbitrary sets of formulas Z and Z_1:*
(i) $Z \subset C(Z)$,
(ii)   *if* $Z \subset Z_1$,   *then* $C(Z) \subset C(Z_1)$,
(iii) $C(C(Z)) = C(Z)$.

The easy proof is left to the reader.   □

Let L be an algorithmic language and C the consequence operation defined above.

The pair $\langle L, C \rangle$ will be called the *deductive system of algorithmic logic* or *algorithmic logic* for short.

If a formula α has a proof from the empty set of formulas, i.e. $\vdash \alpha$ then we shall say that α is a *theorem of algorithmic logic.*

Let A be a set of formulas. By a *formalized algorithmic theory* we shall understand the system $\langle L, C, A \rangle$. The set A will be called the *set of non-logical axioms* or *specific axioms* of the theory.

If a formula α has a proof from the set A, then α is a *theorem of the algorithmic theory* $\langle L, C, A \rangle$.

EXAMPLE 5.2. As an example of a theorem of the algorithmic logic we shall consider the formula

(1)   **begin if** γ **then** M' **else** M'' **fi**; M **end** α
           $\equiv$ **if** γ **then** M'; M **else** M''; M **fi** α,

where $\gamma \in F_0$, M', $M'' \in \Pi$, $\alpha \in F$.

Before we present, in Figure 5.3, the formal proof of formula (1) let us mention two auxiliary facts:

Fact 1. For all formulas β, β_1, δ, if $\vdash (\beta \Rightarrow \beta_1)$ then $\vdash ((\delta \wedge \beta) \Rightarrow (\delta \wedge \beta_1))$.

Fact 2. For all formulas β, β_1, δ, if $\vdash (\beta \Rightarrow \beta_1)$ then $\vdash ((\delta \vee \beta) \Rightarrow (\delta \vee \beta_1))$.

The formal proofs of both facts makes use of classical axioms only, thus we shall present the proof of one of them as an example (Figure 5.2).

Let us introduce the notations used in Figure 5.3

$$M'(M\alpha) = \beta', \qquad\qquad M''(M\alpha) = \beta'',$$
$$\textbf{begin } M'; M \textbf{ end } \alpha = \beta'_1, \qquad \textbf{begin } M''; M \textbf{ end } \alpha = \beta''_1.$$

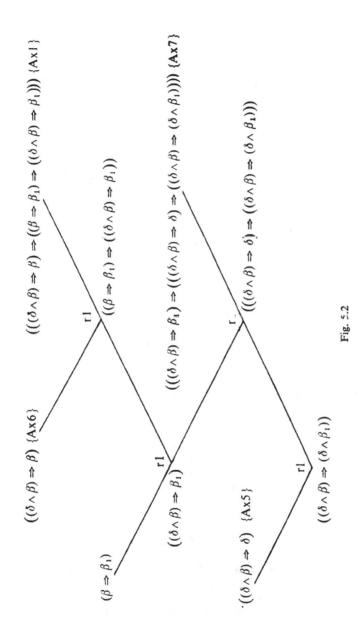

Fig. 5.2

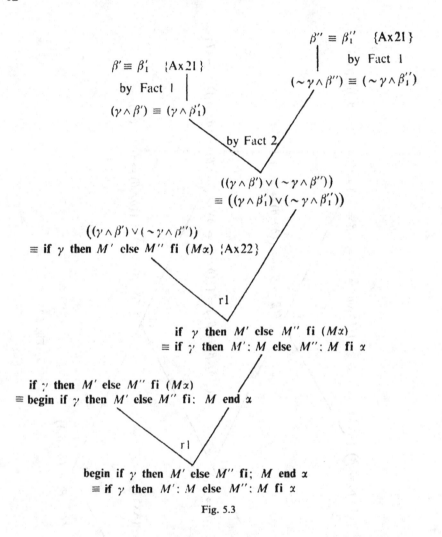

Fig. 5.3

EXAMPLE 5.3. The following formula is a theorem of algorithmic logic:

(2)   **while** γ **do** M **od** $\alpha \equiv \bigcup$ **if** γ **then** M **fi** $(\sim\gamma\wedge\alpha)$

where γ is an open formula, M is a program and α is a formula.

First we prove that for every natural number i and for every formula α,

(3) $\vdash (M^i\alpha \Rightarrow \bigcup M\alpha)$

The proof is by induction with respect to the number of iterations i.
For $i = 0$ we have

(f1) $\quad \vdash \left(\left(\alpha \vee \bigcup M(M\alpha)\right) \Rightarrow \bigcup M\alpha\right),$ $\hspace{2cm}$ {Ax16}

(f2) $\quad \vdash \left(\alpha \Rightarrow \left(\alpha \vee \bigcup M(M\alpha)\right)\right),$ $\hspace{2.5cm}$ {Ax2}

(f3) $\quad \vdash \left(\left(\alpha \Rightarrow \left(\alpha \vee \bigcup M(M\alpha)\right)\right) \Rightarrow \left(\left(\left(\alpha \vee \bigcup M(M\alpha)\right) \Rightarrow \bigcup M\alpha\right)\right.\right.$
$\hspace{5cm} \left.\left. \Rightarrow \left(\alpha \Rightarrow \bigcup M\alpha\right)\right)\right),$ $\hspace{1cm}$ {Ax1}

(f4) $\quad \vdash \left(\left(\left(\alpha \vee \bigcup M(M\alpha)\right) \Rightarrow \bigcup M\alpha\right) \Rightarrow \left(\alpha \Rightarrow \bigcup M\alpha\right)\right),$ {r1, f3, f2}

(f5) $\quad \vdash \left(\alpha \Rightarrow \bigcup M\alpha\right).$ $\hspace{3.5cm}$ {r1, f1, f4}

Assume that for a fixed natural number i and arbitrary formula α,

(f6) $\quad \vdash \left(M^i\alpha \Rightarrow \bigcup M\alpha\right).$

Below we shall prove that $\left(M^{i+1}\alpha \Rightarrow \bigcup M\alpha\right)$ is a theorem of algorithmic
logic.

(f7) $\quad \vdash \left(M^{i+1}\alpha \Rightarrow M^i(M\alpha)\right),$ $\hspace{2.5cm}$ {Ax21}

(f8) $\quad \vdash \left(M^{i+1}\alpha \Rightarrow \bigcup M(M\alpha)\right),$ $\hspace{2cm}$ {r1, f6, f7}

(f9) $\quad \vdash \left(\bigcup M(M\alpha) \Rightarrow \left(\alpha \vee \bigcup M(M\alpha)\right)\right),$ $\hspace{1cm}$ {Ax2}

(f10) $\quad \vdash \left(M^{i+1}\alpha \Rightarrow \left(\alpha \vee \bigcup M(M\alpha)\right)\right),$ $\hspace{1cm}$ {Ax1, f8, f9, r1}

(f11) $\quad \vdash \left(\left(\alpha \vee \bigcup M(M\alpha)\right) \Rightarrow \bigcup M\alpha\right),$ $\hspace{1.5cm}$ {Ax16}

(f12) $\quad \vdash \left(M^{i+1}\alpha \Rightarrow \bigcup M\alpha\right).$ $\hspace{2cm}$ {f10, f11, Ax1, r1}.

Hence by the principle of induction for every i,

$\qquad \vdash \left(M^i\alpha \Rightarrow \bigcup M\alpha\right).$

In particular,

$\qquad \vdash \left(\left(\text{if } \gamma \text{ then } M \text{ fi}\right)^i(\sim\gamma \wedge \alpha) \Rightarrow \bigcup \text{if } \gamma \text{ then } M \text{ fi } (\sim\gamma \wedge \alpha)\right).$

Hence by the ω-rule r6

(f13) $\quad \vdash \left(\text{while } \gamma \text{ do } M \text{ od } \alpha \Rightarrow \bigcup \text{if } \gamma \text{ then } M \text{ fi } (\sim\gamma \wedge \alpha)\right).$

We shall prove

(4) $\quad \vdash \left(\left(\text{if } \gamma \text{ then } M \text{ fi}\right)^i(\sim\gamma \wedge \alpha) \Rightarrow \text{while } \gamma \text{ do } M \text{ od } \alpha\right)$

analogously by induction with respect to the number of iterations i.
For $i = 0$ we have

(f14) $\quad \vdash \left(\left(\sim\gamma \wedge \alpha\right) \Rightarrow \left(\left(\sim\gamma \wedge \alpha\right) \vee \left(\gamma \wedge M \text{ (while } \gamma \text{ do } M \text{ od } \alpha)\right)\right)\right)$
$\hspace{9cm}$ {Ax2}

$\qquad \vdash \left(\left(\sim\gamma \wedge \alpha\right) \Rightarrow \text{while } \gamma \text{ do } M \text{ od } \alpha\right)$ {Ax1, r1, Ax23, f14}

Assume that for a fixed natural number i,

(f15) $\quad \vdash \left(\left(\text{if } \gamma \text{ then } M \text{ fi}\right)^i(\sim\gamma \wedge \alpha) \Rightarrow \text{while } \gamma \text{ do } M \text{ od } \alpha\right).$

We shall prove (4) for the natural number $(i+1)$.

(f16) $\quad \vdash \left(\left(\text{if } \gamma \text{ then } M \text{ fi}\right)^{i+1}(\sim\gamma \wedge \alpha)\right.$
$\hspace{1.5cm} \Rightarrow \left(\left(\gamma \wedge M \text{ (while } \gamma \text{ do } M \text{ od } \alpha)\right) \vee\right.$
$\hspace{1.5cm} \left.\left. \vee \left(\sim\gamma \wedge \left(\text{if } \gamma \text{ then } M \text{ fi}\right)^i(\sim\gamma \wedge \alpha)\right)\right)\right)$ $\hspace{1cm}$ {f15, Ax22}

(f17) $\vdash ((\textbf{if } \gamma \textbf{ then } M \textbf{ fi})^{i+1}(\sim\gamma \wedge \alpha)$
$\Rightarrow (\gamma \wedge M(\textbf{while } \gamma \textbf{ do } M \textbf{ od } \alpha))\vee(\sim\gamma \wedge \alpha)))$,   {f16}
$\vdash ((\textbf{if } \gamma \textbf{ then } M \textbf{ fi})^{i+1}(\sim\gamma \wedge \alpha)$
$\Rightarrow \textbf{while } \gamma \textbf{ do } M \textbf{ od } \alpha)$.   {Ax23, f17}

Thus by the principle of induction (4) is proved.

By the ω-rule r4 we have

(f18) $\vdash (\bigcup \textbf{if } \gamma \textbf{ then } M \textbf{ fi } (\sim\gamma \wedge \alpha) \Rightarrow \textbf{while } \gamma \textbf{ do } M \textbf{ od } \alpha)$.

By (f13) and (f18)

$\vdash (\textbf{while } \gamma \textbf{ do } M \textbf{ od } \alpha \equiv \bigcup \textbf{if } \gamma \textbf{ then } M \textbf{ fi } (\sim\gamma \wedge \alpha))$. $\square$

EXAMPLE 5.4. For every program M, the formula

(5) $\sim M \textbf{ false}$

is a theorem of algorithmic logic.

First of all, we shall prove by induction with respect to the length of the program M, that

(f1) $\vdash (M \textbf{ false} \Rightarrow \textbf{false})$.

If M is an assignment instruction then (f1) follows immediately from Ax12. As an inductive assumption let us suppose that

(f2) $\vdash (M' \textbf{ false} \Rightarrow \textbf{false})$,

and

(f3) $\vdash (M'' \textbf{ false} \Rightarrow \textbf{false})$

for all programs M', M'' shorter than M.

Let us consider the program M of the form $\textbf{begin } M'; M'' \textbf{ end}$.

(f4) $\vdash (\textbf{begin } M'; M'' \textbf{ end false} \Rightarrow M'(M'' \textbf{ false}))$,   {Ax21}
(f5) $\vdash (M'(M'' \textbf{ false}) \Rightarrow M' \textbf{ false})$,   {f3, r2}
(f6) $\vdash (M'(M'' \textbf{ false}) \Rightarrow \textbf{false})$,   {f2, f5, Ax1, r1}
$\vdash (\textbf{begin } M'; M'' \textbf{ end false} \Rightarrow \textbf{false})$.   {f4, f6, Ax1, r1}

Let M be of the form $\textbf{if } \gamma \textbf{ then } M' \textbf{ else } M'' \textbf{ fi}$.

(f7) $\vdash (\textbf{if } \gamma \textbf{ then } M' \textbf{ else } M'' \textbf{ fi false} \Rightarrow ((\gamma \wedge M' \textbf{ false}) \vee$
$\vee (\sim\gamma \wedge M'' \textbf{ false})))$,   {Ax22}
(f8) $\vdash ((\gamma \wedge M' \textbf{ false}) \Rightarrow (\gamma \wedge \textbf{false}))$,   {f2, Ax1–Ax11}
(f9) $\vdash ((\sim\gamma \wedge M'' \textbf{ false}) \Rightarrow (\sim\gamma \Rightarrow \textbf{false}))$,   {f3, Ax1–Ax11}
(f10) $\vdash ((\sim\gamma \wedge \textbf{false}) \Rightarrow \textbf{false})$,   {Ax5}
$\vdash ((\textbf{if } \gamma \textbf{ then } M' \textbf{ else } M'' \textbf{ fi false}) \Rightarrow \textbf{false})$. {f7, f8, f9, f10}

Let M be of the form $\textbf{while } \gamma \textbf{ do } M' \textbf{ od}$. By the above proof we have

$\vdash ((\textbf{if } \gamma \textbf{ then } M' \textbf{ fi})^{i} \textbf{ false} \Rightarrow \textbf{false})$ for every $i \in N$.

Hence by the ω-rule r6 and (f10)

$$\vdash (\textbf{while } \gamma \textbf{ do } M' \textbf{ od false} \Rightarrow \textbf{false}).$$

Hence we shall prove the formula (f1) for every program M. Formula (5) follows from (f1) by Ax10 and rl.   □

6. MODELS AND CONSISTENCY

In this section we shall prove that the syntactic consequence operation is equivalent to the semantic consequence operation defined in § 4 of Chapter II. More strictly, we shall prove that for every set of formulas Z, $C(Z) \subset \text{Cn}(Z)$.

The procedure consists of two steps. Firstly, it will be proved that all axioms are tautologies; and secondly, that every rule of inference leads from valid premises to a valid conclusion. Both facts assure us that the set of all valid formulas in any data structure is closed with respect to the syntactic consequence operation.

As a corollary, we observe that an algorithmic theory which possesses a model is consistent.

LEMMA 6.1. *All axioms of the algorithmic logic* AL *are tautologies.*

PROOF. We shall not verify the axioms of classical propositional calculus Ax1–Ax11 or the axioms of classical predicate calculus Ax19 and Ax20.

The formulas Ax12, Ax14, Ax15, Ax21 and Ax22 are tautologies by Lemma 2.1, Example 2.4 and Lemma 3.4.

Let $\mathfrak{A}$ be an arbitrary data structure for the algorithmic language L and let v be an arbitrary valuation in $\mathfrak{A}$.

Ax13. Consider the formula

$$(s \sim \alpha \equiv \sim s\alpha)$$

where s is a substitution. By the definition of semantics we have,

$$\mathfrak{A}, v \models s \sim \alpha \quad \text{iff} \quad \mathfrak{A}, s_{\mathfrak{A}}(v) \models \sim \alpha \quad \text{iff}$$

$$\text{non } \mathfrak{A}, s_{\mathfrak{A}}(v) \models \alpha \quad \text{iff}$$

$$\text{non } \mathfrak{A}, v \models s\alpha \quad \text{iff} \quad \mathfrak{A}, v \models \sim s\alpha.$$

Ax17. Consider the formula $(\bigcap M\alpha \equiv (\alpha \wedge \bigcap M(M\alpha)))$. By the definition of semantics we have:

$$\mathfrak{A}, v \models \bigcap M\alpha \quad \text{iff} \quad \underset{i \in N}{\text{g.l.b.}}(M^i\alpha)_{\mathfrak{A}}(v) = 1 \quad \text{iff}$$

for every $i \in N$, $(M^i \alpha)_{\mathfrak{A}}(v) = 1$ iff

$\alpha_{\mathfrak{A}}(v) = 1$ and $\underset{i \in N}{\text{g.l.b.}} \ (M^i(M\alpha))_{\mathfrak{A}}(v) = 1$ iff

$\mathfrak{A}, v \models \alpha$ and $\mathfrak{A}, v \models \bigcap M(M\alpha)$ iff

$\mathfrak{A}, v \models (\alpha \wedge \bigcap M(M\alpha))$.

Ax18. Sentences (1)–(7) below are equivalent.

(1) $\mathfrak{A}, v \models s(\exists x)\alpha(x)$.

(2) $\mathfrak{A}, s_{\mathfrak{A}}(v) \models (\exists x)\alpha(x)$.

(3)   There exists $a \in \mathfrak{A}$ such that $\mathfrak{A}, \bar{v}_a^x \models \alpha(x)$, where $\bar{v} = s_{\mathfrak{A}}(v)$.

(4)   There exists $a \in \mathfrak{A}$ such that $\mathfrak{A}, \bar{v}_a^y \models (x := y)\alpha(x)$ where $\bar{v} = s_{\mathfrak{A}}(v)$ and y does not occur in s or in α.

(5)   There exists $a \in \mathfrak{A}$ such that
$\mathfrak{A}, s_{\mathfrak{A}}(v_a^y) \models (x := y)\alpha(x)$,   where $y \notin V(s\alpha)$.

(6)   There exists $a \in \mathfrak{A}$,
$\mathfrak{A}, v_a^y \models s((x := y)\alpha(x))$,   where $y \notin V(s\alpha)$.

(7) $\mathfrak{A}, v \models (\exists y)s((x := y)\alpha(x))$.

Hence,

$$\mathfrak{A}, v \models (s(\exists x)\alpha(x) \equiv (\exists y)(s(x := y)\alpha))$$

if y does not occur in s or in α.

Ax23. Sentences (8)–(12) below are equivalent.

(8) $\mathfrak{A}, v \models \text{while } \gamma \text{ do } M \text{ od } \alpha$.

(9)   There exists $i \in N$ such that $\mathfrak{A}, v \models M^j\gamma$ for $j < i$, $M_{\mathfrak{A}}^i(v)$ is defined and $\mathfrak{A}, v \models M^i(\sim\gamma \wedge \alpha)$.

(10)   Either $i = 0$ and $\mathfrak{A}, v \models (\sim\gamma \wedge \alpha)$ or $i \neq 0$ and $\mathfrak{A}, v \models M(M^j\gamma)$
for $j < i-1$ and $\mathfrak{A}, v \models M(M^{i-1}(\sim\gamma \wedge \alpha))$ for $i \in N$.

(11) $\mathfrak{A}, v \models (\sim\gamma \wedge \alpha)$ or $\mathfrak{A}, v \models \gamma$ and $\mathfrak{A}, v' \models \text{while } \gamma \text{ do } M \text{ od } \alpha$,
where $M_{\mathfrak{A}}(v)$ is defined and $v' = M_{\mathfrak{A}}(v)$.

(12) $\mathfrak{A}, v \models ((\sim\gamma \wedge \alpha) \vee (\gamma \wedge M (\text{while } \gamma \text{ do } M \text{ od } \alpha)))$.
Hence

$\mathfrak{A}, v \models \text{while } \gamma \text{ do } M \text{ od } \alpha$
$\equiv ((\sim\gamma \wedge \alpha) \vee (\gamma \wedge M (\text{while } \gamma \text{ do } M \text{ od } \alpha)))$.   □

LEMMA 6.2. *For every inference rule of AL, if the premises of the rule are valid in a data structure* $\mathfrak{A}$, *then the conclusion of the rule is also valid in* $\mathfrak{A}$.

PROOF. We shall consider only three inference rules in order to show the method of the proof. The rule r2 is proved in Lemma 3.4.

Let $\mathfrak{A}$ be an arbitrary data structure.

r3. $\dfrac{(M((x := y)\alpha) \Rightarrow \beta)}{(M(\exists x)\alpha(x) \Rightarrow \beta)}$ where y does not occur in M, α or β.

Suppose that

(13) $\mathfrak{A} \models (M(x := y)\alpha \Rightarrow \beta)$ and $y \notin V(M) \cup V(\alpha) \cup V(\beta)$

and for a fixed valuation v in $\mathfrak{A}$

(14) $\mathfrak{A}, v \models M((\exists x)\alpha(x))$

and

(15)   non $\mathfrak{A}, v \models \beta$.

Hence by (14), $M_{\mathfrak{A}}(v)$ is defined and for $\bar{v} = M_{\mathfrak{A}}(v)$, $\mathfrak{A}, \bar{v} \models (\exists x)\alpha(x)$. By the definition of semantics, there exists an element $a \in \mathfrak{A}$ such that

$$\mathfrak{A}, \bar{v}_a^x \models \alpha(x).$$

Since $y \notin V(\alpha)$, then $\mathfrak{A}, \bar{v}_a^y \models (x := y)\alpha(x)$. Since $y \notin V(\beta)$, then, by (15), non $\mathfrak{A}, v_a^y \models \beta$. Since $y \notin V(M)$ then, $\bar{v}_a^y = M_{\mathfrak{A}}(v_a^y)$. Thus $\mathfrak{A}, v_a^y \models M((x := y)\alpha(x))$ and non $\mathfrak{A}, v_a^y \models \beta$.

As a consequence non $\mathfrak{A}, v_a^y \models M((x := y)\alpha \Rightarrow \beta)$, which contradicts (13).

r5. $\dfrac{\{(\beta \Rightarrow M'(M^i\alpha))\}_{i \in N}}{(\beta \Rightarrow (M' \bigcap M\alpha))}$.

Assume

(16) $\mathfrak{A} \models (\beta \Rightarrow M'(M^i\alpha))$ for every natural number i.

Let v be an arbitrary valuation such that

(17) $\mathfrak{A}, v \models \beta$.

By (16) we have $\mathfrak{A}, v \models M'(M^i\alpha)$ for every $i \in N$. Thus, $M'_{\mathfrak{A}}(v)$ is defined and for $v' = M_{\mathfrak{A}}(v)$ and all $i \in N$

$$\mathfrak{A}, v' \models M^i\alpha.$$

By the definition of semantics, it follows that

$$\mathfrak{A}, v' \models \bigcap M\alpha, \quad \text{i.e.} \quad \mathfrak{A}, v \models M' \bigcap M\alpha.$$

Hence by (17), $\mathfrak{A}, v \models (\beta \Rightarrow M' \bigcap M\alpha)$.

r6. $\dfrac{\{(M'(\textbf{if } \gamma \textbf{ then } M \textbf{ fi})^i(\sim\gamma \wedge \alpha) \Rightarrow \beta)\}_{i \in N}}{(M'(\textbf{while } \gamma \textbf{ do } M \textbf{ od } \alpha) \Rightarrow \beta)}$.

Assume

(18) $\mathfrak{A} \models (M'((\textbf{if } \gamma \textbf{ then } M \textbf{ fi})^i(\sim\gamma \wedge \alpha)) \Rightarrow \beta)$ for every i

and suppose that for a valuation v,

$$\mathfrak{A}, v \models \sim (M'(\text{while } \gamma \text{ do } M \text{ od } \alpha) \Rightarrow \beta).$$

Hence

(19) $\mathfrak{A}, v \models M'(\text{while } \gamma \text{ do } M \text{ od } \alpha)$

and

(20) $\mathfrak{A}, v \models \sim \beta.$

By (19) and the definition of semantics

$$M'_{\mathfrak{A}}(v) \text{ is defined and for } v' = M'_{\mathfrak{A}}(v)$$

there exists a natural number i_0 such that

$$\mathfrak{A}, v' \models M^j \gamma \quad \text{for } j < i_0 \quad \text{and} \quad \mathfrak{A}, v' \models M^{i_0}(\sim \gamma \wedge \alpha).$$

Thus

$$\mathfrak{A}, v' \models (\text{if } \gamma \text{ then } M \text{ fi})^{i_0}(\sim \gamma \wedge \alpha).$$

Since $v' = M'_{\mathfrak{A}}(v)$, then by (20)

$$\mathfrak{A}, v \models \sim (M'((\text{if } \gamma \text{ then } M \text{ fi})^{i_0}(\sim \gamma \wedge \alpha)) \Rightarrow \beta)$$
—a   contradiction of (18).

Hence

$$\mathfrak{A} \models (M'(\text{while } \gamma \text{ do } M \text{ od } \alpha) \Rightarrow \beta).$$

The fact proved above allows us to say that the inference rules r1–r6 are sound.   □

COROLLARY 6.1. *For every inference rule of* AL *if the premises of the rule are tautologies, then the conclusion of the rule is a tautology.*   □

THEOREM 6.1. *For every formula α and every set of formulas Z, if $Z \vdash \alpha$, then $Z \models \alpha$.*

In other words, the set of syntactic consequences of a set Z is contained in the set of all semantic consequences of the set Z.

PROOF. Let Z be a set of formulas. Assume that $Z \vdash \alpha$. Hence there exists a formal proof $\langle D, d \rangle$, of the formula α from the set Z. We shall proceed by induction on the level of the tree D to show that for every $c \in D$, $Z \models d(c)$.

If $\mathfrak{A}$ is a model of Z, then for every leaf c in D, $d(c)$ is valid in $\mathfrak{A}$.

Consider an internal node c of the tree D and assume that the induction assumption holds for all sons of c, i.e. $Z \models d(c_i)$ for every son c_i of c.

The formula $d(c)$ is a conclusion of an inference rule for the premises $d(c_i)$. By Lemma 6.2 we infer that $Z \models d(c)$. Hence $Z \models \alpha$. $\qquad \square$

DEFINITION 6.1. *Let* T *be an algorithmic theory,* $T = \langle L, C, A \rangle$. *By a model of* T *we shall understand any data structure* $\mathfrak{A}$ *for the language* L *such that* $\mathfrak{A} \models A$. $\qquad \square$

As an immediate consequence of Theorem 6.1 we have the following corollaries:

COROLLARY 6.2. *For every formula* α:
(i) *if* α *is a theorem of the theory* T, *then* α *is valid in every model of* T,
(ii) *if the formula* α *is a theorem of* AL, *then* α *is a tautology.* $\qquad \square$

DEFINITION 6.2. *An algorithmic theory* $T = \langle L, C, A \rangle$ *is consistent iff there exists a formula which is not a theorem of* T. $\qquad \square$

COROLLARY 6.3.
(i) *The algorithmic logic* AL *is consistent.*
(ii) *If a theory* T *has a model, then it is consistent.*

PROOF. It is sufficient to prove property (ii).
Let $\mathfrak{A}$ be a model of a theory $T = \langle L, C, A \rangle$ and let every formula α be a theorem of T. By Corollary 6.1, for an arbitrary valuation v we have

$$\mathfrak{A}, v \models \alpha \quad \text{and} \quad \mathfrak{A}, v \models \sim \alpha,$$

which is a contradiction. $\qquad \square$

7. USEFUL TAUTOLOGIES AND INFERENCE RULES

This section presents the tautologies and inference rules which we consider useful in proving properties of programs.

The proofs in this section are not formal. We have omitted many steps related to classical propositional calculus in order to underline axioms and inference rules specific to algorithmic logic.

In all the formulas below α, β are arbitrary formulas, M, M' are arbitrary programs, γ, γ' are open formulas and Z is a set of formulas.

(1) $\qquad \vdash M \sim \alpha \Rightarrow \sim M\alpha$.

PROOF.

$$\vdash \sim M \text{ false}, \hspace{4cm} \{\text{Example } 5.4\}$$
$$\vdash \sim M(\sim \alpha \wedge \alpha),$$
$$\vdash (\sim M \sim \alpha \vee \sim M\alpha), \hspace{3cm} \{\text{Ax14}\}$$
$$\vdash (M \sim \alpha \Rightarrow \sim M\alpha). \hspace{4cm} \square$$

(2) $\vdash (M \text{ true} \Rightarrow (\sim M\alpha \Rightarrow M \sim \alpha)).$

PROOF.

$$\vdash (\sim M \text{ true} \vee M(\alpha \vee \sim \alpha)), \hspace{2.5cm} \{\text{Ax11}\}$$
$$\vdash (\sim M \text{ true} \vee (M\alpha \vee M \sim \alpha)), \hspace{2cm} \{\text{Ax15}\}$$
$$\vdash ((M \text{ true} \wedge \sim M\alpha) \Rightarrow M \sim \alpha),$$
$$\vdash (M \text{ true} \Rightarrow (\sim M\alpha \Rightarrow M \sim \alpha)). \hspace{1.5cm} \{\text{Ax8}\} \ \square$$

(2') $\vdash (M \text{ true} \Rightarrow (M \sim \alpha \equiv \sim M\alpha)). \hspace{1.5cm} \{(1), (2)\}$

(3) $\vdash (M(\alpha \Rightarrow \beta) \Rightarrow (M\alpha \Rightarrow M\beta)).$

PROOF.

$$\vdash ((\alpha \Rightarrow \beta) \Rightarrow (\sim \alpha \vee \beta)),$$
$$\vdash (M(\alpha \Rightarrow \beta) \Rightarrow M(\sim \alpha \vee \beta)), \hspace{2.5cm} \{\text{r2}\}$$
$$\vdash (M(\alpha \Rightarrow \beta) \Rightarrow (M \sim \alpha \vee M\beta)). \hspace{2cm} \{\text{Ax15}\}$$
$$\vdash (M(\alpha \Rightarrow \beta) \Rightarrow (\sim M\alpha \vee M\beta)), \hspace{2.5cm} \{(1)\}$$
$$\vdash (M(\alpha \Rightarrow \beta) \Rightarrow (M\alpha \Rightarrow M\beta)). \hspace{3cm} \square$$

(4) $\vdash (M \text{ true} \Rightarrow ((M\alpha \Rightarrow M\beta) \Rightarrow M(\alpha \Rightarrow \beta))).$

PROOF.

$$\vdash (M \text{ true} \Rightarrow (\sim M\alpha \Rightarrow M \sim \alpha)), \hspace{2.5cm} \{(2)\}$$
$$\vdash ((M \text{ true} \wedge \sim M\alpha) \Rightarrow M \sim \alpha), \hspace{2.5cm} \{\text{Ax8}\}$$
$$\vdash (((M \text{ true} \wedge \sim M\alpha) \vee M\beta) \Rightarrow (M \sim \alpha \vee M\beta)),$$
$$\vdash ((M \text{ true} \vee M\beta) \Rightarrow ((\sim M\alpha \vee M\beta) \Rightarrow M(\sim \alpha \vee \beta))),$$
$$\hspace{6cm} \{\text{Ax8, Ax15}\}$$
$$\vdash ((M \text{ true} \vee M\beta) \Rightarrow ((M\alpha \Rightarrow M\beta) \Rightarrow M(\alpha \Rightarrow \beta))),$$
$$\vdash (M \text{ true} \Rightarrow ((M\alpha \Rightarrow M\beta) \Rightarrow M(\alpha \Rightarrow \beta))). \hspace{1cm} \square$$

(4') $\vdash (M \text{ true} \Rightarrow ((M\alpha \Rightarrow M\beta) \equiv M(\alpha \Rightarrow \beta))). \hspace{0.5cm} \{(3), (4)\}.$

(5)   For every natural number i,
$$\vdash (M^i\alpha \Rightarrow \bigcup M\alpha).$$

(6)   For every natural number i,
$$\vdash (\bigcap M\alpha \Rightarrow M^i\alpha).$$

(7)   For every natural number i,

$\vdash ((\textbf{if } \gamma \textbf{ then } M \textbf{ fi})^i (\sim \gamma \wedge \alpha) \Rightarrow \textbf{while } \gamma \textbf{ do } M \textbf{ od } \alpha)$.

For the proofs of (5), (6), (7) see Example 5.3.

(8) $\vdash ((\alpha \wedge \bigcap M(\alpha \Rightarrow M\alpha)) \Rightarrow \bigcap M\alpha)$.

PROOF.

$\vdash (\bigcap M(\alpha \Rightarrow M\alpha) \Rightarrow M^i(\alpha \Rightarrow M\alpha))$ for every $i \in N$, $\{(5)\}$

(f1) $\vdash (\bigcap M(\alpha \Rightarrow M\alpha) \Rightarrow (M^i\alpha \Rightarrow M^{i+1}\alpha))$ for every $i \in N$,

$\{(3)\}$

$\vdash ((\alpha \wedge \bigcap M(\alpha \Rightarrow M\alpha)) \Rightarrow \alpha)$. $\{Ax5\}$

Assume that for a natural number i,

(f2) $\vdash ((\alpha \wedge \bigcap M(\alpha \Rightarrow M\alpha)) \Rightarrow M^i\alpha)$,

$\vdash ((\alpha \wedge \bigcap M(\alpha \Rightarrow M\alpha)) \Rightarrow (M^i\alpha \wedge (M^i\alpha \Rightarrow M^{i+1}\alpha)))$,

$\{f1, f2\}$

$\vdash ((\alpha \wedge \bigcap M(\alpha \Rightarrow M\alpha)) \Rightarrow (M^{i+1}\alpha \wedge (M^i\alpha \Rightarrow M^{i+1}\alpha)))$.

Hence by the principle of induction

$\vdash ((\alpha \wedge \bigcap M(\alpha \Rightarrow M\alpha)) \Rightarrow M^i\alpha)$ for every $i \in N$,

$\vdash ((\alpha \wedge \bigcap M(\alpha \Rightarrow M\alpha)) \Rightarrow \bigcap M\alpha)$. $\{\omega\text{-rule } r5\}$ □

(9) $\vdash (\bigcap M(M\alpha) \Rightarrow M \bigcap M\alpha)$.

PROOF.

$\vdash (\bigcap M(M\alpha) \Rightarrow M^i(M\alpha))$ for every $i \in N$, $\{(6)\}$

$\vdash (\bigcap M(M\alpha) \Rightarrow M(M^i\alpha))$ for every $i \in N$, $\{Ax21\}$

(f1) $\vdash (\bigcap M(M\alpha) \Rightarrow M \bigcap M\alpha)$, $\{r5\}$

$\vdash (\bigcap M\alpha \Rightarrow M^i\alpha)$ for every $i \in N$, $\{(6)\}$

$\vdash (M \bigcap M\alpha \Rightarrow M(M^i\alpha))$ for every $i \in N$, $\{r2\}$

$\vdash (M \bigcap M\alpha \Rightarrow M^i(M\alpha))$ for every $i \in N$, $\{Ax21\}$

(f2) $\vdash (M \bigcap M\alpha \Rightarrow \bigcap M(M\alpha))$, $\{r5\}$

$\vdash (M \bigcap M\alpha \equiv \bigcap M(M\alpha))$. $\{f1, f2\}$ □.

(10) $\vdash \bigcup M(M\alpha) \equiv M \bigcup M\alpha$.

The proof analogous to the previous formula is omitted.   □

(11) $\dfrac{(\alpha \Rightarrow \beta)}{(\bigcup M\alpha \Rightarrow \bigcup M\beta)}$.

PROOF. Assume that for an arbitrary fixed set of formulas Z,

$Z \vdash (\alpha \Rightarrow \beta)$.

Then for every natural number $i \in N$

$$Z \vdash (M^i \alpha \Rightarrow M^i \beta). \hspace{3cm} \{r2\}$$

Hence

$$Z \vdash (M^i \alpha \Rightarrow \bigcup M\beta) \quad \text{for every } i \in N, \hspace{2cm} \{(5)\}$$
$$Z \vdash (\bigcup M\alpha \Rightarrow \bigcup M\beta). \hspace{3cm} \{r4\} \quad \square$$

(12) $\qquad \dfrac{(\alpha \Rightarrow \beta)}{(\bigcap M\alpha \Rightarrow \bigcap M\beta)}$.

The proof is analogous to the previous one.

(13) $\qquad \dfrac{\alpha, \ M \text{ true}}{M\alpha}$.

For the proof see Example 5.1.

(14) $\qquad \dfrac{(\gamma \Rightarrow \sim M \sim \gamma)}{(\gamma \Rightarrow \sim \text{while } \gamma \text{ do } M \text{ od true})}$.

PROOF. Let Z be an arbitrary set of formulas

(f1) $\quad Z \vdash (\gamma \Rightarrow \sim M \sim \gamma)$, $\hspace{3cm}$ {assumption}

$\qquad Z \vdash (((\gamma \wedge M \sim \gamma) \vee \sim \gamma) \Rightarrow \sim \gamma)$, $\hspace{1.5cm}$ {Ax1–Ax11}

$\qquad Z \vdash (\text{if } \gamma \text{ then } M \text{ fi } \sim \gamma \Rightarrow \sim \gamma)$. $\hspace{2cm}$ {Ax22}

Assume that for a natural number i

(f2) $\quad Z \vdash ((\text{if } \gamma \text{ then } M \text{ fi})^i \sim \gamma \Rightarrow \sim \gamma)$, {inductive assumption}

$\qquad Z \vdash ((\text{if } \gamma \text{ then } M \text{ fi})^{i+1} \sim \gamma \Rightarrow \text{if } \gamma \text{ then } M \text{ fi } \sim \gamma)$,

$\hspace{9cm} \{r2, f2\}$

$\qquad Z \vdash ((\text{if } \gamma \text{ then } M \text{ fi})^{i+1} \sim \gamma \Rightarrow ((\gamma \wedge M \sim \gamma) \vee \sim \gamma))$,

$\hspace{9cm} \{Ax22\}$

$\qquad Z \vdash ((\text{if } \gamma \text{ then } M \text{ fi})^{i+1} \sim \gamma \Rightarrow \sim \gamma)$. $\hspace{2.5cm} \{f1\}$

Hence by the principle of induction

$\qquad Z \vdash ((\text{if } \gamma \text{ then } M \text{ fi})^i \sim \gamma \Rightarrow \sim \gamma) \quad \text{for every } i \in N.$

Thus by rule r6,

$\qquad Z \vdash (\text{while } \gamma \text{ do } M \text{ od true} \Rightarrow \sim \gamma)$,

$\qquad Z \vdash (\gamma \Rightarrow \sim \text{while } \gamma \text{ do } M \text{ od true})$. $\hspace{3cm} \square$

(15) $\qquad \dfrac{(\gamma \Rightarrow M\gamma)}{(\gamma \Rightarrow \sim \text{while } \gamma \text{ do } M \text{ od true})}$.

PROOF. Let Z be an arbitrary set of formulas.

$\qquad Z \vdash (\gamma \Rightarrow M\gamma)$, $\hspace{4cm}$ {assumption}

$\qquad Z \vdash (M\gamma \Rightarrow \sim M \sim \gamma)$, $\hspace{4cm}$ {(1)}

$\qquad Z \vdash (\gamma \Rightarrow \sim M \sim \gamma)$,

$\qquad Z \vdash (\gamma \Rightarrow \sim \text{while } \gamma \text{ do } M \text{ od true})$. $\hspace{1.5cm}$ {rule (14)} $\quad \square$

(16) $$\frac{\gamma}{\sim \text{while } \gamma \text{ do } M \text{ od true}}\,.$$

The proof follows immediately from rule (14).   □

(17) $$\frac{(\gamma \Rightarrow \gamma')}{(\text{while } \gamma' \text{ do } M \text{ od true} \Rightarrow \text{while } \gamma \text{ do } M \text{ od true})}\,.$$

PROOF.

(f1) $Z \vdash (\gamma \Rightarrow \gamma')$,   {assumption}

$Z \vdash (\sim \gamma' \Rightarrow \text{while } \gamma \text{ do } M \text{ od true})$.   {Ax23}

Suppose that for the natural number i,

$Z \vdash ((\text{if } \gamma' \text{ then } M \text{ fi})^{i} \sim \gamma' \Rightarrow \text{while } \gamma \text{ do } M \text{ od true})$,
        {inductive assumption}

$Z \vdash ((\text{if } \gamma' \text{ then } M \text{ fi})^{i+1} \sim \gamma' \Rightarrow \text{if } \gamma' \text{ then } M \text{ fi } (\text{while } \gamma$
        $\text{do } M \text{ od true}))$.   {r2, inductive assumption}

(f2) $Z \vdash ((\text{if } \gamma' \text{ then } M \text{ fi})^{i+1} \sim \gamma' \Rightarrow ((\gamma' \wedge M \text{ (while } \gamma \text{ do } M \text{ od}$
        $\text{true})) \vee (\sim \gamma' \wedge \text{while } \gamma \text{ do } M \text{ od true})))$,   {Ax22}

$Z \vdash ((\sim \gamma' \wedge \text{while } \gamma \text{ do } M \text{ od true}) \Rightarrow \sim \gamma)$,   {Ax23, f1}

$\vdash (((\gamma' \wedge M \text{ while } \gamma \text{ do } M \text{ od true}) \vee \sim \gamma)$
        $\Rightarrow (\sim \gamma \vee (\gamma' \wedge \gamma \wedge M \text{ (while } \gamma \text{ do } M \text{ od true})) \vee (\gamma' \wedge \sim \gamma \wedge$
        $\wedge M(\text{while } \gamma \text{ do } M \text{ od true}))))$,   {Ax1–Ax11}

$Z \vdash ((\gamma' \wedge M(\text{while } \gamma \text{ do } M \text{ od true}) \vee \sim \gamma)$
        $\Rightarrow (\gamma \wedge M(\text{while } \gamma \text{ do } M \text{ od true}) \vee \sim \gamma))$,

$Z \vdash ((\text{if } \gamma' \text{ then } M \text{ fi})^{i+1} \sim \gamma' \Rightarrow (\gamma \wedge M(\text{while } \gamma \text{ do } M \text{ od}$
        $\text{true}) \vee \sim \gamma))$,   {f2}

$Z \vdash ((\text{if } \gamma' \text{ then } M \text{ fi})^{i+1} \sim \gamma' \Rightarrow \text{while } \gamma \text{ do } M \text{ od true})$,
        {Ax23}

$Z \vdash ((\text{if } \gamma' \text{ then } M \text{ fi})^{i} \sim \gamma' \Rightarrow \text{while } \gamma \text{ do } M \text{ od true})$
        for every $i \in N$,   {principle of induction}

$Z \vdash (\text{while } \gamma' \text{ do } M \text{ od true} \Rightarrow \text{while } \gamma \text{ do } M \text{ od true})$.
        {r6}   □

(18)   If $V(M) \cap V(\alpha) = \varnothing$,   then $\vdash (M \text{ true} \Rightarrow (M\alpha \equiv \alpha.))$

The formal proof of (18) is very long. It goes by induction on the complexity of the expressions M and α. Another proof which is of semantic character will be given in the following chapter.   □

(19) $$\frac{M' \text{ true}}{(\text{while } \gamma \text{ do } M \text{ od true} \Rightarrow \text{while } \gamma \text{ do } M; M' \text{ od true})}\,,$$

where $V(M') \cap V(\text{while } \gamma \text{ do } M \text{ od}) = \varnothing$.

PROOF.

$$Z \vdash (\sim\gamma \Rightarrow \textbf{while } \gamma \textbf{ do } M; M' \textbf{ od true}). \qquad \{\text{Ax23}\}$$

Assume (the inductive assumption) that for a natural number i,

$$Z \vdash ((\textbf{if } \gamma \textbf{ then } M \textbf{ fi})^i \sim\gamma \Rightarrow \textbf{while } \gamma \textbf{ do } M; M' \textbf{ od true}),$$
$$Z \vdash ((\textbf{if } \gamma \textbf{ then } M \textbf{ fi})^i \sim\gamma \Rightarrow M'(\textbf{if } \gamma \textbf{ then } M \textbf{ fi})^i \sim\gamma). \quad \{(18)\}$$

Thus

$$Z \vdash ((\textbf{if } \gamma \textbf{ then } M \textbf{ fi})^{i+1}\gamma \Rightarrow (\gamma \wedge M\,(M'\,(\textbf{if } \gamma \textbf{ then } M \textbf{ fi})^i \sim\gamma) \vee$$
$$\vee \sim \gamma \wedge (\textbf{if } \gamma \textbf{ then } M \textbf{ fi})^i \sim\gamma).$$

By the inductive assumption we have

$$Z \vdash ((\textbf{if } \gamma \textbf{ then } M \textbf{ fi})^{i+1} \sim\gamma \Rightarrow (\gamma \wedge M(M' \textbf{ while } \gamma \textbf{ do } M; M'$$
$$\textbf{od true}) \vee \sim \gamma \wedge \textbf{while } \gamma \textbf{ do } M; M' \textbf{ od true})).$$

Hence

$$Z \vdash ((\textbf{if } \gamma \textbf{ then } M \textbf{ fi})^{i+1} \sim\gamma \Rightarrow \textbf{while } \gamma \textbf{ do } M; M' \textbf{ od true}).$$
$$\{\text{Ax23}\}$$

By the principle of induction, for every $i \in N$

$$Z \vdash ((\textbf{if } \gamma \textbf{ then } M \textbf{ fi})^i \sim\gamma \Rightarrow \textbf{while } \gamma \textbf{ do } M; M' \textbf{ od true}).$$

By rule r6

$$Z \vdash (\textbf{while } \gamma \textbf{ do } M \textbf{ od true} \Rightarrow \textbf{while } \gamma \textbf{ do } M; M' \textbf{ od true}). \quad \square$$

(20) $\vdash ((\gamma \wedge \bigcap M(\gamma \Rightarrow \sim M\sim\gamma)) \Rightarrow \sim \textbf{while } \gamma \textbf{ do } M \textbf{ od true}).$

PROOF.

$$\vdash (\sim\gamma \Rightarrow (\sim\gamma \vee \sim \bigcap M(\gamma \Rightarrow \sim M\sim\gamma))), \qquad (\text{Ax2})$$
$$\vdash ((\textbf{if } \gamma \textbf{ then } M \textbf{ fi})^i \sim\gamma \Rightarrow (\sim\gamma \vee \sim \bigcap M(\gamma \Rightarrow \sim M\sim\gamma)))$$
$$\text{for a natural number } i, \qquad \{\text{inductive assumption}\}$$
$$\vdash ((\textbf{if } \gamma \textbf{ then } M \textbf{ fi})^{i+1} \sim\gamma \Rightarrow (\gamma \wedge M(\sim \gamma \vee \sim \bigcap M(\gamma$$
$$\Rightarrow \sim M\sim\gamma))) \vee (\sim\gamma \wedge (\sim\gamma \vee \sim \bigcap M(\gamma \Rightarrow \sim M\sim\gamma)))),$$
$$\{\text{inductive assumption, Ax22}\}.$$
$$\vdash (\sim\gamma \wedge (\sim\gamma \vee \sim \bigcap M(\gamma \Rightarrow \sim M\sim\gamma))) \equiv \sim\gamma,$$
$$\{\text{Ax1–Ax11}\}$$

$$\vdash ((\textbf{if } \gamma \textbf{ then } M\textbf{fi})^{i+1} \sim\gamma \Rightarrow (((\gamma \wedge M\sim\gamma) \vee$$
$$\vee (\gamma \wedge M\sim \bigcap M(\gamma \Rightarrow \sim M\sim\gamma))) \vee \sim\gamma)), \qquad \{\text{Ax15}\}$$
$$\vdash ((\textbf{if } \gamma \textbf{ then } M\textbf{fi})^{i+1} \sim\gamma \Rightarrow ((\sim(\gamma \Rightarrow \sim M\sim\gamma) \vee$$
$$\vee \sim (M\bigcap M(\gamma \Rightarrow \sim M\sim\gamma))) \vee \sim\gamma)), \qquad \{(1)\}$$
$$\vdash ((\textbf{if } \gamma \textbf{ then } M\textbf{fi})^{i+1} \sim\gamma \Rightarrow (\sim\gamma \vee \sim \bigcap M(\gamma \Rightarrow \sim M\sim\gamma))).$$

By the principle of induction, for every natural number i

$$\vdash ((\textbf{if } \gamma \textbf{ then } M \textbf{ fi})^i \sim\gamma \Rightarrow \sim(\gamma \wedge \bigcap M(\gamma \Rightarrow \sim M\sim\gamma))).$$

Hence by the rule r6

$$\vdash (\textbf{while } \gamma \textbf{ do } M \textbf{ od true} \Rightarrow \sim(\gamma \wedge \bigcap M(\gamma \Rightarrow \sim M \sim \gamma))),$$
$$\vdash ((\gamma \wedge \bigcap M(\gamma \Rightarrow \sim M \sim \gamma)) \Rightarrow \textbf{while } \gamma \textbf{ do } M \textbf{ od true}). \qquad \square$$

8. AN EXAMPLE OF A CORRECTNESS PROOF

In this section we shall present a proof, almost formal, of the statement that the bisection algorithm correctly computes an approximation of a zero of a continuous function in an Archimedean field.

Assume that f is a function defined on an interval $[a, b]$ such that $f(a) \cdot f(b) < 0$.

THEOREM 8.1. *The program K of the form*

> **while** $(b-a) > \varepsilon$ **do**
> $\quad x := (a+b)/2;$
> $\quad$ **if** $f(a) \cdot f(x) \leqslant 0$ **then** $b := x$ **else** $a := x$ **fi**
> **od**

is correct with respect to the precondition

$$\xi: \big(f(a) \cdot f(b) < 0 \wedge (b-a) > \varepsilon > 0\big)$$

and the postcondition

$$\delta: \big(f(a) \cdot f(b) \leqslant 0 \wedge (b-a) \leqslant \varepsilon\big).$$

More strictly we shall prove that the formula $(\xi \Rightarrow K\delta)$ is provable in the theory of Archimedean fields (cf. Chapter IV).

Let us assume the following abbreviations:

> M: **begin**
> $\quad x := (a+b)/2;$
> $\quad$ **if** $f(a) \cdot f(x) \leqslant 0$ **then** $b := x$ **else** $a := x$ **fi**
> **end**,

$$\delta_i : (f(a) \cdot f(b) \leqslant 0 \wedge (b-a) = k/2^i) \quad \text{for } k > 0 \text{ and } i \in N.$$

We shall prove a few lemmas in order to illustrate the role of axioms and inference rules.

LEMMA 8.2. *The following formula is provable in the theory of fields*:
$$(\delta_0 \Rightarrow M\delta_1).$$

PROOF. Observe that the following two formulas are theorems in the theory of fields:

$$((b-a) = k \Rightarrow ((a+b)/2-a = k/2 \wedge b-(a+b)/2 = k/2)),$$
$$(\forall d)(f(a) \cdot f(b) \leqslant 0 \Rightarrow (d \cdot f(a) \leqslant 0 \vee d \cdot f(a) > 0 \wedge d \cdot f(b) \leqslant 0)).$$

Substitute $d = f((a+b)/2)$. By propositional calculus we have

$$(\delta_0 \Rightarrow ((f((a+b)/2) \cdot f(a) \leqslant 0 \wedge ((a+b)/2-a) = k/2) \vee$$
$$\vee (f((a+b)/2) \cdot f(a) > 0 \wedge f((a+b)/2) \cdot f(b) \leqslant 0 \wedge$$
$$\wedge (b-(a+b)/2) = k/2))).$$

Applying the axiom of assignment Ax12 twice

$$(z := \tau)\gamma \equiv \gamma(z/\tau),$$

we obtain

$$(\delta_0 \Rightarrow ((x := (a+b)/2)(f(x) \cdot f(a) \leqslant 0 \wedge (x-a) = k/2) \vee$$
$$\vee (x := (a+b)/2)(f(x) \cdot f(a) > 0 \wedge f(x) \cdot f(b) \leqslant 0 \wedge$$
$$\wedge (b-x) = k/2))).$$

By axiom Ax15

$$M(\alpha \vee \beta) \equiv (M\alpha \vee M\beta),$$

we have

$$(\delta_0 \Rightarrow (x := (a+b)/2)(f(x) \cdot f(a) \leqslant 0 \wedge (x-a) = k/2 \vee$$
$$\vee f(x) \cdot f(a) > 0 \wedge f(x) \cdot f(b) \leqslant 0 \wedge (b-x) = k/2)),$$

which is equivalent by Ax12 to

$$(\delta_0 \Rightarrow (x := (a+b)/2)(f(x) \cdot f(a) \leqslant 0 \wedge (b := x)\delta_1 \vee$$
$$\vee f(x) \cdot f(a) > 0 \wedge (a := x)\delta_1)).$$

By axiom Ax22

$$\textbf{if } \gamma \textbf{ then } M \textbf{ else } N \textbf{ fi } \alpha \equiv ((\gamma \wedge M\alpha) \vee (\sim\gamma \wedge N\alpha)),$$

we prove

$$(\delta_0 = (x := (a+b)/2)(\textbf{if } f(x) \cdot f(a) \leqslant 0 \textbf{ then } b := x$$
$$\textbf{else } a := x \textbf{ fi } \delta_1)).$$

Hence $(\delta_0 \Rightarrow M\delta_1)$ is provable in the theory of ordered fields.   □

As a consequence of the above lemma we have

LEMMA 8.3. *For every natural number i the following formula is provable in the theory of ordered fields*

$$(\delta_0 \Rightarrow M^i \delta_1).$$ $\qquad\qquad\qquad$ □

THE PROOF OF THEOREM 8.1. For every natural number $j > 0$ we can prove by Lemma 8.3 the following formula:

(1) $\qquad ((k > \varepsilon > 0 \wedge \delta_0 \wedge \varepsilon \cdot j = k)$
$\qquad\qquad \Rightarrow M^j(f(a) \cdot f(b) \leqslant 0 \wedge b - a = k/2^j \wedge k/j \leqslant \varepsilon)).$

By axioms of fields and axioms Ax12 and Ax23 of algorithmic logic the following two formulas are provable:

$$((z := \varepsilon)(z := z + \varepsilon)^j z \geqslant k \Rightarrow \varepsilon \cdot j \geqslant k),$$
$$(M^j(f(a) \cdot f(b) \leqslant 0 \wedge b - a = k/2^j \wedge k/j \leqslant \varepsilon)$$
$$\Rightarrow \textbf{while } b - a > \varepsilon \textbf{ do } M \textbf{ od } \delta).$$

Hence using propositional calculus and (1) we have proved

$$((z := \varepsilon)(z := z + \varepsilon)^j \ z \geqslant k \Rightarrow ((k > \varepsilon > 0 \wedge \delta_0) \Rightarrow K\delta))$$

for every $j \geqslant 0$. By ω-rule r6 of algorithmic logic we obtain

(2) $\qquad ((z := \varepsilon) \ (\textbf{while } z < k \textbf{ do } z := z + \varepsilon \textbf{ od true})$
$\qquad\qquad \Rightarrow (k > \varepsilon > 0 \wedge \delta_0 \Rightarrow K\delta)).$

Making use of the following form of Archimedean axiom:

$$(k > \varepsilon > 0 \Rightarrow (z := \varepsilon) \ (\textbf{while } z < k \textbf{ do } z := z + \varepsilon \textbf{ od true})),$$

we obtain, by (2), the following theorem:

$$((\delta_0 \wedge k > \varepsilon > 0) \Rightarrow K\delta).$$

Thus the formula $(\xi \Rightarrow K\delta)$ is also provable. $\qquad\qquad\qquad$ □

BIBLIOGRAPHIC REMARKS

The algorithmic languages discussed here were introduced by Salwicki (1970). The role of semantical properties of programs (termination, partial correctness) and certain formalisms were first presented by Engeler (1967), Floyd (1967) and Hoare (1969). The origins of the theory of programs go back to Turing (1949), Yanov (1959) and McCarthy (1961). The first deductive system for proving equivalence of program schemes was constructed by Yanov (1959). Another system for a combination of first-order logic and λ-calculus was elaborated by Thiele (1966).

The program of research into algorithmic logic was first formulated by Salwicki (1970). The axiomatization and the completeness theorem of algorithmic logic were given by Mirkowska (1971). Kreczmar (1974) studied effectivity problems in algorithmic logic. Algorithmic logic can be also called a logic of the weakest precondition, cf. Dijkstra (1976); the strongest postcondition was studied by Banachowski (1977).

Many authors have studied algorithmic logic using mathematical tools in addition to those mentioned above, e.g. Rasiowa (1975), Grabowski (1981), Dańko (1980), Perkowska (1972) and many others.

CHAPTER III

METAMATHEMATICAL INVESTIGATIONS OF ALGORITHMIC LOGIC

We have seen in the preceding chapter that the axioms of algorithmic logic (AL) are tautologies, and that the inference rules are sound. We have proved that for any algorithmic theory the theorems of the theory are valid in all its models. In this chapter we shall prove the inverse implication, which will be referred to as the *Completeness Theorem*. It shows that semantic and syntactic methods of proving properties of programs are equivalent. The Completeness Theorem allows us to prove many properties of algorithmic logic, e.g., inessentiality of definitions which have a straightforward interpretation in computer science, namely that subroutines (i.e., non-recursive procedures) can be eliminated. Another important corollary which follows from the Completeness Theorem states that axiomatization of AL characterizes the semantics of program connectives in a unique way.

The chapter contains also another axiomatization of AL, which is constructed in a way similar to Gentzen's axiomatization.

Bearing in mind the future use of AL in algorithmic theories of data structures, we provide various extensions of the main result on completeness to the cases of data structures with partial operations and of many-sorted data structures.

1. LINDENBAUM ALGEBRA

Let T denote an algorithmic theory $\langle L, C, A \rangle$, where L is an algorithmic language (cf. Chapter I, § 1), C is the syntactic consequence operation and A is a set of specific axioms.

DEFINITION 1.1. *By $\approx$ we shall denote the equivalence relation in the set of all formulas of the language L such that for arbitrary formulas α, β*

$$\alpha \approx \beta \quad \textit{iff } A \vdash (\alpha \Rightarrow \beta) \quad \textit{and } A \vdash (\beta \Rightarrow \alpha). \qquad \square$$

The following lemma is an extension of the classical fact that the relation $\approx$ is a congruence in the algebra of formulas (cf. Chapter I, § 1).

LEMMA 1.1. *For every formulas* α, β, α', β' *and every program* M, *if* $\beta \approx \beta'$ *and* $\alpha \approx \alpha'$ *then*

$$(\alpha \vee \beta) \approx (\alpha' \vee \beta'), \quad (\alpha \Rightarrow \beta) \approx (\alpha' \Rightarrow \beta'),$$

$$(\alpha \wedge \beta) \approx (\alpha' \wedge \beta'), \quad \sim\alpha \approx \sim\alpha',$$

$$M\alpha \approx M\alpha',$$

$$\bigcup M\alpha \approx \bigcup M\alpha', \quad \bigcap M\alpha \approx \bigcap M\alpha',$$

$$(\exists x)\alpha(x) \approx (\exists x)\alpha'(x), \quad (\forall x)\alpha(x) \approx (\forall x)\alpha'(x).$$

PROOF. The first four equivalences follow by classical propositional calculus.

The equivalence $M\alpha \approx M\alpha'$ follows immediately from the assumption $\alpha \approx \alpha'$ by the rule of inference r2.

The equivalences $\bigcup M\alpha \approx \bigcup M\alpha'$ and $\bigcap M\alpha \approx \bigcap M\alpha'$ follow from the assumption $\alpha \approx \alpha'$ by rules (10) and (11) from Chapter II, § 7.

Let $\alpha(x) \approx \alpha'(x)$ and let x be an individual variable free in α and α'. Then by r2,

$$(x := \tau)\alpha(x) \approx (x := \tau)\alpha'(x) \quad \text{for every term } \tau.$$

Hence by Ax19,

$$(x := \tau)\alpha(x) \Rightarrow (\exists x)\alpha'(x) \quad \text{and}$$

$$(x := \tau)\alpha'(x) \Rightarrow (\exists x)\alpha(x).$$

Let us take as τ an individual variable y such that $y \notin V(\alpha) \cup V(\alpha')$. Then by r3

$$(\exists x)\alpha(x) \Rightarrow (\exists x)\alpha'(x) \quad \text{and} \quad (\exists x)\alpha'(x) \Rightarrow (\exists x)\alpha(x).$$

From the above and Ax20 it follows that

$$(\forall x)\alpha(x) \equiv (\forall x)\alpha'(x). \qquad \square$$

Let $F/\approx$ be the set of all equivalence classes with respect to the relation $\approx$. By $\|\alpha\|$ we shall denote the set of all formulas $\beta \in F$ such that $\alpha \approx \beta$.

As a consequence of Lemma 1.1 we can consider a quotient algebra

$$\langle F/\approx, \cup, \cap, \sim, \Rightarrow \rangle,$$

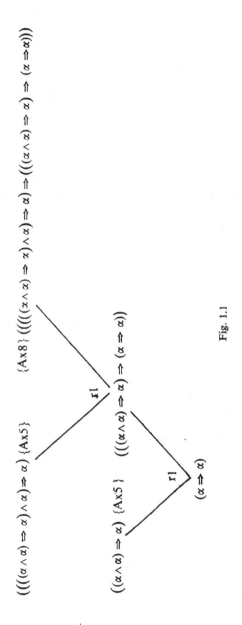

Fig. 1.1

where the operations $\cup, \cap, >, \sim$ are defined as follows. For every $\alpha, \beta \in F$,

$$||\alpha|| \cup ||\beta|| = ||(\alpha \vee \beta)||, \quad ||\alpha|| \Rightarrow ||\beta|| = ||(\alpha \Rightarrow \beta)||,$$

$$||\alpha|| \cap ||\beta|| = ||(\alpha \wedge \beta)||, \quad \sim ||\alpha|| = ||\sim \alpha||.$$

We shall call this algebra the *Lindenbaum algebra* of the theory $T = \langle L, C, A \rangle$.

Observe that the relation $\leqslant$ such that for every $\alpha, \beta \in F$,

(1) $||\alpha|| \leqslant ||\beta||$ iff $A \vdash (\alpha \Rightarrow \beta)$

defines an ordering relation in the Lindenbaum algebra.

And indeed, for every formula $\alpha \in F$, the formula $(\alpha \Rightarrow \alpha)$ is a theorem in T, as can be seen from Figure 1.1 (p. 81).

Hence $||\alpha|| \leqslant ||\alpha||$.

If, for some formulas $\alpha, \beta, \delta \in F$, $||\alpha|| \leqslant ||\beta||$ and $||\beta|| \leqslant ||\delta||$ then $A \vdash (\alpha \Rightarrow \beta)$ and $A \vdash (\beta \Rightarrow \delta)$ and therefore by Ax1, $||\alpha|| \leqslant ||\delta||$.

Finally, suppose $||\alpha|| \leqslant ||\beta||$ and $||\beta|| \leqslant ||\alpha||$. It is the case that $A \vdash (\alpha \Rightarrow \beta)$ and $A \vdash (\beta \Rightarrow \alpha)$. Hence $\alpha \approx \beta$ and as a consequence $||\alpha|| = ||\beta||$.

LEMMA 1.2. *The Lindenbaum algebra of a theory* $T = \langle L, C, A \rangle$ *is a Boolean algebra and for every formula* $\alpha \in F$:

(i) $||\alpha|| = 1$ *iff* α *is a theorem in the theory* T,

(ii) $||\alpha|| \neq 0$ *iff* $\sim \alpha$ *is not a theorem in* T,

where 1 *is the unit-element and* 0 *is the zero-element in the Boolean algebra.*

PROOF. By axioms Ax2 and Ax3, for arbitrary formulas α, β, $||\alpha|| \leqslant ||(\alpha \vee \beta)||$ and $||\beta|| \leqslant ||(\alpha \vee \beta)||$.

By axiom Ax4, for every formula δ, if $||\alpha|| \leqslant ||\delta||$ and $||\beta|| \leqslant ||\delta||$, then $||(\alpha \vee \beta)|| \leqslant ||\delta||$, cf. Figure 1.2. Hence l.u.b. $\{||\alpha||, ||\beta||\} = ||\alpha|| \cup \cup ||\beta||$.

Analogously, by Ax5 and Ax6, $||(\alpha \wedge \beta)|| \leqslant ||\alpha||$ and $||(\alpha \wedge \beta)|| \leqslant ||\beta||$. By the proof indicated in Figure 1.3, for arbitrary formula δ such that $||\delta|| \leqslant ||\alpha||$ and $||\delta|| \leqslant ||\beta||$ we have $||\delta|| \leqslant ||\alpha|| \cap ||\beta||$. Hence g.l.b.$\{||\alpha||, ||\beta||\} = ||\alpha|| \cap ||\beta||$. Thus we shall prove that the Lindenbaum algebra is a lattice (cf. Appendix A).

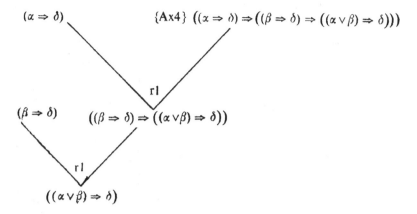

Fig. 1.2

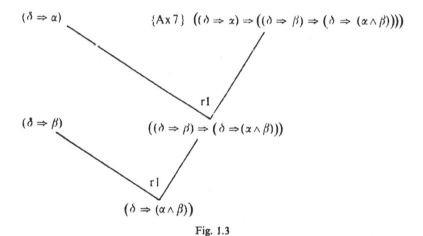

Fig. 1.3

Figures 1.4 (p. 84) and 1.5 (p. 85), where

$$\beta_1 = ((\alpha \wedge \delta) \Rightarrow \alpha) \quad \{Ax5\}, \qquad \beta_6 = ((\beta \wedge \delta) \Rightarrow \beta_8),$$
$$\beta_2 = (\alpha \Rightarrow (\alpha \vee \beta)) \quad \{Ax2\}, \qquad \beta_7 = ((\alpha \wedge \delta) \vee (\beta \wedge \delta)),$$
$$\beta_3 = ((\alpha \wedge \delta) \Rightarrow (\alpha \vee \beta)), \qquad \beta_8 = ((\alpha \vee \beta) \wedge \delta),$$
$$\beta_4 = ((\alpha \wedge \delta) \Rightarrow \delta) \quad \{Ax6\}, \qquad \beta_9 = ((\beta \wedge \delta) \Rightarrow \delta) \quad \{Ax6\},$$
$$\beta_5 = ((\alpha \wedge \delta) \Rightarrow \beta_8), \qquad \beta_{10} = ((\beta \wedge \delta) \Rightarrow (\alpha \vee \beta)),$$

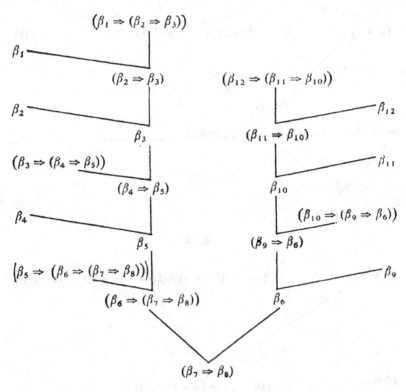

Fig. 1.4

$$\beta_{11} = (\beta \Rightarrow (\alpha \vee \beta)) \ \{\text{Ax3}\}, \quad \beta_{13} = (\delta \Rightarrow \beta_7),$$
$$\beta_{12} = ((\beta \wedge \delta) \Rightarrow \beta) \ \{\text{Ax5}\}, \quad \beta_{14} = ((\alpha \vee \beta) \Rightarrow \beta_{13}),$$

are formal proofs, which show that $(||\alpha|| \cup ||\beta||) \cap ||\delta|| = ((||\alpha|| \cap ||\delta||) \vee (||\beta|| \cap ||\delta||))$ for all $\alpha, \beta, \delta \in F$.

By Ax9, $|\textbf{false}|| = \mathbf{0}$ and by Ax11, $||\textbf{true}|| = \mathbf{1}$. To prove that $(||\alpha|| \cap \cap \sim ||\alpha||) \cup ||\beta|| = ||\beta||$ and $(||\alpha|| \cup \sim ||\alpha||) \cap ||\beta|| = ||\beta||$, see Figure 1.6. (p. 86). $- ||\alpha||$ is hence a complement of $||\alpha||$ for every $\alpha \in F$.

To prove (i) let us note that if $||\alpha|| = \mathbf{1}$, then for every $\beta \in F$, $||\beta|| \leqslant ||\alpha||$. In particular, $||(\alpha \vee \sim \alpha)|| \leqslant ||\alpha||$. Hence by Ax11, $A \vdash \alpha$.

Conversely, if $A \vdash \alpha$, then $A \vdash (\beta \Rightarrow \alpha)$ as can be seen from Figure 1.7 (p. 87). Thus for every $\beta \in F$, $||\beta|| \leqslant ||\alpha||$, i.e. $||\alpha|| = \mathbf{1}$.

To prove (ii) observe that if $||\alpha|| = \mathbf{0}$, then for every β, $||\alpha|| \leqslant ||\beta||$, i.e. $A \vdash (\alpha \Rightarrow \beta)$. Conversely, if $A \vdash \sim \alpha$, then Figure 1.8 (p. 87) is a proof

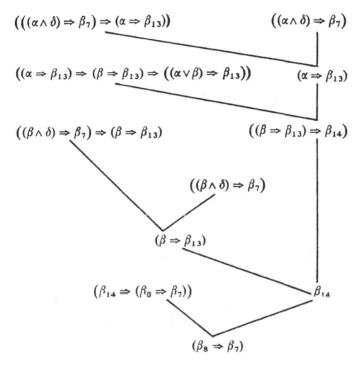

Fig. 1.5

of the formula $(\alpha \Rightarrow \beta)$ for every $\beta \in F$. Thus $||\alpha|| \leqslant ||\beta||$ for every $\beta \in F$, i.e. $||\alpha|| = \mathbf{0}$. $\square$

COROLLARY 1.1. *If a theory T is consistent then the Lindenbaum algebra of this theory is a non-degenerate Boolean algebra.* $\square$

The Lindenbaum algebra can be treated as an algebra with additional operations induced by programs. More strictly, for every program $M \in \Pi$, M can be treated as a one-argument operation in the Lindenbaum algebra such that

$$M(||\alpha||) = ||M\alpha||$$

for every formula $\alpha \in F$.

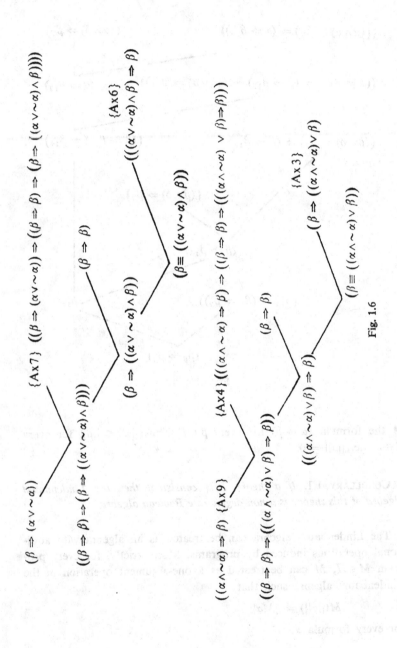

Fig. 1.6

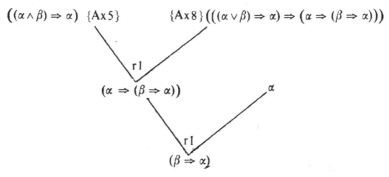

Fig. 1.7

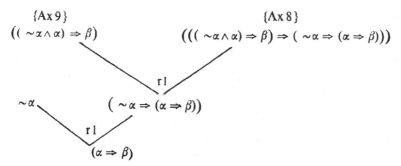

Fig. 1.8

LEMMA 1.3. *For every formula* α *and all programs M and M' the following properties hold*:

(i) $||M'\bigcup M\alpha|| = \underset{i \in N}{\text{l.u.b.}}||M'(M^i\alpha)||$,

(ii) $||M'\bigcap M\alpha|| = \underset{i \in N}{\text{g.l.b.}}||M'(M^i\alpha)||$,

(iii) $||M'(\exists x)\alpha(x)|| = \underset{\tau \in T}{\text{l.u.b.}}||M'(x := \tau)\alpha(x)||$,

(iv) $||M'(\forall x)\alpha(x)|| = \underset{\tau \in T}{\text{g.l.b.}}||M'(x := \tau)\alpha(x)||$.

PROOF. By properties (5) and (6) from Chapter II, § 7, it follows that

$$\vdash (M'(M^i\alpha) \Rightarrow M'\bigcup M\alpha) \text{ and } \vdash (M'\bigcap M\alpha \Rightarrow M'(M^i\alpha))$$

for every natural number i. Hence, as a result of application of rules r4 and r5 we have

$$||M'(M^i\alpha)|| \leqslant ||M'\bigcup M\alpha|| \text{ and } ||M'\bigcap M\alpha|| \leqslant ||M'(M^i\alpha)||.$$

Let us suppose that there are formulas δ, δ' such that for every natural number i,

$$||M'(M^i\alpha)|| \leqslant ||\delta|| \quad \text{and} \quad ||\delta'|| \leqslant ||M'(M^i\alpha)||.$$

By inference rules r4 and r5,

$$||M'\bigcup M\alpha|| \leqslant ||\delta|| \quad \text{and} \quad ||\delta'|| \leqslant ||M'\bigcap M\alpha||.$$

This implies (i) and (ii).

To prove (iii), note that by Ax19

$$||M'(x := \tau)\alpha(x)|| \leqslant ||M'(\exists x)\alpha(x)||$$

for every term $\tau \in T$. Suppose for some $\beta \in F$ that

$$||M'(x := \tau)\alpha(x)|| \leqslant ||\beta|| \quad \text{for all} \quad \tau \in T.$$

In particular

$$A \vdash (M'(x := y)\alpha(x) \Rightarrow \beta)$$

for an arbitrary individual variable y occurring neither in $M'\alpha$ nor in β. Thus by rule r3, $A \vdash (M'(\exists x)\alpha(x) \Rightarrow \beta)$, i.e. $||M'(\exists x)\alpha(x)|| \leqslant ||\beta||$. Hence (iii) holds.

The analogous proof of (iv) is omitted.   □

Let Q denote the following set of elements of the Lindenbaum algebra:

$$\text{l.u.b.} ||M'(M^i\alpha)_i||, \quad \text{g.l.b.} ||M'(M^i\alpha)||,$$
$$_{i \in N} _{i \in N}$$
$$\text{l.u.b.} ||M(x := \tau)\alpha||, \quad \text{g.l.b.} ||M(x := \tau)\alpha||.$$
$$_{\tau \in T} _{\tau \in T}$$

where M', M are programs and α is a formula.

The set Q is denumerable since the alphabet of the algorithmic language is a denumerable set.

Let us recall that a Q-filter in the Lindenbaum algebra is a non-empty proper subset V of $F/\approx$ with the following properties:

(2) $a \cap b \in V$ iff $a \in V$ and $b \in V$,

(3)   if $a \in V$ and $a \leqslant b$, then $b \in V$,

(4)   if $a \cup b \in V$, then $a \in V$ or $b \in V$,

(5)   for every element $a = \text{l.u.b } a_j, \; a \in Q$,
$$_{j \in J}$$
         if $a \in V$ then there exists $j \in J$ such that $a_j \in V$,

(6)   for every element $b = \text{g.l.b. } b_j, \; b \in Q$,
$$_{j \in J}$$
         if for all $j \in J$, $b_j \in V$ then $b \in V$.

LEMMA 1.4. *For every non-zero element a_0 of the Lindenbaum algebra of theory T there exists a Q-filter V such that $a_0 \in V$.*

For the proof see Appendix B.   □

2. THE COMPLETENESS THEOREM

The present section is devoted to a comparison of the syntactic and the semantic consequence operations. It has been proved (cf. Chapter II, § 6) that every theorem of an algorithmic theory is valid in any model of the theory. Now we shall prove that every formula valid in any model of an algorithmic theory is a theorem in this theory.

Let $T = \langle L, C, A \rangle$ be a consistent (cf. Chapter II. § 6) algorithmic theory. According to Corollary 1.1 and Lemma 1.4 there exists a Q-filter in the Lindenbaum algebra of the theory.

DEFINITION 2.1. *By a canonical data structure determined by a Q-filter V we shall mean the relational system*

$$\mathfrak{A}_V = \langle T, \{\varphi_{\mathfrak{A}_V}\}_{\varphi \in \Phi}, \{\varrho_{\mathfrak{A}_V}\}_{\varrho \in P} \rangle,$$

where $\langle T, \{\varphi_{\mathfrak{A}_V}\}_{\varphi \in \Phi} \rangle$ is an algebra of terms (cf. Chapter II., § 1), i.e.

$$\varphi_{\mathfrak{A}_V}(\tau_1, \ldots, \tau_n) = \varphi(\tau_1, \ldots, \tau_n) \quad \text{for } \varphi \in \Phi,$$

and where for every predicate $\varrho \in P$,

$$\varrho_{\mathfrak{A}_V}(\tau_1, \ldots, \tau_n) = \mathbf{1} \quad \text{iff} \quad ||\varrho(\tau_1, \ldots, \tau_n)|| \in V. \qquad □$$

Let us denote by v_V a valuation in a canonical data structure $\mathfrak{A}_V$ such that

$$v_V(x) = x \quad \text{for every individual variable } x \in V_i,$$
$$v_V(p) = \mathbf{1} \text{ iff } ||p|| \in V, \quad \text{for every propositional variable } p \in V_0.$$

The following lemma is crucial for further considerations.

LEMMA 2.1. *For every formula α in theory T,*

(1) $\mathfrak{A}_V, v_V \models \alpha \quad \text{iff} \quad ||\alpha|| \in V.$

The proof of Lemma 2.1 will proceed by induction with respect to the complexity of formulas. The ordering relation we need should be adequate to reflect the evaluation of a value of a formula.

Let Z denote a set which contains the following pairs:

$$\langle \overline{s\varrho(\tau_1, ..., \tau_n)}, \ s\varrho(\tau_1, ..., \tau_n) \rangle \quad \text{for} \quad \varrho \in P, \ \tau_i \in T, \ i \leqslant n,$$

$$\langle \beta, \sim\beta \rangle, \quad \langle \alpha, (\alpha \vee \beta) \rangle, \quad \langle \beta, (\alpha \vee \beta) \rangle, \quad \langle \alpha, (\alpha \Rightarrow \beta) \rangle,$$

$$\langle \beta, (\alpha \Rightarrow \beta) \rangle, \quad \langle \alpha, (\alpha \wedge \beta) \rangle, \quad \langle \beta, (\alpha \wedge \beta) \rangle,$$

$$\langle M^i\alpha, \textstyle\bigcup M\alpha \rangle, \quad \langle M^i\alpha, \textstyle\bigcap M\alpha \rangle \quad \text{for every } i \in N,$$

$$\langle M(M'\alpha), \ \textbf{begin } M; \ M' \textbf{ end } \alpha \rangle,$$

$$\langle (\gamma \wedge M\alpha), \ \textbf{if } \gamma \textbf{ then } M \textbf{ else } M' \textbf{ fi } \alpha \rangle,$$

$$\langle (\sim\gamma \wedge M'\alpha), \ \textbf{if } \gamma \textbf{ then } M \textbf{ else } M' \textbf{ fi } \alpha \rangle,$$

$$\langle (\textbf{if } \gamma \textbf{ then } M \textbf{ fi})^i (\sim\gamma \wedge \alpha), \ \textbf{while } \gamma \textbf{ do } M \textbf{ od } \alpha \rangle \quad \text{for all } i \in N,$$

$$\langle (x := \tau)\alpha(x), \ (\exists x)\alpha(x) \rangle, \quad \langle (x := \tau)\alpha(x), \ (\forall x)\alpha(x) \rangle,$$

and is closed with respect to the following rules:

(i) if $\langle \alpha, \beta \rangle \in Z$, then for every $s \in S$, $\langle s\alpha, s\beta \rangle \in Z$,

(ii) if $\langle \alpha, \beta \rangle \in Z$ and $\langle \beta, \delta \rangle \in Z$, then $\langle \alpha, \delta \rangle \in Z$.

where α, β are arbitrary formulas, M, M' are programs, γ is an open formula and τ is a term.

DEFINITION 2.2. *We shall say that a formula α is submitted to a formula β, in short $\alpha \prec \beta$, if and only if $\langle \alpha, \beta \rangle$ belongs to Z.* □

LEMMA 2.2. *For every set of formulas Z there exists a formula α which is the minimal element in Z with respect to the relation $\prec$.*

For the proof see Appendix B. □

PROOF OF LEMMA 2.1. Clearly Lemma 2.1 holds for all propositional variables and for all elementary formulas.

Assume that Lemma 2.1 holds for all formulas which are submitted to the formula α

(2) $\mathfrak{A}_\mathcal{V}, v_\mathcal{V} \models \beta$ iff $||\beta|| \in V$,   for all $\beta \prec \alpha$.

Below we shall consider the different forms of the formula α.

1. Let α be of the form $s\varrho(\tau_1, ..., \tau_n)$. By Lemma 2.1 in Chapter II it then follows that

$$\mathfrak{A}_\mathcal{V}, v_\mathcal{V} \models s\varrho(\tau_1, ..., \tau_n) \quad \text{iff} \quad \mathfrak{A}_\mathcal{V}, v_\nabla \models \overline{s\varrho(\tau_1, ..., \tau_n)}.$$

Since $\overline{s\varrho(\tau_1, ..., \tau_n)} \prec s\varrho(\tau_1, ..., \tau_n)$, by the inductive assumption, (1) holds.

Let us assume that throughout this proof s denotes a sequence of assignment instructions.

2. Let α be of the form $s(\beta \vee \delta)$. Since both formulas $s\beta$ and $s\delta$ are submitted to $s(\beta \vee \delta)$, and by (2),

$$\mathfrak{A}_{\overline{V}}, v_{\overline{V}} \models s\beta \quad \text{iff} \quad \|s\beta\| \in \overline{V}, \,$$

$$\mathfrak{A}_{\overline{V}}, v_{\overline{V}} \models s\delta \quad \text{iff} \quad \|s\delta\| \in \overline{V}.$$

Hence by properties (3) and (4) of Q-filters (cf. § 1)

$$\|s\beta\| \vee \|s\delta\| \in \overline{V} \quad \text{iff} \quad \mathfrak{A}_{\overline{V}}, v_{\overline{V}} \models s\beta \text{ or } \mathfrak{A}_{\overline{V}}, v_{\overline{V}} \models s\delta.$$

Thus by the definition of semantics and Lemma 3.4 from Chapter II

$$\mathfrak{A}_{\overline{V}}, v_{\overline{V}} \models s(\beta \vee \delta) \quad \text{iff} \quad \|s(\beta \vee \delta)\| \in \overline{V}.$$

The similar proofs for the formulas $s(\beta \vee \delta)$, $s(\beta \Rightarrow \delta)$ and $s \sim \beta$ are omitted.

3. Let α be of the form $s(M\beta)$, where M is not an assignment instruction. We shall consider three different forms of the program M.

3a. $\qquad M - $ **begin** M'; M'' **end**.

By Lemma 3.4 from Chapter II and since

$$s\big(M'(M''\beta)\big) \prec s(\textbf{begin } M'; \ M'' \textbf{ end } \beta)$$
$$\mathfrak{A}_{\overline{V}}, v_{\overline{V}} \models \alpha \quad \text{iff} \quad \|s\big(M'(M''\beta)\big)\| \in \overline{V}.$$

From axiom Ax21 and the formula just proved (1) follows.

3b. $\qquad M = $ **if** γ **then** M' **else** M'' **fi**.

As a consequence of the definition of semantics and inductive assumption,

$$\mathfrak{A}_{\overline{V}}, v_{\overline{V}} \models \alpha \quad \text{iff} \quad \|(\gamma \wedge M'\alpha)\| \in \overline{V} \quad \text{or} \quad \|(\sim\gamma \wedge M''\alpha)\| \in \overline{V}.$$

Hence by properties (3) and (4) of Q-filters (cf. § 1)

$$\mathfrak{A}_{\overline{V}}, v_{\overline{V}} \models \alpha \quad \text{iff} \quad \|(\gamma \wedge M'\alpha) \vee (\sim\gamma \wedge M''\alpha)\| \in \overline{V}.$$

Property (1) follows from the above equivalence by Ax22.

3c. $\qquad M = $ **while** γ **do** M' **od**.

By Example 5.4, from Chapter II

$$\mathfrak{A}_{\overline{V}}, v_{\overline{V}} \models \alpha$$

iff there exists a natural number i such that

$$\mathfrak{A}_{\overline{V}}, v_{\overline{V}} \models s \, (\textbf{if } \gamma \textbf{ then } M' \textbf{ fi})^i \, (\sim\gamma \wedge \alpha).$$

From Definition 2.2 it follows that the formula $s \, (\textbf{if } \gamma \textbf{ then } M' \textbf{ fi})^i \, (\sim\gamma \wedge \alpha)$

is submitted to formula α for every natural number i. Hence by the inductive assumption,

$$\mathfrak{A}_P, v_P \models \alpha$$

iff there exists a natural number i such that

$$||s(\text{if } \gamma \text{ then } M' \text{ fi})^i(\sim\gamma\wedge\alpha)|| \in V.$$

By property (5), § 1, and Lemma 1.3 we obtain

$$\mathfrak{A}_P, v_P \models \alpha \quad \text{iff} \quad ||s\bigcup\text{if } \gamma \text{ then } M' \text{ fi } (\sim\gamma\wedge\beta)|| \in V.$$

Finally, since the formula

$$s\bigcup \text{if } \gamma \text{ then } M' \text{ fi } (\sim\gamma\wedge\beta) \equiv s (\text{while } \gamma \text{ do } M' \text{ od } \beta)$$

is an algorithmic tautology (cf. Chapter II, Example 5.3), then

$$\mathfrak{A}_P, v_P \models \alpha \quad \text{iff} \quad ||s(\text{while } \gamma \text{ do } M' \text{ od } \beta)|| \in V.$$

4. Let α be of the form $s\bigcap M'\beta$. By the definition of semantics

$$\mathfrak{A}_P, v_P \models \alpha \quad \text{iff for every } i \in N, \mathfrak{A}_P, v_P \models s(M'^i\beta).$$

However, $s(M'^i\beta) \prec s\bigcap M'\beta$ for all $i \in N$. Thus by inductive assumption (2)

$$\mathfrak{A}_P, v_P \models s(M'^i\beta) \quad \text{iff} \quad ||s(M'^i\beta)|| \in V.$$

From property (6), § 1, and Lemma 1.3 we have

$$||s(M'^i\beta)|| \in V \text{ for every } i \in N \quad \text{iff} \quad ||s\bigcap M'\beta|| \in V.$$

Clearly the previous two equivalences imply (1).

5. Let α be of the form $s((\exists x)\beta)$. From the definition of semantics it follows that

$$\mathfrak{A}_P, v_P \models \alpha$$

iff there exists a term τ such that

$$\mathfrak{A}_P, s_{\mathfrak{A}_P}(v_P) \models (x := \tau)\beta(x).$$

Applying the inductive assumption and by property (5), § 1, and Lemma 1.3 we obtain

$$||s((x := \tau)\beta(x))|| \in V \text{ for some } \tau \in T, \text{ iff } ||s(\exists x)\alpha(x)|| \in V.$$

Hence

$$\mathfrak{A}_P, v_P \models \alpha \quad \text{iff} \quad ||s((\exists x)\beta(x))|| \in V.$$

In an entirely similar manner we can prove the property (1) for formulas of the form $s\bigcup M'\beta$ and $s(\forall x)\beta(x)$.

This completes the proof of Lemma 2.1.   □

LEMMA 2.3. *For every valuation v in the data structure $\mathfrak{A}_V$ and for every formula α, there exists a program M such that*

$$\alpha_{\mathfrak{A}_V}(v) = (M\alpha)_{\mathfrak{A}_V}(v_V).$$

PROOF. Let $x_1, \ldots, x_n$ be the sequence of all individual variables that occur in α and let v be a valuation such that $v(x_i) = \tau_i$, for $i = 1, \ldots, n$.

Let $p_1, \ldots, p_m$ be the sequence of all propositional variables that occur in α and let us assume that

$$\alpha_j = \begin{cases} \text{true,} & \text{if } v(p_j) = 1, \\ \text{false,} & \text{if } v(p_j) = 0 \text{ for } j = 1, \ldots, m. \end{cases}$$

Consider a program M of the form

begin $x_1 := \tau_1 ; \ldots ; x_n := \tau_n ; p_1 := \alpha_1 ; \ldots ; p_m := \alpha_m$ **end.**

Observe that $M_{\mathfrak{A}_V}(v_V)$ is a valuation v' such that

$$v'(x_i) = \tau_i(v_V) = \tau_i = v(x_i), \quad \text{for } 1 \leqslant i \leqslant n,$$
$$v'(p_j) = \alpha_j(v_V) = 1 \quad \text{iff} \quad ||\alpha_j|| \in V \quad \text{iff} \quad v(p_j) = 1,$$
$$\text{for } 1 \leqslant j \leqslant m,$$

and $v'(z) = v(z)$ for all remaining variables. Hence

$$\mathfrak{A}_V, v_V \models M\alpha \quad \text{iff} \quad \mathfrak{A}_V, v' \models \alpha \quad \text{iff} \quad \mathfrak{A}_V, v \models \alpha. \qquad \square$$

THEOREM 2.4 (Model Existence Theorem). *Every consistent algorithmic theory has a model.*

PROOF. Let T be a consistent theory, $T = \langle L, C, A \rangle$. Hence (cf. Chapter II, Definition 6.2) there exists a formula α_0 such that non $A \models \alpha_0$. By Lemma 1.2, $||\sim\alpha_0|| \neq 0$ and therefore by Lemma 1.4, there exists a Q-filter V such that $||\sim\alpha_0|| \in V$.

Let us consider the canonical structure $\mathfrak{A}_V$ determined by this filter. We shall prove that $\mathfrak{A}_V$ is a model of the set A.

Consider a formula α and a valuation v in $\mathfrak{A}_V$. By Lemma 2.3 there exists a program M such that $\mathfrak{A}_V, v \models \alpha$ iff $\mathfrak{A}_V, v_V \models M\alpha$ and $\vdash M$ **true**. As a consequence of the auxiliary rule

$$\frac{\alpha, M \text{ true}}{M\alpha} \quad \text{(cf. Example 5.1 in Chapter II)}$$

we have $A \vdash M\alpha$. Hence by Lemma 1.2 $||M\alpha|| = 1$ and therefore $||M\alpha|| \in V$.

From Lemmas 2.1 and 2.3 it follows that

$$\mathfrak{A}_V, v \models \alpha \quad \text{iff} \quad \mathfrak{A}_V, v_\Gamma \models M\alpha \quad \text{iff} \quad ||M\alpha|| \in V.$$

Hence $\mathfrak{A}_V, v \models \alpha$ for every $\alpha \in A$ and every valuation v in $\mathfrak{A}_V$, i.e. $\mathfrak{A}_V$ is a model of theory T.   □

THEOREM 2.5 (Completeness Theorem). *For every formula α in a consistent theory T the following conditions are equivalent:*

(i) *α is a theorem in T,*

(ii) *α is valid in every model of the theory T.*

PROOF. By Theorem 6.1 (Chapter II) (i) implies (ii).

Suppose α is not a theorem in T. As a consequence of Lemma 1.2, $||\sim\alpha|| \neq 0$ and therefore there exists a Q-filter V such that $||\sim\alpha|| \in V$ (cf. Lemma 1.4). It follows from Lemma 2.1 that $\mathfrak{A}_V, v_V \models \sim\alpha$. Applying Theorem 2.4 we obtain the conclusion that α is not valid in every model of T.   □

Theorem 2.5 asserts that the syntactic consequence operation and the semantic consequence operation determine the same sets of formulas, i.e., $C(Z) = \text{Cn}(Z)$ for every set of formulas Z.

THEOREM 2.6. *For every formula α, $\vdash \alpha$ iff $\models \alpha$, i.e. the algorithmic logic is complete.*

This theorem follows directly from the previous one.   □

Theorem 2.5 indicated that the semantic and the syntactic methods can be used exchangeably. To prove a theorem we can construct a formal proof or discuss its validity. In most examples the second method is easier than the first.

EXAMPLE. For every formula α and every program M, if $V(\alpha) \cap \cap V(M) = \varnothing$, then the formula

$$(M \text{ true} \Rightarrow (M\alpha \equiv \alpha))$$

is a theorem of algorithmic logic.

PROOF. If $V(\alpha) \cap V(M) = \varnothing$, then for every data structure $\mathfrak{A}$ and valuation v such that $M_\mathfrak{A}(v)$ is defined

$$\mathfrak{A}, v \models \alpha \quad \text{iff} \quad \mathfrak{A}, M_\mathfrak{A}(v) \models \alpha,$$

since the value of a formula depends only on the variables that occur in it.

Hence for every $\mathfrak{A}$ and v

$$\mathfrak{A}, v \models (M \text{ true} \Rightarrow (\alpha \equiv M\alpha))$$

and, as a consequence of the Completeness Theorem, the formula $(M \text{ true} \Rightarrow (\alpha \equiv M\alpha))$ is a theorem of algorithmic logic.   □

3. TWO COROLLARIES OF THE COMPLETENESS THEOREM

Let $T = \langle L, C, A \rangle$ be an algorithmic theory. In constructing proofs we frequently make use of the following important fact.

THEOREM 3.1 (Deduction Theorem). *Let α be a formula without free variables. A formula β is a theorem of the theory $T' = \langle L, C, A \cup \{\alpha\}\rangle$ iff the formula $(\alpha \Rightarrow \beta)$ is a theorem of the theory T, i.e. $A \cup \{\alpha\} \vdash \beta$ iff $A \vdash (\alpha \Rightarrow \beta)$.*

PROOF. Assume that $A \cup \{\alpha\} \vdash \beta$. By completeness it follows that

(1) $A \cup \{\alpha\} \models \beta$.

Let us suppose that there is a model $\mathfrak{M}$ of the set A such that $(\alpha \Rightarrow \beta)$ is not valid in it. Thus

(2) $\mathfrak{M}, v \models \alpha$ and   non $\mathfrak{M}, v \models \beta$,

for some valuation v in $\mathfrak{M}$. Since the value of the formula α does not depend on any valuation, then $\mathfrak{M}$ is a model of the formula α. Hence $\mathfrak{M} \models A \cup \{\alpha\}$ and, as a consequence of (1), $\mathfrak{M} \models \beta$, which contradicts (2). Hence $A \models (\alpha \Rightarrow \beta)$. By completeness, $A \vdash (\alpha \Rightarrow \beta)$.

Conversely, if $A \vdash (\alpha \Rightarrow \beta)$, then $A \cup \{\alpha\} \vdash (\alpha \Rightarrow \beta)$. Since $A \cup \{\alpha\} \vdash \alpha$, then by rule r1 (*modus ponens*), $A \cup \{\alpha\} \vdash \beta$.   □

The above theorem can be strengthened if the syntactic assumption is replaced by the semantic one.

DEFINITION 3.1. *We shall say that a formula α is closed iff the value of α does not depend on any valuation in any data structure.*   □

For example, every formula which has no free occurrences of any variable is a closed formula and the expression $(q := \text{true}) \ (q \Rightarrow \sim q)$ is also a closed formula.

THEOREM 3.2. *Let α be a closed formula of a theory $T = \langle L, C, A \rangle$. For every formula β of T, $A \vdash (\alpha \Rightarrow \beta)$ iff $A \cup \{\alpha\} \vdash \beta$.*   □

Let us note that the Deduction Theorem does not hold if α is not required to be closed.

In view of Chapter II, § 4, the Upward Skolem–Löwenheim Theorem of classical logic fails to hold in algorithmic logic. However, it can be easily proved that the downward theorem holds.

THEOREM 3.3 (Downward Skolem–Löwenheim Theorem). *If an algorithmic theory has an infinite model, then it has a denumerable model.*

PROOF. Let $T = \langle L, C, A \rangle$ be an algorithmic theory and let $\mathfrak{M}$ be its infinite model. From Corollary 6.3 of Chapter II it follows that T is consistent. As a consequence of Theorem 2.4 we find that T has a denumerable model in the set of all terms.   □

The third theorem of this section is analogous to the Herbrand theorem in classical logic.

THEOREM 3.4. *Let K and M be arbitrary programs without a **while**-operation and let α be an open formula.*

A formula $M \bigcup K\alpha$ is a theorem of AL iff there exists a natural number m such that the formula $M \bigvee_{i \leqslant m} K^i \alpha$ is a theorem of AL.

For a proof see Chapter VI, § 5.   □

As a simple generalization of Theorem 3.4 we obtain the following lemma.

LEMMA 3.5. *Let α be a formula of the form*

$$\bigcap K_1 \ldots \bigcap K_m (M_1 \bigcup K_{1+m} (\ldots M_n \bigcup K_{n+m} \ \beta) \ldots),$$

*where β is an open formula and $K_1, \ldots, K_{n+m}, M_1, \ldots, M_n$ are programs without a **while**-operation. The formula α is a theorem of algorithmic logic iff there exists a sequence $i_1, \ldots, i_n$ of natural numbers such that the formula*

$$M_1 \bigvee_{i \leqslant i_1} K^i_{m+1} (\ldots (M_n \bigvee_{j \leqslant i_n} K^j_{m+n} \beta) \ldots)$$

is a theorem of AL.   □

We shall now present a simple application of Theorem 3.4 in the theory of programs.

LEMMA 3.6. *Let K be a program of the form*

begin M_1; **while** γ **do** M_2 **od end**,

where M_1, M_2 are programs without a **while** *operation. Let* $\models K$ **true**. *Then there exists a natural number n such that the length of every computation of K is less than n.*

PROOF. By Example 5.3 of Chapter II the formula

$$M_1 \bigcup \text{if } \gamma \text{ then } M_2 \text{ fi } (\sim\gamma \wedge \alpha)$$

is a theorem of algorithmic logic. Thus by the Completeness Theorem and Theorem 3.4 there exists a natural number m such that

$$\models M_1 \bigvee_{i \leq m} (\text{if } \gamma \text{ then } M_2 \text{ fi})^i (\sim\gamma \wedge \alpha).$$

We shall prove that the length of any computation of K is proportional to m, i.e. that the number of iterations of M_2 in any computation is bounded by m.

Suppose $\mathfrak{A}$ is a data structure and v is a valuation such that

$$K_{\mathfrak{A}}(v) = M_{1\mathfrak{A}}(M_{2\mathfrak{A}}^j(v)) \quad \text{and} \quad j > m.$$

Hence by the definition of semantics

$$\mathfrak{A}, v \models M_1 (\text{if } \gamma \text{ then } M_2 \text{ fi})^i \gamma \quad \text{for all } i < j.$$

Thus

$$\text{non } \mathfrak{A}, v \models \bigvee_{i \leq m} M_1 (\text{if } \gamma \text{ then } M_2 \text{ fi})^i (\sim\gamma \wedge \alpha),$$

contrary to the assumption. □

4. THE STANDARD EXECUTION METHOD IS IMPLICITLY DEFINED BY THE AXIOMATIZATION OF ALGORITHMIC LOGIC

It has been shown in the preceding sections that our knowledge of the semantics of a chosen programming language is sufficiently complete since there exists a proof of every algorithmic property which is semantically valid. Here we give a deeper insight into this.

The semantics of an algorithmic language L consists of three elements:

(i) an interpretation of functors and predicates,

(ii) an execution method for programs,

(iii) a satisfiability relation.

The execution method defined in Chapter II, § 2 is based on the notion of computation. We shall call this the *standard execution method*. This definition of execution method is by no means a unique one: there are other possible definitions. In general, by the *execution method* we shall mean a function which to every program of the language L assigns a binary relation in the set of all valuations in a given data structure.

DEFINITION 4.1. *We shall say that the execution method for programs is proper for* AL *iff the satisfiability relation which is based on it allows the soundness of* AL *axiomatization to be proved.*   □

Obviously the standard execution method is proper for AL. The question naturally arises as to whether there are other different execution methods proper for AL.

The program execution method is strictly connected to the problem of implementation. Can we treat our axiomatic system as a criterion for the correctness of implementation?

The main conclusion of this section is that all conceivable proper execution methods of programs are similar in the sense that they induce the same input-output relations.

The completeness theorem can be then interpreted in a way which shows that the notion of computation is the one natural execution method for programs.

Now we shall formulate the thesis of this section more strictly.

DEFINITION 4.2. *By a semantic structure for L we shall mean the triple* $\langle \mathfrak{A}, I, \models \rangle$ *where* $\mathfrak{A}$ *is a data structure for L, I is an execution method, and* $\models$ *is a satisfiability relation.*   □

In what follows we shall restrict our considerations to the class of semantic structures $\langle \mathfrak{A}, I, \models \rangle$ such that

(1) a data structure $\mathfrak{A}$ is normalized, i.e. for arbitrary valuations v_1, v_2 in $\mathfrak{A}$,

$$v_1 \neq v_2 \quad \text{iff} \quad (\exists \beta) \ (\mathfrak{A}, v_1 \models \beta \text{ and non } \mathfrak{A}, v_2 \models \beta)$$

(different valuations can be distinguished by means of a formula in the language L) and

(2) the satisfiability relation $\models$ is such that

$$\mathfrak{A}, v \models (\alpha \vee \beta) \quad \text{iff} \quad \mathfrak{A}, v \models \alpha \quad \text{or} \quad \mathfrak{A}, v \models \beta,$$

$$\mathfrak{A}, v \models (\alpha \wedge \beta) \quad \text{iff} \quad \mathfrak{A}, v \models \alpha \quad \text{and} \quad \mathfrak{A}, v \models \beta,$$

$$\mathfrak{A}, v \models \sim \alpha \quad \text{iff} \quad \text{non } \mathfrak{A}, v \models \alpha,$$

$$\mathfrak{A}, v \models M\alpha \quad \text{iff} \quad (\exists v')(v, v') \in I(M) \quad \text{and} \quad \mathfrak{A}, v' \models \alpha$$

for arbitrary formulas α, β program M and arbitrary valuation v. Let $\langle \mathfrak{A}, I, \models \rangle$ be a fixed semantic structure of the above defined class.

LEMMA 4.1. *For every program M, if property (3) holds for arbitrary formulas α, β, where*

(3) $\mathfrak{A} \models M(\alpha \wedge \beta) \equiv (M\alpha \wedge M\beta),$

then $I(M)$ is a partial function.

PROOF. Suppose $(v, v_1) \in I(M)$ and $(v, v_2) \in I(M)$ and $v_1 \neq v_2$. It follows from assumption (1) that there exists a formula α such that $\mathfrak{A}, v_1 \models \alpha$ and $\mathfrak{A}, v_2 \models \sim \alpha$. Hence, $\mathfrak{A}, v \models M\alpha$ and $\mathfrak{A}, v \models M \sim \alpha$. However, non $\mathfrak{A}, v \models M(\alpha \wedge \sim \alpha)$, contrary to (3).   □

LEMMA 4.2. *Let K, M be arbitrary programs. If properties (3), (4) hold for K and M, where*

(4) $\mathfrak{A} \models \textbf{begin } K; M \textbf{ end } \alpha \equiv K(M\alpha) \quad \text{for every } \alpha \in F,$

then $I(\textbf{begin } K; M \textbf{ end}) = I(K) \circ I(M)$.

PROOF. Let $(v_1, v_2) \in I(\textbf{begin } K; M \textbf{ end})$ and let α be a formula such that $\mathfrak{A}, v_2 \models \alpha$. Hence by (2)

$$\mathfrak{A}, v_1 \models \textbf{begin } K; M \textbf{ end } \alpha.$$

It follows from (4) that $\mathfrak{A}, v_1 \models K(M\alpha)$. As a consequence of (2), there exists a valuation v' and a valuation v'' such that $(v_1, v') \in I(K)$, $(v', v'') \in I(M)$ and $\mathfrak{A}, v'' \models \alpha$. Since there exists at most one valuation v' and one valuation v'' with the above property, we have obtained for every formula α, if $\mathfrak{A}, v_2 \models \alpha$, then $\mathfrak{A}, v'' \models \alpha$. Thus from (1), $v_2 = v''$, and therefore

$$(v_1, v') \in I(K) \text{ and } (v', v_2) \in I(M) \text{ for some } v'.$$

Hence $(v_1, v_2) \in I(K) \circ I(M)$.

Let us suppose conversely that $(v_1, v_2) \in I(K) \circ I(M)$. By the definition of the composition of relations, there exists a valuation v' such that

$$(v_1, v') \in I(K) \quad \text{and} \quad (v', v_2) \in I(M).$$

Let us suppose that $\mathfrak{A}, v_2 \models \alpha$ for an arbitrary, fixed formula α. It follows that $\mathfrak{A}, v' \models M\alpha$ and, moreover by (4), that $\mathfrak{A}, v_1 \models K(M\alpha)$ and $\mathfrak{A}, v_1 \models$ **begin** K; M **end** α. Hence there exists a valuation v'' such that

$$(v_1, v'') \in I(\text{begin } K; M \text{ end}) \quad \text{and} \quad \mathfrak{A}, v'' \models \alpha.$$

The valuation v'' is unique since we have assumed property (3). Hence for every formula α,

$$\mathfrak{A}, v_2 \models \alpha \quad \text{implies} \quad \mathfrak{A}, v'' \models \alpha.$$

It follows by (1) that $v_2 = v''$, i.e.

$$(v_1, v_2) \in I(\text{begin } K; M \text{ end}). \qquad \square$$

For any open formula γ, let $\text{id}(\gamma)$ denote the set $\{(v, v) : \mathfrak{A}, v \models \gamma\}$.

LEMMA 4.3. *Let K, M be arbitrary programs. If properties (3) and (5) hold for K, M and for the arbitrary formulas $\alpha \in F$, $\gamma \in F_0$ where,*

(5) $\mathfrak{A} \models$ **if** γ **then** K **else** M **fi** $\alpha \equiv ((\gamma \wedge K\alpha) \vee (\sim\gamma \wedge M\alpha))$,

then

$$I(\text{if } \gamma \text{ then } K \text{ else } M \text{ fi}) = (\text{id}(\gamma) \circ I(K)) \cup (\text{id}(\sim\gamma) \circ I(M)).$$

The proof is similar to the proof of Lemma 4.2 and is therefore omitted.

$$\square$$

LEMMA 4.4. *If for every formula α and every open formula γ properties (3)–(6) hold, where*

(6) $\mathfrak{A} \models$ **while** γ **do** M **od** α

$$\equiv ((\sim\gamma \wedge \alpha) \vee (\gamma \wedge M \text{ while } \gamma \text{ do } M \text{ od } \alpha)),$$

then

$$I(\text{while } \gamma \text{ do } M \text{ od}) \supset \bigcup_{i \in N} I(\text{if } \gamma \text{ then } M \text{ fi})^i \circ \text{id}(\sim\gamma).$$

PROOF. Suppose $(v_1, v_2) \in \bigcup_{i \in N} I(\text{if } \gamma \text{ then } M \text{ fi})^i \circ \text{id}(\sim\gamma)$. Hence there exists a natural number m such that

$$(v_1, v_2) \in I(\text{if } \gamma \text{ then } M \text{ fi})^m \quad \text{and} \quad \mathfrak{A}, v_2 \models \sim\gamma,$$

by Lemma 4.2.

Let us assume that for some formula α, $\mathfrak{A}, v_2 \models \alpha$. It follows from the above properties that

$$\mathfrak{A}, v_1 \models (\text{if } \gamma \text{ then } M \text{ fi})^m (\sim\gamma \wedge \alpha).$$

As a consequence of property (6) we find that for every valuation v,

if $\quad \mathfrak{A}, v \models (\text{if } \gamma \text{ then } M \text{ fi})^m \ (\sim\gamma \wedge \alpha)$

then $\quad \mathfrak{A}, v \models \text{while } \gamma \text{ do } M \text{ od } \alpha$.

Hence $\mathfrak{A}, v_1 \models \text{while } \gamma \text{ do } M \text{ od } \alpha$ and there exists a valuation v' such that

$$(v_1, v') \in I \,(\text{while } \gamma \text{ do } M \text{ od}) \quad \text{and} \quad \mathfrak{A}, v' \models \alpha.$$

Thus $v' = v_2$ by (1), since for every formula α,

$$\mathfrak{A}, v_2 \models \alpha \quad \text{implies} \quad \mathfrak{A}, v' \models \alpha.$$

Therefore $(v_1, v_2) \in I(\text{while } \gamma \text{ do } M \text{ od})$. $\qquad\qquad\square$

Let us assume that the algorithmic language L contains the binary predicate $=$. Moreover, let us assume that the semantic structure $\langle \mathfrak{A}, I, \models \rangle$ is such that $=$ is interpreted as identity relation in $\mathfrak{A}$ and for every element a of the data structure $\mathfrak{A}$ there exists a term τ_a such that for an arbitrary valuation v, $a = \tau_{a\mathfrak{A}}(v)$. We shall call such semantic structure a *Herbrand structure*.

LEMMA 4.5. *If $\mathfrak{A}$ is a Herbrand structure and properties (3)–(7) hold for program M, every open formula γ and arbitrary formulas α, β, where*

(7) $\quad \mathfrak{A} \models ((\text{if } \gamma \text{ then } M \text{ fi})^i (\sim\gamma \wedge \alpha) \Rightarrow \beta) \quad$ *for all $i \in N$ implies*
$\quad\quad \mathfrak{A} \models (\text{while } \gamma \text{ do } M \text{ od } \alpha \Rightarrow \beta),$

then

$$I(\text{while } \gamma \text{ do } M \text{ od}) = \bigcup_{i \in N} I(\text{if } \gamma \text{ then } M \text{ fi})^i \circ \text{id}(\sim\gamma).$$

PROOF. Suppose $(v_1, v_2) \in I(\text{while } \gamma \text{ do } M \text{ od})$ and $\mathfrak{A}, v_2 \models \alpha_1$. Let us assume that β_1 is a formula which describes the valuation v_1 with respect to all variables occurring in **while** γ **do** M **od** α_1, i.e.

(8) $\quad \beta_1 \equiv (x_1 = \tau_{v_1(x_1)} \wedge \ldots \wedge x_n = \tau_{v_1(x_n)} \wedge q_1 \equiv c_1 \wedge \ldots$
$\quad\quad \ldots \wedge q_m \equiv c_m),$

where $x_1, \ldots, x_n$ are all individual variables and $q_1, \ldots, q_m$ are all propositional variables occurring in **while** γ **do** M **od** α_1 and

$$c_j = \begin{cases} \textbf{true} & \text{iff } v_1(q_j) = 1, \\ \textbf{false} & \text{iff } v_1(q_j) = 0. \end{cases}$$

Hence

$\quad \mathfrak{A}, v_1 \models \text{while } \gamma \text{ do } M \text{ od } \alpha_1 \quad \text{and} \quad \mathfrak{A}, v_1 \models \beta_1$, i.e.
$\quad \text{non } \mathfrak{A} \models (\text{while } \gamma \text{ do } M \text{ od } \alpha_1 \Rightarrow \sim\beta_1).$

By property (7) there exists a natural number m such that

$$\text{non } \mathfrak{A} \models ((\text{if } \gamma \text{ then } M \text{ fi})^m (\sim\gamma \wedge \alpha_1) \Rightarrow \sim\beta_1).$$

As a consequence there exists a valuation v' such that

$$\mathfrak{A}, v' \models (\text{if } \gamma \text{ then } M \text{ fi})^m (\sim\gamma \wedge \alpha_1) \quad \text{and} \quad \mathfrak{A}, v' \models \beta_1.$$

By the last property and assumption (8) we have

(9) $\mathfrak{A}, v_1 \models (\text{if } \gamma \text{ then } M \text{ fi})^m (\sim\gamma \wedge \alpha_1).$

Assume that m is the minimal natural number with such property. By (9) there exists a valuation v_2' such that

$$(v_1, v_2') \in I(\text{if } \gamma \text{ then } M \text{ fi})^m \quad \text{and} \quad \mathfrak{A}, v_2' \models (\sim\gamma \wedge \alpha_1).$$

Let us consider an arbitrary formula α_2 and let $\mathfrak{A}, v_2 \models \alpha_2$. We shall prove that $\mathfrak{A}, v_2' \models (\sim\gamma \wedge \alpha_2)$.

Following considerations presented above we have

$$\mathfrak{A}, v_1 \models (\text{if } \gamma \text{ then } M \text{ fi})^j (\sim\gamma \wedge \alpha_2)$$

for some natural number j. Suppose $j < m$. By property (5)

$$\mathfrak{A} \models (\text{if } \gamma \text{ then } M \text{fi})^m (\sim\gamma \wedge \alpha_1) \equiv (\text{if } \gamma \text{ then } M \text{fi})^j (\sim\gamma \wedge \alpha_1),$$

and by (9)

$$\mathfrak{A}, v_1 \models (\text{if } \gamma \text{ then } M \text{ fi})^j (\sim\gamma \wedge \alpha_1),$$

contrary to the assumption that m is the smallest natural number with such property. Thus $j \geqslant m$ and therefore

$$\mathfrak{A}, v_1 \models (\text{if } \gamma \text{ then } M \text{ fi})^m (\sim\gamma \wedge \alpha_2).$$

Hence by property (3) $\mathfrak{A}, v_2' \models (\sim\gamma \wedge \alpha_2)$. Thus, there exists a natural number m and a valuation v_2' such that for an arbitrary formula α, $\mathfrak{A}, v_2 \models \alpha$ implies $(v_1, v_2') \in I(\text{if } \gamma \text{ then } M \text{ fi})^m$ and $\mathfrak{A}, v_2' \models (\alpha \wedge \sim\gamma)$. Hence $v_2' = v_2$ and consequently

$$(v_1, v_2) \in I(\text{if } \gamma \text{ then } M \text{ fi})^m \circ \text{id}(\sim\gamma). \qquad \square$$

LEMMA 4.6. *If $\mathfrak{A}$ is a Herbrand structure and property* (10) *holds for an assignment instruction $s = (x := w)$ and every open formula γ, where*

(10) $\mathfrak{A} \models s\gamma \equiv \overline{s}\overline{\gamma},$

then

$$I(x := w) = \{(v_1, v_2) \colon v_1(z) = v_2(z) \text{ for } z \neq x \text{ and } v_2(x)$$
$$= w_{\mathfrak{A}}(v_1)\}.$$

PROOF. Let us assume that x is an individual variable and w is a term. Let $(v_1, v_2) \in I(x := w)$. Suppose $v_2(x) = a$. Thus $\mathfrak{A}, v_2 \models (x = \tau_a)$. Hence $\mathfrak{A}, v_1 \models (x := w)(x = \tau_a)$ and therefore by (10), $w_{\mathfrak{A}}(v_1) = \tau_a(v_1) = v_2(x)$. Let y be an individual variable and $y \neq x$. Suppose $v_2(y) = b$, then $\mathfrak{A}, v_2 \models (y = \tau_b)$. It follows that $\mathfrak{A}, v_1 \models (x := w)(y = \tau_b)$ and therefore $v_2(y) = v_1(y)$. Let q be a propositional variable. If $\mathfrak{A}, v_2 \models q$, then $\mathfrak{A}, v_1 \models (x := w)q$ and by (10) $\mathfrak{A}, v_1 \models q$.

Hence $v_2(z) = v_1(z)$ for $z \neq x$ and $v_2(x) = w_{\mathfrak{A}}(v_1)$.

The discussion is similar in the case where $s = (q := \gamma)$, q is a propositional variable and γ is an open formula.   □

As a straightforward consequence of the above lemmas we obtain the main result of this section.

THEOREM 4.7. *Algorithmic logic determines the unique execution method for programs. More strictly, for every semantic Herbrand structure* $\langle \mathfrak{A}, I, \models \rangle$ *if all axioms of algorithmic logic are valid and all inference rules are sound, then the execution method I satisfies the following equalities*:

$$I(\text{begin } K; M \text{ end}) = I(K) \circ I(M),$$

$$I(\text{if } \gamma \text{ then } M \text{ else } K \text{ fi}) = \text{id}(\gamma) \circ I(M) \cup \text{id}(\sim\gamma) \circ I(K),$$

$$I(\text{while } \gamma \text{ do } M \text{ od}) = \bigcup_{i \in N} I(\text{if } \gamma \text{ then } M \text{ fi})^i \circ \text{id}(\sim\gamma),$$

$$I(x := w) = \{(v_1, v_2): v_1(z) = v_2(z) \text{ for } z \neq x \text{ and } v_2(x) = w_{\mathfrak{A}}(v_1)\}$$

for arbitrary programs K, M, every open formula γ and an arbitrary assignment instruction $(x := w)$.   □

5. GENTZEN TYPE AXIOMATIZATION

In the preceding sections an axiomatic system for reasoning about algorithmic formulas has been presented and studied. There have been many examples of formal proofs, but no algorithm has been given for their mechanical construction.

In this section we shall discuss a deductive system in which proof of a formula is determined by the formula itself. Informally, the process of deduction will consists of decomposition of a formula into parts. Each step of decomposition will be determined univocally. This kind of deductive system seems appropriate for the automatization of the process of proving and is called *Gentzen type axiomatization*.

Let L be an algorithmic language.

Throughout this section Γ and Δ (with indices if necessary) will denote finite sequences of formulas. Any expression of the form $\Gamma \to \Delta$ will be called a *sequent*.

DEFINITION 5.1. *A sequent $\Gamma \to \Delta$ is called indecomposable iff every formula that occurs in $\Gamma \cup \Delta$ is either a propositional variable or is an elementary formula. We shall call such formulas indecomposable.* □

DEFINITION 5.2. *A sequent $\Delta \to \Gamma$ is said to be axiom-sequent iff $\Gamma \cap \Delta \neq \varnothing$.* □

Let $\Gamma = \{\alpha_1, ..., \alpha_n\}$ and $\Delta = \{\beta_1, ..., \beta_m\}$ for some $n, m \in N$. We shall use $(\bigwedge \Gamma \Rightarrow \bigvee \Delta)$ as the shortened form of the formula $(\bigwedge_{i \leqslant n} \alpha_i \Rightarrow \bigvee_{j \leqslant m} \beta_j)$. If $\Gamma = \varnothing$, then $\bigwedge \Gamma \equiv$ **true**; if $\Delta = \varnothing$, then $\bigvee \Delta \equiv$ **false**.

LEMMA 5.1. *For every axiom-sequent $\Gamma \to \Delta$ and every data structure $\mathfrak{A}$ for L, $\mathfrak{A} \models (\bigwedge \Gamma \Rightarrow \bigvee \Delta)$.* □

The rules of decomposition are listed below.

$$1A \quad \frac{\Gamma_1, \Gamma_2, s_1 \ldots s_{k-1}\overline{s_k\gamma} \to \Delta}{\Gamma_1, s_1 \ldots s_k\gamma, \Gamma_2 \to \Delta}, \qquad 1B \quad \frac{\Gamma \to \Delta_1, \Delta_2, s_1 \ldots \overline{s_k\gamma}}{\Gamma \to \Delta_1, s_1 \ldots s_k\gamma, \Delta_2},$$

$$2A \quad \frac{\Gamma_1, \Gamma_2 \to \Delta, s\gamma}{\Gamma_1, s\sim\gamma, \Gamma_2 \to \Delta}, \qquad 2B \quad \frac{\Gamma, s\gamma \to \Delta_1, \Delta_2}{\Gamma \to \Delta_1, s\sim\gamma, \Delta_2},$$

$$3A \quad \frac{\Gamma_1, \Gamma_2, s\alpha, s\beta \to \Delta}{\Gamma_1, s(\alpha \wedge \beta), \Gamma_2 \to \Delta}, \qquad 3B \quad \frac{\Gamma \to \Delta_1, \Delta_2, s\alpha; \Gamma \to \Delta_1, \Delta_2, s\beta}{\Gamma \to \Delta_1, s(\alpha \wedge \beta), \Delta_2},$$

$$4A \quad \frac{\Gamma_1, \Gamma_2, s\alpha \to \Delta; \Gamma_1, \Gamma_2, s\beta \to \Delta}{\Gamma_1, s(\alpha \vee \beta), \Gamma_2 \to \Delta}, \qquad 4B \quad \frac{\Gamma \to \Delta_1, \Delta_2, s\alpha, s\beta}{\Gamma \to \Delta_1, s(\alpha \vee \beta), \Delta_2},$$

$$5A \quad \frac{\Gamma_1, \Gamma_2 \to \Delta, s\alpha; \Gamma_1, \Gamma_2, s\beta \to \Delta}{\Gamma_1, s(\alpha \Rightarrow \beta), \Gamma_2 \to \Delta}, \qquad 5B \quad \frac{\Gamma, s\alpha \to \Delta_1, \Delta_2, s\beta}{\Gamma \to \Delta_1, s(\alpha \Rightarrow \beta), \Delta_2},$$

$$6A \quad \frac{\Gamma_1, \Gamma_2, s(K(M\alpha)) \to \Delta}{\Gamma_1, s \text{ begin } K; M \text{ end } \alpha, \Gamma_2 \to \Delta},$$

$$6B \quad \frac{\Gamma \to \Delta_1, \Delta_2, s(K(M\alpha))}{\Gamma \to \Delta_1, s \text{ begin } K; M \text{ end } \alpha, \Delta_2},$$

$$7A \quad \frac{\Gamma_1, \Gamma_2, s(\gamma \wedge K\alpha) \to \Delta; \Gamma_1, \Gamma_2, s(\sim\gamma \wedge M\alpha)}{\Gamma_1, s \text{ if } \gamma \text{ then } K \text{ else } M \text{ fi } \alpha, \Gamma_2 \to \Delta},$$

$$7B \quad \frac{\Gamma \to \Delta_1, \Delta_2, s(\gamma \wedge K\alpha), s(\sim\gamma \wedge M\alpha)}{\Gamma \to \Delta_1, s \text{ if } \gamma \text{ then } K \text{ else } M \text{ fi } \alpha, \Delta_2},$$

$$8A \quad \frac{\{\Gamma_1, \Gamma_2, s(\text{if } \gamma \text{ then } M \text{ fi})^i(\sim\gamma \wedge \alpha) \to \Delta\}_{i\in N}}{\Gamma_1, s \text{ while } \gamma \text{ do } M \text{ od } \alpha, \Gamma_2 \to \Delta},$$

$$8B \quad \frac{\Gamma \to \Delta_1, \Delta_2, s(\sim\gamma \wedge \alpha), s(\gamma \wedge M \text{ while } \gamma \text{ do } M \text{ od } \alpha)}{\Gamma \to \Delta_1, s \text{ while } \gamma \text{ do } M \text{ od } \alpha, \Delta_2},$$

$$9A \quad \frac{\{\Gamma_1, \Gamma_2, s(M^i\alpha) \to \Delta\}_{i\in N}}{\Gamma_1, s\bigcup M\alpha, \Gamma_2 \to \Delta}, \qquad 9B \quad \frac{\Gamma \to \Delta_1, \Delta_2, s\alpha, s\bigcup M(M\alpha)}{\Gamma \to \Delta_1, s\bigcup M\alpha, \Delta_2},$$

$$10A \quad \frac{\Gamma_1, \Gamma_2, s\alpha, s\bigcap M(M\alpha) \to \Delta}{\Gamma_1, s\bigcap M\alpha, \Gamma_2 \to \Delta}, \qquad 10B \quad \frac{\{\Gamma \to \Delta_1, \Delta_2, s(M^i\alpha)\}_{i\in N}}{\Gamma \to \Delta_1, s\bigcap M\alpha, \Delta_2},$$

$$11A \quad \frac{\Gamma_1, \Gamma_2, s(x := \tau)\alpha(x), s(\forall x)\alpha \to \Delta}{\Gamma_1, s(\forall x)\alpha(x), \Gamma_2 \to \Delta},$$

$$11B \quad \frac{\Gamma \to \Delta_1, \Delta_2, s(x := y)\alpha}{\Gamma \to \Delta_1, s(\forall x)\alpha(x), \Delta_2},$$

where y is an individual variable which does not appear in s and α,

$$12A \quad \frac{\Gamma_1, \Gamma_2 \to \Delta, s(\forall x)\sim\alpha}{\Gamma_1, s(\exists x)\alpha, \Gamma_2 \to \Delta}, \qquad 12B \quad \frac{\Gamma, s(\forall x)\sim\alpha \to \Delta_1, \Delta_2}{\Gamma \to \Delta_1, s(\exists x)\alpha, \Delta_2}.$$

In all the above schemes Γ_1 and Δ_1 denote sequences of indecomposable formulas and $\Gamma_2, \Delta_2, \Gamma, \Delta$ are arbitrary sequents of formulas; s denotes a sequence of assignment instructions; α, β arbitrary formulas; γ denotes an open formula; M, M' denote arbitrary programs; τ is a term.

Observe that the rules of decomposition reflect the axioms and rules of Hilbert style axiomatization (cf. Chapter II, § 5). Rule r1 (*modus ponens*) has no counterpart among decomposition rules.

DEFINITION 5.3. *By a diagram of a formula α we shall understand an ordered pair $\langle D, d \rangle$, where D is a tree (cf. Definition 5.1 from Chapter II) and d is a mapping which assigns a certain non-empty sequent to every element of the tree. The mapping d and the tree D are defined by induction on level l of D as follows:*

1. For $l = 0$, the only vertex on this level is the empty sequence denoted by $\varnothing$, the root of the tree, and $d(\varnothing)$ is of the form $\to \alpha$.

2. Suppose we have defined the elements of the tree D and the function d on them up to level l not higher than n.

Let $c = (i_1, \ldots, i_n)$ be a vertex on the level n. If $d(c)$ is indecomposable or $d(c)$ is an axiom then c is a leaf of the tree and $d(c)$ is called a leaf-

sequent. In the opposite case let us assume $d(c)$ is of the form $\Gamma \to \Delta$.
CASE A. n is an odd number. The unique son of vertex c is of the form

$$(i_1, \ldots, i_n, 0) \quad and \quad d(i_1, \ldots i_n, 0) = d(c)$$

only if the sequence Δ contains indecomposable formulas.

In the opposite case, if the sequent $d(c)$ is the conclusion in the rule of decomposition of group A,

$$\frac{\{\Gamma_j \to \Delta_j\}_{j \in J}}{\Gamma \to \Delta},$$

then $(i_1, \ldots, i_n, j) \in D$ and $d(i_1, \ldots, i_n, j) = \Gamma_j \to \Delta_j$ for all $j \in J$.
CASE B. n is an even number. The unique son of vertex c is

$$(i_1, \ldots, i_n, 0) \quad and \quad d(i_1, \ldots, i_n, 0) = d(c),$$

only if Γ is a sequence which consists of indecomposable formulas.

In the opposite case, if the sequent $d(c)$ is the conclusion in a rule of decomposition of group B,

$$\frac{\{\Gamma_j \to \Delta_j\}_{j \in J}}{\Gamma \to \Delta},$$

then $(i_1, \ldots, i_n, j) \in D$ and $d(i_1, \ldots, i_n, j) = \Gamma_j \to \Delta_j$ for all $j \in J$. $\square$

REMARK. Let $\langle D, d \rangle$ be a diagram of a formula. If α is an indecomposable formula such that $\alpha \in d(c)$ for some $c = (i_1, \ldots, i_n) \in D$, then for every $c' = (i_1, \ldots, i_n, i_{n+1}, \ldots, i_m)$ if $c' \in D$, then $\alpha \in d(c')$. In other words, if α appears in a vertex, then it also appears in all successors of this vertex. $\square$

LEMMA 5.2. *For every data structure $\mathfrak{A}$ for the language L, for every valuation v in $\mathfrak{A}$ and for every rule of decomposition of the form*

$$\frac{\{\Gamma_j \to \Delta_j\}_{j \in J}}{\Gamma \to \Delta}$$

the following condition holds

$$(\bigwedge \Gamma \Rightarrow \bigvee \Delta)_{\mathfrak{A}}(v) = \operatorname*{g.l.b.}_{j \in J}(\bigwedge \Gamma_j \Rightarrow \bigvee \Delta_j)_{\mathfrak{A}}(v).$$

The proof follows immediately from Lemma 5.1 of Chapter II. $\square$

As a consequence of Lemma 5.2 and Lemma 5.1 we obtain the following fact:

LEMMA 5.3. *If the diagram of a formula* α *is a finite path tree and all leaf-sequents are axioms, then* α *is a tautology.*   □

LEMMA 5.4. *If* α *is a tautology and the diagram of the formula* α *is a finite-path tree, then all leaf-sequents are axioms.*

PROOF. Let $\langle D, d \rangle$ be a diagram of a tautology α, where D is a finite--path tree.

Suppose that there exists a leaf $c \in D$ such that the leaf-sequent $d(c)$ is not an axiom. From the definition of a diagram it follows that $d(c)$ is indecomposable. Let $d(c)$ be of the form $\Gamma \rightarrow \Delta$.

We shall define a data structure $\mathfrak{A}$ for the language L such that

$$\mathfrak{A} = \langle T, \{\varphi_{\mathfrak{A}}\}_{\varphi \in \Phi}, \{\varrho_{\mathfrak{A}}\}_{\varrho \in P} \rangle,$$

where T is the set of all terms in the language L and for arbitrary terms $\tau_1, \ldots, \tau_n$

$$\varphi_{\mathfrak{A}}(\tau_1, \ldots, \tau_n) = \varphi(\tau_1, \ldots, \tau_n), \quad \text{for any functor } \varphi,$$

$$\varrho_{\mathfrak{A}}(\tau_1, \ldots, \tau_n) = 1 \text{ iff } \varrho(\tau_1, \ldots, \tau_n) \in \Gamma, \text{ for any predicate } \varrho.$$

Observe that the last definition is proper since we have assumed $\Gamma \cap \Delta = \varnothing$.

Let v_0 be a valuation in $\mathfrak{A}$ such that $v_0(x) = x$ for all individual variables x, and $v_0(p) = 1$ iff $p \in \Gamma$ for all propositional variables p.

By the definition of a data structure $\mathfrak{A}$ we immediately have $\mathfrak{A}, v_0 \models \sim (\bigwedge \Gamma \Rightarrow \bigvee \Delta)$. Suppose that for some $c' \in D$, $d(c') = \Gamma' \rightarrow \Delta'$ and non $\mathfrak{A}, v_0 \models (\bigwedge \Gamma' \Rightarrow \bigvee \Delta')$. By Lemma 5.2 there exists c'' such that c' is a son of c'', $d(c'') = \Gamma'' \rightarrow \Delta''$ and non $\mathfrak{A}, v_0 \models (\bigwedge \Gamma'' \Rightarrow \bigvee \Delta'')$. Hence there exists a finite path $c_0, \ldots, c_n$, such that $c_0 = \varnothing$, $c_n = c$ and such that for every vertex c_j from this path, if $d(c_j) = \Gamma_j \rightarrow \Delta_j$, then non $\mathfrak{A}, v_0 \models (\bigwedge \Gamma_j \Rightarrow \bigvee \Delta_j)$. In particular, for $j = 0$ we have

$$\mathfrak{A}, v_0 \models \sim (\textbf{true} \Rightarrow \alpha), \quad \text{i.e. } \mathfrak{A}, v_0 \models \sim \alpha.$$

As a consequence, α is not a tautology, contrary to the assumption. □

LEMMA 5.5. *If the diagram of the formula* α_0 *has an infinite path, then* α_0 *is not a tautology.*

PROOF. Let $\langle D, d \rangle$ be the diagram of the formula α_0 and let Path $= \{c_i\}_{i \in N}$ be an infinite path in D. Assume that $d(c_i) = \Gamma_i \rightarrow \Delta_i$, for $i \in N$.

To prove the lemma we shall construct a data structure $\mathfrak{A}$ in the set of all terms T and a valuation v_0 such that

$$\mathfrak{A}, v_0 \models \sim \alpha_0.$$

Denote $F_S = \bigcup_{i \in N} \Delta_i$ and $F_P = \bigcup_{i \in N} \Gamma_i$.

Note that if γ is an indecomposable formula such that $\gamma \in d(c_{i_0})$ for some i_0, then $\gamma \in d(c_i)$ for every $i \geqslant i_0$.

Since for every $i \in N$, $\Gamma_i \cap \Delta_i = \emptyset$, then $\gamma \in F_S - F_P$ or $\gamma \in F_P - F_S$.

Let us consider a data structure $\mathfrak{A}$

$$\mathfrak{A} = \langle T, \{\varphi_{\mathfrak{A}}\}_{\varphi \in \Phi}, \{\varrho_{\mathfrak{A}}\}_{\varrho \in P} \rangle,$$

such that for arbitrary terms $\tau_1, \ldots, \tau_n$

$$\varphi_{\mathfrak{A}}(\tau_1, \ldots, \tau_n) = \varphi(\tau_1, \ldots, \tau_n) \quad \text{for all } \varphi \in \Phi,$$

$$\varrho_{\mathfrak{A}}(\tau_1, \ldots, \tau_n) = 1 \quad \text{iff} \quad \varrho(\tau_1, \ldots, \tau_n) \in F_P \quad \text{for } \varrho \in P,$$

and let

$$v_0(p) = 1 \quad \text{iff} \quad p \in F_P \quad \text{for every propositional variable } p \in P,$$

$$v_0(x) = x \quad \text{for every individual variable } x.$$

We shall prove by induction with respect to the relation $\prec$ (cf. Definition 2.2) that for every formula α

(1)   if $\alpha \in F_S$, then $\mathfrak{A}, v_0 \models \sim \alpha$,
        if $\alpha \in F_P$, then $\mathfrak{A}, v_0 \models \alpha$.

By the definition of the structure $\mathfrak{A}$, property (1) holds for all indecomposable formulas.

Suppose that property (1) holds for all formulas that are submitted to the formula α and let $\alpha \in F_S \cup F_P$.

If α is a decomposable formula then it appears also as a first formula in a certain sequent of the infinite path and therefore it will be decomposed. In particular, if $s \bigcup M(M^i \beta) \in F_S$, then the formulas $s \bigcup M(M^{i+1} \beta)$ and $sM^i \beta$ are in F_S as a consequence of rule (9B). Thus, if $s \bigcup M\beta \in F_S$ then all formulas of the form $s(M^i \beta)$ for $i \in N$ are in F_S. By the inductive assumption $\mathfrak{A}, v_0 \models s(M^i \beta)$ for all $i \in N$ since $s(M^i \beta) \prec \alpha$ and therefore non $\mathfrak{A}, v_0 \models \alpha$.

By Definition 5.3, if the formula α is of the form $\sim \beta$ and $\alpha \in F_S$ then $\beta \in F_P$. By the inductive assumption $\mathfrak{A}, v_0 \models \beta$ since $\beta \prec \alpha$ and therefore non $\mathfrak{A}, v_0 \models \alpha$.

If the formula α is of the form $(\delta \vee \beta)$, $(\delta \wedge \beta)$, $s\gamma$, s **begin** K; M **end** β, s **if** γ **then** K **else** M **fi** β, s **while** γ **do** M **od** β, $s \bigcup M\beta$, $s \bigcap M\beta$ or $s(\forall x)\beta(x)$ and if $\alpha \in F_S$, then by the definition of the diagram there exists a set of formulas $\{\beta_i\}_{i \in J}$ for some set $J \subset N$ such that $\beta_i \in F_S$, $\beta_i \prec \alpha$ and l.u.b. $\beta_{i_{\mathfrak{M}}}(v) = \alpha_{\mathfrak{M}}(v)$ for arbitrary data structure $\mathfrak{A}$ and an arbi-
${\scriptstyle i \in J}$
trary valuation v. By the inductive assumption non $\mathfrak{A}, v_0 \models \beta_i$ for $i \in J$, hence non $\mathfrak{A}, v_0 \models \alpha$.

The case $\alpha \in F_P$ can be discussed in an analogous way. As a result we obtain $\mathfrak{A}, v_0 \models \alpha$.

From the above considerations we have non $\mathfrak{A}, v_0 \models \alpha_0$, since $\alpha_0 \in F_S$, i.e. α_0 is not a tautology. $\qquad \square$

THEOREM 5.1. *The diagram of the formula α is a finite-path tree with all leaf-sequents being axioms iff α is a tautology.* $\qquad \square$

6. THE NORMAL FORM OF PROGRAMS

The aim of this section is to prove that every program can be transformed into a form which contains the single occurrence of the **while**-operation.

We shall start with the auxiliary definitions. Let v, v' be the two valuations and let X be a set of variables. We shall say that $v = v'$ off X if and only if for every $z \notin X, v'(z) = v(z)$.

DEFINITION 6.1. *We shall say that the variable x is inessential for the program M iff the following conditions are valid for arbitrary data structure $\mathfrak{A}$ and for arbitrary valuations v, v' such that $v = v'$ off $(\{x\})$*
 (i) $M_{\mathfrak{A}}(v)$ *is defined iff* $M_{\mathfrak{A}}(v')$ *is defined and*
 (ii) *if* $M_{\mathfrak{A}}(v)$ *and* $M_{\mathfrak{A}}(v')$ *are defined then* $M_{\mathfrak{A}}(v) = M_{\mathfrak{A}}(v')$. $\qquad \square$

Let us consider an example.
$$M: \textbf{begin } u := x+y; \ x := u \cdot z \textbf{ end}.$$
The variable u is then inessential for program M.

DEFINITION 6.2. *Two programs M, M' are equivalent up to a set of variables VAR in symbols $M \sim M'$ off VAR iff for every data structure $\mathfrak{A}$ and every valuation v*

(i) $M_{\mathfrak{A}}(v)$ *is defined iff* $M'_{\mathfrak{A}}(v)$ *is defined and*

(ii) *if both mappings* $M_{\mathfrak{A}}$ *and* $M'_{\mathfrak{A}}$ *are defined at* v, *then*

$$M_{\mathfrak{A}}(v) = M'_{\mathfrak{A}}(v) \text{ off VAR.}$$

In the case where VAR $= \varnothing$ *we shall write* $M \sim M'$.   □

This definition formalizes our intuitive idea of two programs being equivalent iff their results are identical up to the auxiliary variables.

Let Π_0 be a class of programs and VAR a set of variables which are inessential for any M from Π_0. The following properties are then valid for all M, M', M'' from Π_0:

$M \sim M$ off VAR,

if $M \sim M'$ off VAR, then $M' \sim M$ off VAR,

if $M \sim M'$ off VAR   and $M' \sim M''$ off VAR,

then $M \sim M''$ off VAR.

Hence $\sim$ off VAR is an equivalence relation in Π_0.

EXAMPLE 6.1. Let γ be an open formula, M, M' programs and q a propositional variable such that $q \notin V(\gamma) \cup V(M) \cup V(M')$. The following two programs are equivalent up to the set $\{q\}$:

M_1 : **begin**

      while γ **do** M **od**;

      M'

    end,

M_2 : **begin**

      $q := $ **true**;

      while q **do**

          if γ **then** M **else** M'; $q := $ **false fi**

      od

    end.   □

As a consequence of the definition we obtain the following useful lemma:

LEMMA 6.1. *For arbitrary programs* M, M' *and arbitrary set of variables* VAR, *if* $M \sim M'$ *off* VAR, *then for every formula* α *such that* $V(\alpha) \cap$ VAR $= \varnothing$, *for every data structure* $\mathfrak{A}$ *and every valuation* v *in* $\mathfrak{A}$,

$$\mathfrak{A}, v \models M\alpha \quad \textit{iff} \quad \mathfrak{A}, v \models M'\alpha.$$

The proof follows immediately from the fact that the value of every formula depends solely on variables which occur in it.   □

LEMMA 6.2. *For arbitrary programs K, M, K', M' and arbitrary sets of variables* VAR_1, VAR_2, $\text{VAR} = \text{VAR}_1 \cup \text{VAR}_2$, *the following conditions hold*:

(i) *if $K \sim M$ off VAR_1 and $M \sim M'$ off VAR_2, then $K \sim M'$ off VAR*;

(ii) *if $K \sim K'$ off VAR_1 and $M \sim M'$ off VAR_2 and VAR_1 is inessential for M and M' then*

begin K; M **end** $\sim$ **begin** K'; M' **end** *off* VAR;

(iii) *if $K \sim K'$ off VAR_1 and $M \sim M'$ off VAR_2, then for every formula* γ,

if γ **then** K **else** M **fi** $\sim$ **if** γ **then** K' **else** M' **fi** *off* VAR;

(iv) *if $K \sim K'$ off VAR_1 and γ is an open formula such that $V(\gamma) \cap \cap \text{VAR}_1 = \emptyset$, and VAR_1 is inessential for K and K', then*

while γ **do** K **od** $\sim$ **while** γ **do** K' **od** *off* VAR_1.

PROOF. Let $\mathfrak{A}$ be a data structure and v a valuation.
For the proof of (i) let us assume that

$$K \sim M \text{ off } \text{VAR}_1 \quad \text{and} \quad M \sim M' \text{ off } \text{VAR}_2.$$

Suppose $K_\mathfrak{A}$ is defined at v. Hence by the assumption $M_\mathfrak{A}$ is defined at v and finally $M'_\mathfrak{A}$ is defined at v. Analogously, if $M_\mathfrak{A}(v)$ is defined, then $K_\mathfrak{A}(v)$ is defined.

Suppose $\bar{v} = K(v)$ and $\bar{v}' = M'_\mathfrak{A}(v)$. Hence by the assumption $M_\mathfrak{A}(v)$ is also defined and for $\tilde{v} = M_\mathfrak{A}(v)$ we have

$$\bar{v}(z) = \tilde{v}(z) \quad \text{for every } z \notin \text{VAR}_1,$$
$$\bar{v}'(z) = \tilde{v}(z) \quad \text{for every } z \notin \text{VAR}_2.$$

Thus $\bar{v}(z) = v'(z)$ for every $z \notin \text{VAR}_1 \cup \text{VAR}_2$, which completes the proof of (i).
For the proof of (ii) let us assume that

$$K \sim K' \text{ off } \text{VAR}_1 \quad \text{and} \quad M \sim M' \text{ off } \text{VAR}_2.$$

Suppose **begin** K; M **end**$_\mathfrak{A}$ is defined at v. Then $K_\mathfrak{A}$ is defined at v and for $\bar{v} = K_\mathfrak{A}(v)$, $M_\mathfrak{A}(\bar{v})$ is defined. Let $M_\mathfrak{A}(\bar{v}) = \bar{\bar{v}}$. By the assumption $M'_\mathfrak{A}(\bar{v})$ and $K'_\mathfrak{A}(v)$ are defined.

Let $\bar{v}' = K'_{\mathfrak{A}}(v)$ and $\bar{\bar{v}}' = M'_{\mathfrak{A}}(\bar{v})$. Thus by the assumptions

(1) $\bar{v}(z) = \bar{v}'(z)$ for every $z \notin \text{VAR}_1$,

(2) $\bar{\bar{v}}'(z) = \bar{\bar{v}}(z)$ for every $z \notin \text{VAR}_2$.

Since VAR_1 is the set of variables inessential for M and for M', $M'_{\mathfrak{A}}(\bar{v}')$ is defined and for $\bar{\bar{v}}'' = M'_{\mathfrak{A}}(\bar{v}')$ we have by (1),

(3) $\bar{\bar{v}}''(z) = \bar{\bar{v}}'(z)$ for every $z \notin \text{VAR}_1$.

Hence **begin** K'; M' **end**$_{\mathfrak{A}}(v)$ is defined.

Conversely, if **begin** K'; M' **end**$_{\mathfrak{A}}$ (v) is defined, then by the definition of semantics $K'_{\mathfrak{A}}(v)$ is defined and $M_{\mathfrak{A}}(K'_{\mathfrak{A}}(v))$ is defined. By the assumption we have $K_{\mathfrak{A}}(v)$ is defined and $M_{\mathfrak{A}}(K'_{\mathfrak{A}}(v))$ is defined. Since $K'_{\mathfrak{A}}(v)$ and $K'_{\mathfrak{A}}(v)$ differs at most on variables VAR_1 then $M_{\mathfrak{A}}(K_{\mathfrak{A}}(v))$ is also defined.

Moreover by (2) and (3) we have

$$\bar{\bar{v}}(z) = \bar{\bar{v}}''(z)   \text{ for every } z \notin \text{VAR}_1 \cup \text{VAR}_2.$$

As a consequence we obtain

$$\textbf{begin } K; \ M \textbf{ end} \sim \textbf{begin } K'; \ M' \textbf{ end off VAR},$$

which completes the proof of (ii).

The similar proofs of (iii) and (iv) are omitted.   □

DEFINITION 6.2. *A program M is in the normal form iff*

$$M = \textbf{begin } M_1; \textbf{ while } \gamma \textbf{ do } M_2 \textbf{ od end},$$

where M_2 and M_1 are programs without a **while**-*operation*.   □

THEOREM 6.3. *Composition, branching and iteration of programs in the normal form are equivalent to a program in the normal form.*

PROOF. Let K and M be two programs in the normal form

$$K = \textbf{begin } K_1; \textbf{ while } \gamma_1 \textbf{ do } K_2 \textbf{ od end},$$

$$M = \textbf{begin } M_1; \textbf{ while } \gamma_2 \textbf{ do } M_2 \textbf{ od end},$$

and suppose that q does not belong to $V(K)$ and $V(M)$. The theorem follows from equivalences (4), (5), (6).

(4)   **begin** K; M **end** $\sim M'$ off($\{q\}$).

           M': **begin**

                   $q := \textbf{true}$; K_1;

                   while $(q \wedge \gamma_1) \vee (\sim q \wedge \gamma_2)$ **do**

if $(\gamma_1 \wedge q)$ **then** K_2 **else**
  if $(\sim \gamma \wedge q)$ **then**
    M_1 ; $q :=$ **false else** M_2
  fi
fi
od
end

(5) **if** γ **then** K **else** M **fi** $\sim M''$ **off** $(\{q\})$.
M'': **begin**
  $q := \gamma$;
  if q **then** K_1 **else** M_1 **fi**;
  while $(q \wedge \gamma_1) \vee (\sim q \wedge \gamma_2)$ **do**

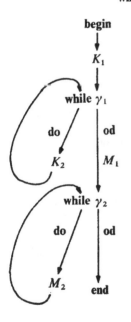

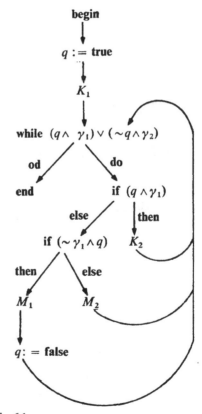

Fig. 6.1

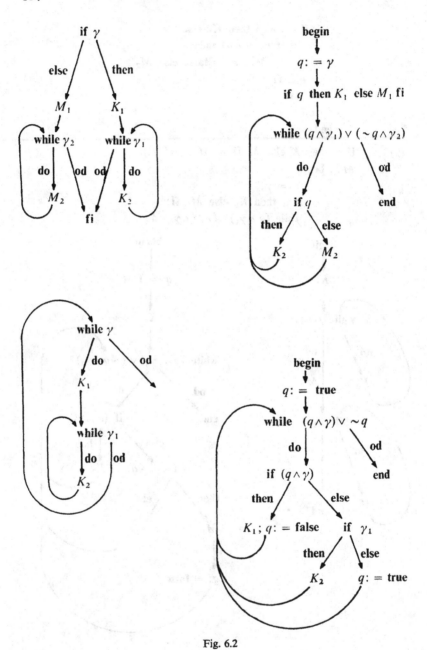

Fig. 6.2

$$\text{if } q \text{ then } K_2 \text{ else } M_2 \text{ fi}$$
$$\textbf{od}$$
$$\textbf{end}$$

(6) $\textbf{while } \gamma \textbf{ do } K \textbf{ od } \sim M''' \text{ off } (\{q\}).$

M''': $\textbf{begin}$

$q := \textbf{true};$

$\textbf{while } ((q \wedge \gamma) \vee \sim q) \textbf{ do}$

$\textbf{if } (q \wedge \gamma) \textbf{ then } K_1; \; q := \textbf{false else}$

$\textbf{if } \gamma_1 \textbf{ then } K_2 \textbf{ else } q := \textbf{true fi}$

$\textbf{fi}$

$\textbf{od}$

$\textbf{end}.$

The lengthy proofs of (4), (5) and (6) are omitted.

We shall illustrate equivalences (4)–(6) by the diagrams shown in Figures 6.1 (p. 113) and 6.2 (p. 114). $\qquad\qquad\qquad\qquad\qquad$ □

THEOREM 6.4. *For every program M there exists a program M' in the normal form such that $V(M') \supset V(M)$ and $M \sim M'$ off $(V(M') - V(M))$. Moreover, all variables from the set $V(M') - V(M)$ are inessential for the program M'.* $\qquad\qquad\qquad\qquad\qquad\qquad\qquad\qquad$ □

7. EQUALITY

In this and the next few sections we shall discuss some extensions of the algorithmic language introduced in Chapter II. The character of these extensions will differ. In this section we extend the alphabet by admitting equality, in § 8 we extend the set of well-formed expressions by generalized terms and parallel substitutions, and in § 9 we extend the notion of data structure in order to discuss partial functions.

In all these extensions the corresponding notion of tautology can be axiomatized and the Completeness Theorem can be proved.

Let us assume that the alphabet of algorithmic language L contains the binary predicate of equality $=$.

DEFINITION 7.1. *We shall say that a data structure $\mathfrak{A}$ for algorithmic language L is proper for equality iff the interpretation of $=$ in the structure $\mathfrak{A}$ is the identity relation.* $\qquad\qquad\qquad\qquad\qquad\qquad$ □

By *algorithmic logic with identity* we shall understand an extension of the axiomatic system described in Chapter II, § 5, by additional axioms characterizing predicate $=$.

(e1) $x = x$,

(e2) $(x = y \Rightarrow y = x)$,

(e3) $((x = y \land y = z) \Rightarrow x = z)$,

(e4) for every n-argument functor $\varphi \in \Phi$,

$$((x_1 = y_1 \land \ ... \ \land x_n = y_n) \Rightarrow \varphi(x_1, ..., x_n) = \varphi(y_1, ..., y_n)),$$

(e5) for every n-argument predicate $\varrho \in P$,

$$((x_1 = y_1 \land \ ... \ \land x_n = y_n) \Rightarrow \varrho(x_1, ..., x_n) \equiv \varrho(y_1, ..., y_n)).$$

In all the above formulas $x, y, z, x_1, ..., x_n, y_1, ..., y_n$ are individual variables.

The first three axioms state that $=$ is an equivalence relation, and the last two concern the extensionality of $=$.

As an immediate consequence of the above axioms we have the following corollary:

COROLLARY. For every term τ and every formula α in the algorithmic language with equality

$$\vdash (\dot{x} = \dot{y} \Rightarrow \tau(\dot{z}/\dot{x}) = \tau(\dot{z}/\dot{y})),$$

$$\vdash (\dot{x} = \dot{y} \Rightarrow \alpha(\dot{z}/\dot{x}) \equiv \alpha(\dot{z}/\dot{y})). \qquad \square$$

Algorithmic logic with identity is obviously consistent. Moreover the following Completeness Theorem is a straightforward consequence of Theorem 2.5 from Chapter III.

THEOREM 7.1. *For every formula α and every set of formulas A*

(i) *α is a theorem in a theory $\langle L, C, A \rangle$ based on algorithmic logic with identity if and only if*

(ii) *α is valid in all models for A which are proper structures for identity.*

The proof is analogous to that of Theorem 2.5 and is therefore omitted. $\square$

In Chapter II, § 3, we have seen properties (e.g. the strongest postcondition) that are expressible in the language with identity. Now we shall mention some others.

LEMMA 7.2. *Let M be a program and let* α *be a formula. Thus the formula*

(1) $\bigcap M\alpha \equiv (\forall \mathring{y})$ (**while** $\sim (\mathring{x} = \mathring{y})$ **do** M **od true**
                    $\Rightarrow \left(\alpha(\mathring{x}/\mathring{y}) \wedge M(\mathring{x}/\mathring{y}) \text{ true}\right))$

is a theorem of algorithmic logic with identity, where $\mathring{x}$ *is a sequence* $x_1, \ldots, x_n$ *of all variables which appear in* $M\alpha$ *and* $\mathring{y}$ *is a copy of* $\mathring{x}$.

PROOF. Let $\mathfrak{A}$ be a data structure proper for identity and let v be a valuation in $\mathfrak{A}$.

Suppose

(2) $\mathfrak{A}, v \models (\forall \mathring{y})$ (**while** $\sim (\mathring{x} = \mathring{y})$ **do** M **od true**
                    $\Rightarrow \left(\alpha(\mathring{x}/\mathring{y}) \wedge M(\mathring{x}/\mathring{y}) \text{ true}\right))$

and

$\mathfrak{A}, v \models \sim \bigcap M\alpha.$

Hence there exists a natural number i such that

(3) $\mathfrak{A}, v \models \sim M^i \alpha$ and $\mathfrak{A}, v \models M^j \alpha$ for $j < i$.

By (2)

(4) $\mathfrak{A}, v_{\mathring{a}}^{\mathring{y}} \models \left($ **while** $\sim (\mathring{x} = \mathring{y})$ **do** M **od true**
                    $\Rightarrow \left(\alpha(\mathring{x}/\mathring{y}) \wedge M(\mathring{x}/\mathring{y}) \text{ true}\right)\right)$

for an arbitrary vector $\mathring{a}$ of elements in $\mathfrak{A}$ which corresponds to $\mathring{y}$, where $v_{\mathring{a}}^{\mathring{y}}$ is a natural extension of the denotation v_a^y (cf. Chapter II, § 2).

Let us take as $\mathring{a}$ a sequence $a_1, \ldots, a_n$ such that

$a_j = M_{\mathfrak{A}}^{i-1}(v)(x_j)$ for $j \leqslant n$.

Thus

$\mathfrak{A}, v_{\mathring{a}}^{\mathring{y}} \models M^{i-1}(\mathring{x} = \mathring{y})$

and therefore $\mathfrak{A}, v_{\mathring{a}}^{\mathring{y}} \models$ **while** $\sim (\mathring{x} = \mathring{y})$ **do** M **od true**. As a consequence of (4) we obtain

$\mathfrak{A}, v_{\mathring{a}}^{\mathring{y}} \models M(\mathring{x}/\mathring{y}) \text{ true}$,   i.e. $\mathfrak{A}, v \models M^i \text{ true}$.

Let us take as $\mathring{a}$ a sequence $a_1, \ldots, a_n$ such that

$a_j = M_{\mathfrak{A}}^i(v)(x_j)$ for all $1 \leqslant j \leqslant n$.

Thus

$$\mathfrak{A}, v_{\bar{a}}^{\bar{x}} \models M^i(\bar{x} = \bar{y})$$

and therefore

$$\mathfrak{A}, v_{\bar{a}}^{\bar{y}} \models \textbf{while } \sim (\bar{x} = \bar{y}) \textbf{ do } M \textbf{ od true}.$$

By (4) we obtain

$$\mathfrak{A}, v_{\bar{a}}^{\bar{x}} \models \alpha(\bar{x}/\bar{y}), \quad \text{i.e. } \mathfrak{A}, v \models M^i\alpha,$$

a contradiction.

The converse implication can be proved analogously. Hence formula (1) is valid for every data structure and every valuation. It follows from the Completeness Theorem that formula (1) is a theorem of algorithmic logic with identity.   □

Note, that Lemma 7.2 allows us to eliminate the iteration quantifier from formulas of algorithmic logic.

The next property we shall discuss is the equivalence of programs. It appears that in the language with identity, equivalence of programs is expressible by a formula.

LEMMA 7.3. *For all programs* M, M' *and every set of variables* VAR M ∼ M' *off* VAR *iff for every formula* α, *such that* $V(\alpha) \cap \text{VAR} = \emptyset$, Mα ≡ M'α *is a theorem of algorithmic logic with identity.*

PROOF. Assume Mα ≡ M'α is a theorem of algorithmic logic with identity for every α such that $V(\alpha) \cap \text{VAR} = \emptyset$. Let $\mathfrak{A}$ be a data structure proper for identity and v a valuation in $\mathfrak{A}$. By the Completeness Theorem and by the assumption $M_{\mathfrak{A}}(v)$ is defined iff $M'_{\mathfrak{A}}(v)$ is defined. Suppose that for some v, $M_{\mathfrak{A}}(v)$, $M'_{\mathfrak{A}}(v)$ are defined but that there exists an individual variable x such that x ∉ VAR and

$$M_{\mathfrak{A}}(v)(x) \neq M'_{\mathfrak{A}}(v)(x).$$

Let us consider the formula x = y, where $y \notin V(M) \cup V(M')$, and let $a = M_{\mathfrak{A}}(v)(x)$.

It follows from the above that

$$\mathfrak{A}, v_a^y \models M(x = y) \quad \text{and} \quad \text{non } \mathfrak{A}, v_a^y \models M'(x = y).$$

Hence Mα ≡ M'α is not valid in $\mathfrak{A}$, a contradiction.

The converse implication has already been proved in Lemma 6.1. □

LEMMA 7.4. *Programs K and M are equivalent up to the set of variables* VAR *if and only if the following formula is a theorem of algorithmic logic with identity.*

$$(5) \qquad \left(\sim (K\ \mathbf{true} \vee M\ \mathbf{true}) \vee (K\ \mathbf{true} \wedge M\ \mathbf{true}) \wedge \bigwedge_{j \leqslant m} (Kq_j \equiv Mq_j) \wedge \right.$$
$$\left. \wedge \bigwedge_{i \leqslant n} K(\vec{x}/\vec{y})\, M(x_{:} = y_i) \right),$$

where

$x_1, ..., x_n$ *are individual variables such that*
$$\{x_1, ..., x_n\} = V_i \cap (V(K) \cup V(M)) - \mathrm{VAR},$$

$q_1, ..., q_m$ *are propositional variables such that*
$$\{q_1, ..., q_m\} = V_0 \cap (V(K) \cup V(M)) - \mathrm{VAR},$$

$y_1, ..., y_n$ *are individual variables such that for* $1 \leqslant j \leqslant n$
$$y_j \notin V(M) \cup V(K) \cup \mathrm{VAR}. \qquad \qquad \square$$

8. GENERALIZED TERMS

Every term defines a total function in a data structure. However an important role is played by partial functions in some situations. How are they to be described in algorithmic language? The solution is based on the notion of generalized term.

DEFINITION 8.1. *By the set of generalized terms we shall understand an extension of the set T by the following rule: if τ is a term and M is a program, then $M\tau$ is a term.* $\qquad \square$

Let $\mathfrak{A}$ be a data structure for the language L. The semantics of generalized terms is as follows: For every τ, $\tau_{\mathfrak{A}}$ is a partial function in the set of all valuations in $\mathfrak{A}$ such that

$$x_{\mathfrak{A}}(v) = v(x);$$

$$\varphi(\tau_1, ..., \tau_n)_{\mathfrak{A}}(v) = \begin{cases} \varphi_{\mathfrak{A}}\big(\tau_{1\mathfrak{A}}(v), ..., \tau_{n\mathfrak{A}}(v)\big), & \text{if} \\ \qquad \tau_{1\mathfrak{A}}(v), ..., \tau_{n\mathfrak{A}}(v) \text{ are defined,} \\ \text{undefined} \quad \text{otherwise;} \end{cases}$$

$$(K\tau)_{\mathfrak{A}}(v) = \begin{cases} \tau_{\mathfrak{A}}(v'), & \text{if } K_{\mathfrak{A}}(v) \text{ is defined and } v' = K_{\mathfrak{A}}(v), \\ \text{undefined} \quad \text{otherwise,} \end{cases}$$

where x is an individual variable; φ is an n-argument functor; $\tau_1, ..., \tau_n$, τ are generalized terms and K is a program.

EXAMPLE 8.1. Let M be the following program:

begin $x := 0$; **while** $x+1 \neq y$ **do** $x := x+1$ **od end**.

The generalized term Mx in the data structure $\mathfrak{N}$ of natural numbers is defined if and only if the value of y is not equal to zero. Moreover, for every valuation v in $\mathfrak{N}$

$$(Mx)_{\mathfrak{N}}(v) = n \quad \text{iff} \quad n+1 = v(y). \qquad \square$$

One of the most important properties of generalized terms is the existence of the normal form. Our considerations are based on the following lemma which is an immediate consequence of the definition of semantics. In this section we shall read equality $\tau_{\mathfrak{A}}(v) = \tau'_{\mathfrak{A}}(v)$ in the following way: $\tau_{\mathfrak{A}}(v)$ is defined iff $\tau'_{\mathfrak{A}}(v)$ is defined and if values of both sides are defined then they are identical.

LEMMA 8.1. *For every data structure and every valuation v*

$$\big(K\varphi(\tau_1, \ldots, \tau_n)\big)_{\mathfrak{A}}(v) = \varphi\big((K\tau_1)_{\mathfrak{A}}(v), \ldots, (K\tau_n)_{\mathfrak{A}}(v)\big),$$

where K is an arbitrary program, $\tau_1, \ldots, \tau_n$ are arbitrary generalized terms and φ is an n-argument functor. $\qquad \square$

DEFINITION 8.2. *We shall say that a generalized term τ is in the normal form iff $\tau = M\eta$, where M is a program and η is a term (classical not generalized!).* $\qquad \square$

LEMMA 8.2. *For every generalized term τ there exists a generalized term in the normal form $M\eta$ such that for every data structure $\mathfrak{A}$ and every valuation v in $\mathfrak{A}$*

$$\tau_{\mathfrak{A}}(v) = (M\eta)_{\mathfrak{A}}(v).$$

PROOF. The proof is by induction on the length of the generalized term.

Lemma 8.2 obviously holds for all classical terms. Let us consider a generalized term $\varphi(\tau_1, \ldots, \tau_n)$ and by the inductive assumption, let $M_i\eta_i$ for $i \leqslant n$ be the generalized terms in the normal form such that

$$(M_i\eta_i)_{\mathfrak{A}}(v) = \tau_{i\mathfrak{A}}(v), \quad 1 \leqslant i \leqslant n,$$

for every data structure $\mathfrak{A}$ and every valuation v. Hence

$$\varphi(\tau_1, \ldots, \tau_n)_{\mathfrak{A}}(v) = \varphi_{\mathfrak{A}}\big((M_1\eta)_{\mathfrak{A}}(v), \ldots, (M_n\eta_n)_{\mathfrak{A}}(v)\big).$$

Let $\dot{x}_i = (x_{i_1}, \ldots, x_{i_m})$ be the sequence of all variables that occur in $M_i \eta_i$ for $i \leqslant n$, and let $\dot{y}_i = (y_{i_1}, \ldots, y_{i_m})$ be a copy of $\dot{x}_i$ such that $\dot{y}_i \subset V - \bigcup V(M_i \eta_i)$ and $\dot{y}_i \cap \dot{y}_j = \emptyset$ for $i \neq j$.

Let s_i denote the program

$$\textbf{begin } y_{i_1} := x_{i_1}; \ldots; y_{i_m} := x_{i_m} \textbf{ end}$$

and $M_i(\dot{x}_i/\dot{y}_i)$, $\eta_i(\dot{x}_i/\dot{y}_i)$ be copies of the program M_i and the term η_i.

The generalized term

$$\tau = s_1 M_1(\dot{y}_1) \ldots s_n M_n(\dot{y}_n) \varphi(\eta_1(\dot{y}_1), \ldots, \eta_n(\dot{y}_n))$$

is in the normal form and for every data structure $\mathfrak{A}$ and valuation v, $\tau_{\mathfrak{A}}(v) = \varphi_{\mathfrak{A}}(\tau_{1\mathfrak{A}}(v), \ldots, \tau_{n\mathfrak{A}}(v))$.

It remains to consider a generalized term of the form $M\tau$. By the inductive assumption there exists a normalized term $K\eta$ for τ such that

$$(K\eta)_{\mathfrak{A}}(v) = \tau_{\mathfrak{A}}(v)$$

for all $\mathfrak{A}$ and v.

Thus $\textbf{begin } M; K \textbf{ end } \eta$ is a generalized term in the normal form such that for all $\mathfrak{A}$ and v,

$$(\textbf{begin } M; K \textbf{ end } \eta)_{\mathfrak{A}}(v) = (M\tau)_{\mathfrak{A}}(v). \qquad \square$$

Let L' be an extension of an algorithmic language L by generalized terms such that if ϱ is an n-argument predicate and $\tau_1, \ldots, \tau_n$ are arbitrary generalized terms then the expression

$$(1) \qquad \varrho(\tau_1, \ldots, \tau_n)$$

is a formula.

We shall assume the following interpretation. For every data structure $\mathfrak{A}$ and every valuation v

$$\varrho(\tau_1, \ldots, \tau_n)_{\mathfrak{A}}(v) = \begin{cases} \varrho(\tau_{1\mathfrak{A}}(v), \ldots, \tau_{n\mathfrak{A}}(v)) & \text{if } \tau_{i\mathfrak{A}}(v) \text{ is} \\ & \text{defined for all } i \leqslant n, \\ 0 & \text{otherwise.} \end{cases}$$

REMARK. The formula of form (1) should be not considered as an elementary one. $\qquad \square$

LEMMA 8.3. *For every formula of the form* (1) *there exists a program* K *and terms* $\eta_1, \ldots, \eta_n$ *such that*

$$\models \varrho(\tau_1, \ldots, \tau_n) \equiv K\varrho(\eta_1, \ldots, \eta_n).$$

The proof is similar to the proof of Lemma 8.2. $\qquad \square$

The result of Lemma 8.3 can be generalized to the set of all formulas.

LEMMA 8.4. *For every formula α of the language L' there exists a formula $\chi(\alpha)$ of the language L such that*

$$\models \alpha \equiv \chi(\alpha).$$

The details of mapping χ can be found in Mirkowska, 1975. □

Lemma 8.4 states that the extension L' of an algorithmic language is not essential.

Let us now consider the problem of axiomatization of algorithmic logic with generalized terms. Obviously, all the axioms and rules mentioned in Chapter II, § 5, are still valid. However, to obtain a complete characterization of the set of tautologies of the language L' it is necessary to characterize the behaviour of generalized terms.

The following theorem gives a solution to the problem.

THEOREM 8.5. *For every set of formulas A and every formula α of the language L' the following conditions are equivalent:*

(i) α *is valid in every model of the set A;*

(ii) α *has a formal proof from the set A extended by the formulas of the form*

$$M\varrho(\tau_1, ..., \tau_n) \equiv \varrho(M\tau_1, ..., M\tau_n),$$
$$\varrho(\tau_1, ..., \tau_n) \equiv \chi\big(\varrho(\tau_1, ..., \tau_n)\big),$$

where ϱ is an n-argument predicate, M is a program and $\tau_1, ..., \tau_n$ are arbitrary generalized terms. □

9. PARTIAL FUNCTIONS

We have so far discussed data structures for the algorithmic language in which functors have been interpreted as total functions. In this section we shall extend the notion of data structure to the class of relational systems with partial operations.

Let L be an algorithmic language of the type $\langle \{n_\varphi\}_{\varphi \in \Phi}, \{m_\varrho\}_{\varrho \in P}\rangle$. By a partial data structure for L we shall understand a relational system

$$\mathfrak{A} = \langle A, \{\psi_{\mathfrak{A}}\}_{\psi \in \Phi}, \{\varrho_{\mathfrak{A}}\}_{\varrho \in P}\rangle$$

such that

(i) for every m_ϱ-argument predicate ϱ of L, $\varrho_{\mathfrak{A}}$ is an m_ϱ-argument relation in A,

(ii) for every n_ψ-argument functor ψ of L, $\psi_\mathfrak{A}$ is an n_ψ-argument partial operation in A,

(iii) for every n_ψ-argument functor $\psi \in \Phi$ there exists an n_ψ-argument relation $\varrho_\psi \in P$ such that for arbitrary elements $a_1, \dots, a_{n_\psi}$, $\psi_\mathfrak{A}(a_1, \dots, a_{n_\psi})$ is defined iff $(a_1, \dots, a_{n_\psi}) \in \varrho_{\psi\mathfrak{A}}$.

For a given data structure $\mathfrak{A}$ and valuation v we shall define the semantics of terms and open formulas as in Chapter II, § 2, with some exceptions:

$$\psi(\tau_1, \dots, \tau_n)_\mathfrak{A}(v) = \begin{cases} \psi_\mathfrak{A}\big(\tau_{1\mathfrak{A}}(v), \dots, \tau_{n\mathfrak{A}}(v)\big) & \text{if } \tau_{i\mathfrak{A}}(v) \text{ is} \\ & \text{defined for all } i \leqslant n, \\ \text{undefined} & \text{otherwise,} \end{cases}$$

$$\varrho(\tau_1, \dots, \tau_n)_\mathfrak{A}(v) = \begin{cases} \varrho_\mathfrak{A}\big(\tau_{1\mathfrak{A}}(v), \dots, \tau_{n\mathfrak{A}}(v)\big) & \text{if } \tau_{i\mathfrak{A}}(v) \text{ is} \\ & \text{defined for all } i \leqslant n, \\ \mathbf{0} & \text{otherwise.} \end{cases}$$

Let τ be a term and let $E(\tau)$ be an open formula of L such that

$$E(x) = \mathbf{true} \quad \text{for } x \in V_i,$$
$$E\big(\psi(\tau_1, \dots, \tau_n)\big) = \varrho_\psi(\tau_1, \dots, \tau_n) \quad \text{for } \tau_i \in T \text{ and } \psi \in \Phi.$$

Analogously, for an arbitrary open formula γ we shall define an open formula $E(\gamma)$ of the language L such that

$$E(q) = \mathbf{true} \quad \text{for } q \in V_0,$$
$$E\big(\varrho(\tau_1, \dots, \tau_n)\big) = \bigwedge_{i \leqslant n} E(\tau_i) \quad \text{for } \tau_i \in T, \varrho \in P,$$
$$E(\gamma \wedge \beta) = E(\gamma \vee \beta) = E(\gamma) \wedge E(\beta) \quad \text{for } \beta \in F_0,$$
$$E(\sim\gamma) = E(\gamma).$$

The sense of the formula $E(w)$, where w is a term or an open formula, is given by the following equivalence: for arbitrary data structure $\mathfrak{A}$ and valuation v,

$$\mathfrak{A}, v \models_{\text{pf}} E(w) \quad \text{iff} \quad \text{for every subterm } \varphi(\tau_1, \dots, \tau_n) \text{ of the}$$
expression w the sequence $\big(\tau_{1\mathfrak{A}}(v), \dots, \tau_{n\mathfrak{A}}(v)\big)$ belongs to the domain of the function $\varphi_\mathfrak{A}$.

We shall write $\models_{\text{pf}}$ to underline the fact that the satisfiability relation concerns the class of partial data structures.

The formulas $E(\tau)$ and $E(\gamma)$ play an important role in the definition of the semantics of programs:

$$(x := w)_{\mathfrak{A}}(v) = \begin{cases} v' & \text{if } \mathfrak{A}, v \models_{pf} E(w) \text{ and } v'(x) = w_{\mathfrak{A}}(v) \\ & v'(z) = v(z) \text{ for all } z \neq x, \\ \text{undefined} & \text{otherwise,} \end{cases}$$

$$(\text{if } \gamma \text{ then } M \text{ else } M' \text{ fi})_{\mathfrak{A}}(v) = \begin{cases} M_{\mathfrak{A}}(v) & \text{if } \mathfrak{A}, v \models E(\gamma) \wedge \gamma \\ & \text{and } M_{\mathfrak{A}}(v) \text{ is defined} \\ M'_{\mathfrak{A}}(v) & \text{if } \mathfrak{A}, v \models E(\gamma) \wedge \sim \gamma \\ & \text{and } M'_{\mathfrak{A}}(v) \text{ is defined,} \\ \text{undefined} & \text{otherwise,} \end{cases}$$

$$(\text{begin } K; M \text{ end})_{\mathfrak{A}}(v) = \begin{cases} M_{\mathfrak{A}}(v') & \text{if } K_{\mathfrak{A}}(v) \text{ is defined and} \\ & M_{\mathfrak{A}}(v') \text{ is defined for} \\ & v' = K_{\mathfrak{A}}(v), \\ \text{undefined} & \text{otherwise,} \end{cases}$$

$$(\text{while } \gamma \text{ do } M \text{ od})_{\mathfrak{A}}(v) = \begin{cases} M_{\mathfrak{A}}^i(v) & \text{if } M_{\mathfrak{A}}^j(v) \text{ is defined for} \\ & \text{all } j \leqslant i, \\ & \mathfrak{A}, v \models_{pf} M_{\mathfrak{A}}^j(E(\gamma) \wedge \gamma) \\ & \text{for } j \leqslant i, \\ & \mathfrak{A}, v \models_{pf} M_{\mathfrak{A}}^i(E(\gamma) \vee \sim \gamma), \\ \text{undefined} & \text{otherwise.} \end{cases}$$

In all the above expressions τ is a term, γ is an open formula, and K, M are arbitrary programs.

Let us note that the result of a program is not defined whenever we find an operation whose arguments do not belong to the domain of the operation. This implies the existence of computations which are finite sequences but which have no results. We shall call such computations *unsuccessful* to distinguish them from those finite computations which do possess results and which are termed *successful computations*.

The meaning of the formula $K\alpha$ is now as follows: for every data structure $\mathfrak{A}$ and every valuation v

$\mathfrak{A}, v \models_{pf} K\alpha$ iff   there exists a successful computation of K from the initial valuation v in $\mathfrak{A}$ whose result satisfies the formula α.

In particular $\mathfrak{A}, v \models_{pf} K$ **true** means the program K has a successful computation from the valuation v in $\mathfrak{A}$. Let us consider the negation of the formula K **true**. $\mathfrak{A}, v \models_{pf} \sim K$ **true** if and only if program K does not possesses a successful computation from the valuation v in $\mathfrak{A}$. The last sentence implies that the computation of K is either infinite or unsuccessful. The property "program K has an infinite computation" is expressible in the language L by the formula loop(K) (see Chapter II, § 3). Is it possible to express the other property by a formula? The following lemma provides a positive answer to this question.

LEMMA 9.1. *For every program K of the language L there exists a formula* fail(K) *of L such that for every data structure $\mathfrak{A}$ and every valuation v*

$$\mathfrak{A}, v \models_{pf} \text{fail}(K) \quad \textit{iff} \quad \textit{there exists an unsuccessful computation of } K \textit{ from } v \textit{ in } \mathfrak{A}.$$

PROOF. Let us consider the following recursive definition

$$\text{fail}(x := w) \overset{df}{=} \sim E(w),$$

$$\text{fail}(\textbf{begin } K; M \textbf{ end}) = \text{fail}(K) \vee K \text{ fail}(M),$$

$$\text{fail}(\textbf{if } \gamma \textbf{ then } M \textbf{ else } K \textbf{ fi})$$
$$= \big(E(\gamma) \Rightarrow \big(\gamma \wedge \text{fail}(M) \vee \sim\gamma \wedge \text{fail}(K)\big)\big)$$

$$\text{fail}(\textbf{while } \gamma \textbf{ do } M \textbf{ od})$$
$$= \bigcup \textbf{if } \gamma \textbf{ then } M \textbf{ fi } \big(E(\gamma) \Rightarrow (\gamma \wedge \text{fail}(M))\big).$$

The lemma follows immediately from the definition of semantics and the construction of the formula fail(K). □

To summarize our considerations let us note two tautologies

(1) $\qquad \models_{pf} K \textbf{ true} \equiv \big(\sim\text{loop}(K) \wedge \sim\text{fail}(K)\big)$

(2) $\qquad \models_{pf} \big(\text{loop}(K) \Rightarrow \sim\text{fail}(K)\big),$

We now turn to the problem of axiomatization. It is easy to observe that the set of all formulas valid in any data structure is closed with respect to all the inference rules mentioned in Chapter II, § 5. Moreover, if a formula α is valid in every data structure with partial operations then it is valid in every data structure with total operations,

(3) $\qquad \models_{pf} \alpha \quad \text{implies} \quad \models \alpha.$

The converse is not true. In particular axioms Ax12, Ax13, Ax22 and Ax23 of AL (cf. Chapter II, § 5) are no longer valid.

Let Ax_{pf} be the set of formulas which contains all the axioms of the system AL except for Ax12, Ax13, Ax22, Ax23 and the following schemes

$$(s \textbf{ true} \Rightarrow (s\gamma \equiv \overline{s}\overline{\gamma})), \quad E(\tau) \equiv (x := \tau) \textbf{ true},$$
$$(s \textbf{ true} \Rightarrow (s \sim \alpha \equiv \sim (s\alpha))), \quad E(\gamma) \equiv (q := \gamma) \textbf{ true},$$
$$\textbf{if } \gamma \textbf{ then } M \textbf{ else } M' \textbf{ fi } \alpha \equiv E(\gamma) \wedge (\gamma \wedge M\alpha \vee \sim\gamma \wedge M'\alpha),$$
$$\textbf{while } \gamma \textbf{ do } M \textbf{ od } \alpha$$
$$\equiv E(\gamma) \wedge (\sim\gamma \wedge \alpha \vee \gamma \wedge M(E(\gamma) \wedge \textbf{while } \gamma \textbf{ do } M \textbf{ od } \alpha)),$$

where s is an assignment instruction, M, M' are programs, γ is an open formula and α is an arbitrary formula.

Let C_{pf} be a syntactical consequence operation such that for every set of formulas Z, $C_{pf}(Z)$ is the smallest set containing $Z \cup Ax_{pf}$ and which is closed with respect to the rules of inference r1–r6 (cf. Chapter II, § 5). To denote that $\alpha \in C_{pf}(Z)$ we shall write $Z \vdash_{pf} \alpha$ for short.

LEMMA 9.2. *For an arbitrary formula α and arbitrary set of formulas Z,*

$$Z \vdash_{pf} \alpha \quad implies \quad Z \models_{pf} \alpha.$$

The proof is by verification of all axioms and rules of inference.   □

By the Completeness Theorem for AL and property (3) we obtain

(4) $\vdash_{pf} \alpha$ implies $\vdash \alpha$

for every formula α of the language L.

The logic introduced here is consistent. Furthermore, if a data structure $\mathfrak{A}$ with total operations is a model of a theory $\langle L, C, A \rangle$, then $\mathfrak{A}$ is a model of $\langle L, C_{pf}, A \rangle$. This implies the following lemma:

LEMMA 9.3. *If $\langle L, C, A \rangle$ is a consistent algorithmic theory, then the theory $\langle L, C_{pf}, A \rangle$ is also consistent.*   □

The model existence theorem and the Completeness Theorems are also valid. The method of proof is in both cases similar to that presented in Chapter III, § 2.

THEOREM 9.4.

(i) *If a theory $\langle L, C_{pf}, A \rangle$ is consistent, then it has a model.*

(ii) *For every consistent theory $T = \langle L, C_{pf}, A \rangle$ and for every formula α of L*

$$A \vdash_{pf} \alpha \quad iff \quad \mathfrak{A} \models_{pf} \alpha \quad for \ the \ arbitrary \ partial \ data \ structure \ \mathfrak{A} \ which \ is \ a \ model \ of \ A.$$ □

10. MANY SORTED STRUCTURES

Many-sorted data structures frequently appear in programming, e.g. stacks, dictionaries, etc. (cf. Chapter IV). These structures have functions and relations whose arguments are of different sorts, e.g. the relation "e is a member of stack s" has two arguments: s, which is a stack, and e, which is an element of the stack.

In this section we shall examine an algorithmic language which is convenient for discussing many-sorted data structures. In a way this extends what we did in the previous sections.

Let V be a set of propositional and individual variables, P a set of predicates and Φ a set of functors of a certain algorithmic language. Let SR be a set, its elements will be called *sorts* or *types*. We shall make the following assumptions:

(1) The set of all individual variables consists of disjoint sets V_j for every $j \in SR$; if $x \in V_j$, then j is called the *type* of x.

(2) For every n-argument predicate $\varrho \in P$ we define a type of predicate ϱ as a sequence $(j_1 \times \ldots \times j_n)$ of sorts.

(3) For every n-argument functor $\varphi \in \Phi$ we define a type of functor φ as a sequence $(j_1 \times \ldots \times j_n \to j)$ of sorts and a predicate ϱ_φ of type $(j_1 \times \ldots \times j_n)$.

A *many sorted algorithmic language L_m* is defined like an algorithmic language but there are some natural differences in the definitions of terms, elementary formulas and assignment instructions (which results from assumptions (1), (2) and (3)). In all these expressions we shall take care of the types of variables and the types of functors and predicates.

DEFINITION 10.1. *The set of all terms T_m is the least set of expressions such that*:

(i) *if $x \in V_j$ for $j \in SR$, then x is a term of type j,*

(ii) *if φ is an n-argument functor of type $(j_1 \times \ldots \times j_n \to j)$ and τ_i is a term of type j_i for $i \leqslant n$, then $\varphi(\tau_1, \ldots, \tau_n)$ is a term of type j.* □

DEFINITION 10.2. *The set of all elementary formulas is the least set of expressions such that if ϱ is an n-argument predicate of type $(j_1 \times \ldots \times j_n)$ and $\tau_1, \ldots, \tau_n$ are terms whose types are $j_1, \ldots, j_n$ respectively, then the expression $\varrho(\tau_1, \ldots, \tau_n)$ is an elementary formula.* □

DEFINITION 10.3. *The set of all assignment instructions consists of all expressions of the form* $(q := \gamma)$, *where* q *is a propositional variable and* γ *is an open formula, together with all expressions of the form* $(x := \tau)$, *where* x *is an individual variable and* τ *is a term such that if* $x \in V_j$, *then the type of* τ *is also* j. □

For the rest of this section let L_m be a fixed many sorted algorithmic language and let L_{pf} be a fixed partial function language based on the same alphabet (cf. Chapter III, § 9). It may be easily observed that L_{pf} is an extension of L_m.

DEFINITION 10.4. *By a data structure for the language* L_m *we shall understand a heterogeneous structure*

$$\mathfrak{A} = \langle A, \{\psi_{\mathfrak{A}}\}_{\psi \in \Phi}, \{\varrho_{\mathfrak{A}}\}_{\varrho \in P} \rangle$$

such that

(i) $A = \bigcup\limits_{j \in SR} A_j$ *for some non-empty, disjoint sets* A_j,

(ii) *for every n-argument predicate* ϱ *of type* $(j_1 \times \ldots \times j_n)$,

$$\varrho_{\mathfrak{A}} \subset A_{j_1} \times \ldots \times A_{j_n},$$

(iii) *for every n-argument functor* ψ *of type* $(j_1 \times \ldots \times j_n \to j)$, $\psi_{\mathfrak{A}}$ *is a partial function such that*

$$\psi_{\mathfrak{A}}: A_{j_1} \times \ldots \times A_{j_n} \to A_j$$

and for arbitrary $a_1, \ldots, a_n$, $\psi_{\mathfrak{A}}(a_1, \ldots, a_n)$ *is defined iff* $a_i \in A_{j_i}$ *for* $i \leqslant n$ *and* $(a_1, \ldots, a_n) \in \varrho_{\psi \mathfrak{A}}$.

The structure defined above will be called a many-sorted data structure. □

It follows from the last definition that every partial data structure for the language L_{pf} can be considered as a many sorted data structure for the corresponding many sorted language L_m (cf. Chapter III, § 9). If $\mathfrak{A}_{pf} = \langle A, \{\psi_{\mathfrak{A}_{pf}}\}_{\psi \in \Phi}, \{\varrho_{\mathfrak{A}_{pf}}\}_{\varrho \in P} \rangle$ is a data structure for L_{pf}, then the following structure

(4) $\mathfrak{A} = \langle \bigcup\limits_{j \in SR} A_j, \{\psi_{\mathfrak{A}}\}_{\psi \in \Phi}, \{\varrho_{\mathfrak{A}}\}_{\varrho \in P} \rangle$

where $A_j = A \times \{j\}$ for $j \in SR$ and for every functor ψ of the type $(j_1 \times \ldots \times j_n) \to j$

$$\psi_{\mathfrak{A}}((a_1, j_1), \ldots, (a_n, j_n)) \overset{\text{df}}{=} \begin{cases} (\psi_{\mathfrak{A}_{pf}}(a_1, \ldots, a_n), j) \\ \quad \text{if } \psi_{\mathfrak{A}_{pf}}(a_1, \ldots, a_n) \text{ is defined,} \\ \text{undefined} \quad \text{otherwise} \end{cases}$$

and for every predicate of the type $(j_1 \times \ldots \times j_m)$

$$\varrho_{\mathfrak{A}}\big((a_1, j_1), \ldots, (a_m, j_m)\big) \stackrel{df}{=} \varrho_{\mathfrak{A}_{pf}}(a_1, \ldots, a_n)$$

is a many sorted data structure for the language L_m which corresponds to $\mathfrak{A}_{pf}$.

Conversely, if $\mathfrak{A}_m = \langle \bigcup_{j \in SR} A_j, \{\psi_{\mathfrak{A}_m}\}_{\psi \in \Phi}, \{\varrho_{\mathfrak{A}_m}\}_{\varrho \in P}\rangle$ is a many sorted data structure for the language L_m, then we can define a corresponding partial data structure

(5) $\mathfrak{A} = \langle A, \{\psi_{\mathfrak{A}}\}_{\psi \in \Phi}, \{\varrho_{\mathfrak{A}}\}_{\varrho \in P}\rangle$,

such that:

(a) $A = \bigcup_{j \in SR} A_j$,

(b) for every n-argument predicate ϱ of the type $(j_1 \times \ldots \times j_n)$, $(a_1, \ldots, a_n) \in \varrho_{\mathfrak{A}}$ iff $a_i \in A_{j_i}$ for $i \leqslant n$ and $(a_1, \ldots, a_m) \in \varrho_{\mathfrak{A}_m}$,

(c) for every n-argument functor ψ of the type $(j_1 \times \ldots \times j_n \to j)$, $\psi_{\mathfrak{A}}(a_1, \ldots, a_n)$ is defined iff $(a_1, \ldots, a_n) \in \varrho_{\psi_{\mathfrak{A}_m}}$ and if $\psi_{\mathfrak{A}}(a_1, \ldots, a_n)$ is defined, then $\psi_{\mathfrak{A}}(a_1, \ldots, a_n) \stackrel{df}{=} \psi_{\mathfrak{A}_m}(a_1, \ldots, a_n)$ for arbitrary elements $a_1, \ldots, a_n$.

The semantics of a many sorted algorithmic language is defined in exactly the same way as for the language of algorithmic logic with partial functors. However we shall consider only those valuations of individual variables which are compatible with types. A strict definition follows.

By a valuation in a many sorted data structure for L_m we shall understand a mapping

$$v \colon V \to A \cup \{1, 0\},$$

such that $v(q) \in \{1, 0\}$ for all $q \in V_0$ and for $j \in SR$

(6) $v(x) \in A_j$ iff $x \in V_j$.

Let us denote by $\models_m$ a satisfiability relation for the language L_m.

Let $\mathfrak{A}_{pf}$ be a partial structure for L_{pf} and let $\mathfrak{A}$ be a corresponding structure for the language L_m, defined by (4).

LEMMA 10.1. *For every formula α of L_m and for every valuation v in $\mathfrak{A}$,*

$$\mathfrak{A}_{pf}, v \models_{pf} \alpha \quad iff \quad \mathfrak{A}, v \models_m \alpha.$$

The proof follows directly from the assumed definitions.   □

COROLLARY. *For every formula $\alpha \in L_m$,*

$$\mathfrak{A}_{pf} \models_{pf} \alpha \quad iff \quad \mathfrak{A} \models_m \alpha.$$ □

Let $\mathfrak{A}_m$ be a many sorted data structure for L_m and $\mathfrak{A}$ the corresponding data structure for the language L_{pf} as defined in (5).

LEMMA 10.2. *For every formula α of L_m and for every valuation v which satisfies condition* (6) *the following equivalence holds*

$$\mathfrak{A}_m, v \models_m \alpha \quad \text{iff} \quad \mathfrak{A}, v \models_{pf} \alpha.$$

PROOF. It suffices to determine whether the lemma holds for elementary formulas.

Let ψ be an n-argument functor of the type $(j_1 \times \ldots \times j_n \to j)$ and let $x_i \in V_{j_i}$ for $i \leqslant n$. $\psi(x_1, \ldots, x_n)_{\mathfrak{A}_m}(v)$ is then defined iff $\psi_{\mathfrak{A}_m}(a_1, \ldots, a_n)$ is defined for $a_i = v(x_i)$, where $i \leqslant n$. Hence by assumption (3) and Definition 10.4, $(a_1, \ldots, a_n) \in \varrho_{\psi\mathfrak{A}_m}$ and $v(x_i) \in A_{j_i}$. This is equivalent by (5) to $(a_1, \ldots, a_n) \in \varrho_{\psi\mathfrak{A}}$ and therefore $\psi(x_1, \ldots, x_m)_{\mathfrak{A}}(v)$ is defined.

Thus by induction on the length of term τ we can prove that $\tau_{\mathfrak{A}_m}(v)$ is defined if and only if $\tau_{\mathfrak{A}}(v)$ is defined and, moreover

(7) $\tau_{\mathfrak{A}_m}(v) = \tau_{\mathfrak{A}}(v).$

Let ϱ be a predicate of type $(j_1 \times \ldots \times j_n)$ and let $\tau_1, \ldots, \tau_n$ be terms whose types of results are $j_1, \ldots, j_n$, respectively. $\mathfrak{A}_m, v \models_m \varrho(\tau_1, \ldots, \tau_n)$ if $(a_1, \ldots, a_n) \in \varrho_{\mathfrak{A}_m}$ for $\tau_{i\mathfrak{A}_m}(v)$ defined and equal to a_i, $i \leqslant n$, where $a_i \in A_{j_i}$. Hence by (7) and the definition of the structure $\mathfrak{A}$ we obtain

$$\tau_{i\mathfrak{A}}(v) \text{ is defined,} \quad a_i = \tau_{i\mathfrak{A}}(v) \in A_{j_i} \quad \text{and} \quad (a_1, \ldots, a_n) \in \varrho_{\mathfrak{A}}.$$

This last property is equivalent to $\mathfrak{A}, v \models_{pf} \varrho(\tau_1, \ldots, \tau_n)$. Hence

$$\mathfrak{A}_m, v \models_m \varrho(\tau_1, \ldots, \tau_n) \quad \text{iff} \quad \mathfrak{A}, v \models_{pf} \varrho(\tau_1, \ldots, \tau_n). \qquad \square$$

COROLLARY. *For every formula α of the language L_m,*

(8) $\mathfrak{A} \models_{pf} \alpha \quad \text{implies} \quad \mathfrak{A}_m \models_m \alpha.$ $\square$

As a consequence of Lemmas 10.1 and 10.2 we have for every formula α of the language L_m

$$\models_m \alpha \quad \text{iff} \quad \models_{pf} \alpha,$$

and additionally for arbitrary set of formulas A of L_m

(9) $A \models_m \alpha \quad \text{implies} \quad A \models_{pf} \alpha.$

We can now easily verify that all instances of axioms of algorithmic logic with partial functors which are formulas of L_m are valid in every many sorted data structure. Furthermore, the set of valid formulas of L_m is closed with respect to the inference rules of algorithmic logic with partial functors (see Chapter III, § 9).

This justifies the following definition:

For every set of formulas A of the language L_m

$\alpha \in C_m(A)$ *if and only if α is a formula of L_m and $\alpha \in C_{pf}(A)$.*

It is clear that for an arbitrary set A of formulas of the language L_m,

(10) $\quad A \vdash_m \alpha \quad$ implies $\quad A \models_m \alpha$

and

(11) $\quad A \vdash_{pf} \alpha \quad$ iff $\quad A \vdash_m \alpha$.

Let $T = \langle L_m, C_m, A \rangle$ be a many sorted algorithmic theory. It follows immediately from (11) that if a corresponding theory with partial functors $\langle L_{pf}, C_{pf}, A \rangle$ is consistent then T is also consistent.

We now show that the Completeness Theorem is also valid for many sorted algorithmic theories.

THEOREM 10.3. *For the arbitrary many sorted algorithmic theory*
$$T = \langle L_m, C_m, A \rangle, \quad A \models_m \alpha \quad iff \quad A \vdash_m \alpha.$$

PROOF. Suppose $A \models_m \alpha$, by (9) we then have $A \models_{pf} \alpha$. Hence by the Completeness Theorem for algorithmic theories with partial functors we have $A \vdash_{pf} \alpha$ and by (11) $A \vdash_m \alpha$.

This completes the proof by (10). $\qquad\qquad\qquad\qquad\qquad\square$

11. DEFINABILITY AND PROGRAMMABILITY

Let L be an algorithmic language and let $\mathfrak{A}$ be a data structure for L. Denote by A the universe of the structure $\mathfrak{A}$.

DEFINITION 11.1. *A relation $r \subset A^n$ is algorithmically definable in a data structure $\mathfrak{A}$ iff there exists a formula α of the language L with at least n variables $x_1, ..., x_n$ such that for every valuation v*

$$\big(v(x_1), ..., v(x_n)\big) \in r \quad iff \quad \mathfrak{A}, v \models \alpha(x_1, ..., x_n).$$

We shall also say that the formula α defines the relation r in the structure $\mathfrak{A}$. $\qquad\qquad\qquad\qquad\square$

EXAMPLE 11.1. Let $\mathfrak{N}$ be the data structure of natural numbers with zero-argument operation 0, the two-argument operation $+$ of addition and the two-argument binary relation $=$ of identity.

For arbitrary natural numbers m, n we have

$$m \leqslant n \quad \text{iff} \quad \mathfrak{N}, v \models (z := 0) \bigcup (z := z+1)x + z = y,$$

$$\text{where } v(y) = n \text{ and } v(x) = m.$$

Hence the relation $\leqslant$ is definable in the data structure $\mathfrak{A}$.   □

DEFINITION 11.2. *A relation $r \subset A^n$ is programmable in the data structure $\mathfrak{A}$ iff it is definable by a formula of the form $K\alpha$, where K is a program and α is an open formula.*

We shall say that the relation r is strongly programmable in $\mathfrak{A}$ iff it is programmable by the formula $K\alpha$ and $\mathfrak{A} \models K$ true.   □

EXAMPLE 11.2.
A. The formula

begin $y := x$; **while** $y \neq z$ **do** $y := y \cdot x$ **od end true**

defines the relation r in the data structure of real numbers such that

$$(x, z) \in r \quad \text{iff} \quad (\exists n \in N) x^n = z.$$

B. Every recursive relation is strongly programmable in the data structure of natural numbers.   □

LEMMA 11.1. *If a relation $r \subset A^n$ is strongly programmable in $\mathfrak{A}$ by the formula $K\alpha$, then the relation $A^n - r$ is programmable by the formula $K \sim \alpha$.*

PROOF. The above follows immediately from the tautology

$$\models (K \text{ true} \Rightarrow (\sim K\alpha \equiv K \sim \alpha)).$$ □

REMARK. A relation r is strongly programmable in $\mathfrak{A}$ iff its complement is strongly programmable in $\mathfrak{A}$.   □

The following theorem is an analogue of the Post Theorem in the theory of recursive functions (cf. Rogers 1967).

THEOREM 11.2. *A relation $r \subset A^n$ is strongly programmable in the structure $\mathfrak{A}$ iff both relation r and its complement $A^n - r$ are programmable in $\mathfrak{A}$.*

PROOF. Let us suppose the relations r and $A^n - r$ are programmable in $\mathfrak{A}$. Making use of the normal form theorem for programs (cf. Theorem 6.4) we can assume that the relations are definable by formulas $K\alpha$ and $M\beta$ of the form:

> **begin** K_1 ; **while** γ_1 **do** K_2 **od end** α,
>
> **begin** M_1 ; **while** γ_2 **do** M_2 **od end** β,

where K_1, K_2, M_1, M_2 are **while**-free programs and γ_1, γ_2 are open formulas.

Let $\vec{x} = (x_1, \ldots, x_m)$, be a vector of all variables that occur in $K\alpha$ and let $\vec{z}$ be a copy of $\vec{x}$ such that $\{z_1, \ldots, z_m\} \cap V(M\beta) = \varnothing$. Let $K(\vec{z})\alpha(\vec{z})$ be a copy of the formula $K\alpha$ obtained by the simultaneous replacement of all occurrences of $x_1, \ldots, x_m$ by the corresponding variables $z_1, \ldots, z_m$. Finally let s denote a program **begin** $z_1 := x_1$; $\ldots$ $\ldots$; $z_m := x_m$ **end** and let q be a propositional variable such that

$$q \notin V(K\alpha) \cup V(M\beta).$$

The program M'

> **begin** s ;
> > $K_1(\vec{z})$; M_1 ; $q := $ **true**;
> > **while** $((\gamma_1(\vec{z}) \wedge q) \vee (\gamma_2 \wedge \sim q))$ **do**
> > > **if** q **then** $K_2(\vec{z})$ **else** M_2 **fi**
> > > $q := \sim q$
> > **od**
> **end** ;

simulates the behaviour of both K and M. The programs K and M are executed interchangeably at even and odd passes throughout the loop of M'. The program M' terminates if the formula $\gamma_1(\vec{z})$ or the formula γ_2 holds after a finite number of steps. Note that for every valuation v in $\mathfrak{A}$ either $(v(x_1), \ldots, v(x_n)) \in r$ or $(v(x_1), \ldots, v(x_n)) \in A^n - r$. Hence after a finite number of iterations either $\gamma_1(\vec{z})$ or γ_2 will be falsified. The latter implies that for every valuation v in $\mathfrak{A}$, the program M' terminates, i.e. $\mathfrak{A} \models M'$ **true** and

$$(v(x_1), \ldots, v(x_n)) \in r \quad \text{iff} \quad \mathfrak{A}, v \models M'((\sim q \wedge \alpha(\vec{z})) \vee$$
$$\vee (q \wedge \sim \beta(\vec{x}))).$$

This completes the proof by Lemma 11.1.   □

DEFINITION 11.3. *The function* $f: A^n \to A$ *is algorithmically definable in* $\mathfrak{A}$ *iff there exists a term* τ *with at least n individual variables* $x_1, \ldots$ $\ldots, x_n$ *such that for every valuation v in* $\mathfrak{A}$

$$\tau_{\mathfrak{A}}(v) = a \quad \text{iff} \quad f(v(x_1), \ldots, v(x_n)) = a. \qquad \square$$

Let L be an algorithmic language which allows generalized terms (see Section 7 of this chapter).

DEFINITION 11.4. *The function* $f: A^n \to A$ *is programmable in* $\mathfrak{A}$ *iff* f *is algorithmically definable in* $\mathfrak{A}$ *by a generalized term Ky, where y is an individual variable and K is a program with at least n individual variables* $x_1, \ldots, x_n$. $\qquad \square$

REMARK. If a function f (total) is programmable in $\mathfrak{A}$ by the term Ky, then $\mathfrak{A} \models K$ **true**. $\qquad \square$

EXAMPLE 11.3.

A. Let K be the following program

K: **begin**
    $y := 0; z := 0;$
    while $z \neq x_2$ **do**
        $u := 0;$
        while $u \neq x_1$ **do**
            $y := y+1;$
            $u := u+1$
        od;
        $z := z+1$
    od
end.

The term Ky defines, in the data structure of natural numbers $\mathfrak{N}$, the function $f(x_1, x_2) = x_1 \cdot x_2$, since for every valuation v in $\mathfrak{N}$ we have

$$(Ky)_{\mathfrak{N}}(v) = a \quad \text{iff} \quad v(x_1) \cdot v(x_2) = a.$$

B. Every recursive function is programmable in the data structure of natural numbers with zero and successor. $\qquad \square$

The definition of programmability can be generalized to the class of partial functions.

DEFINITION 11.5. *A partial function* $f: A^n \to A$ *is programmable in* $\mathfrak{A}$ *iff there exists a term* Ky *with free individual variables* $x_1, \ldots, x_n, y$ *such that for every* $a_1, \ldots, a_n, a \in A$ *and for valuation* v *satisfying* $v(x_i) = a_i$:
  (i) *if* $f(a_1, \ldots, a_n)$ *is defined and* $f(a_1, \ldots, a_n) = a$, *then* $(Ky)_{\mathfrak{A}}(v) = a$,
  (ii) *if* $f(a_1, \ldots, a_n)$ *is not defined, then* $K_{\mathfrak{A}}(v)$ *is not defined either.* $\square$

EXAMPLE 11.4. Every partial recursive function is programmable in the structure of natural numbers with zero and successor. $\square$

12. INESSENTIALITY OF DEFINITIONS

The problem of definitions will be now discussed in the formalized theory $T = \langle L, C, A \rangle$.

The general idea is quite typical in mathematics: to form a new notion by admission of a suitable definition. The aim of such a procedure is twofold. It emphasizes and facilitates the investigation of an important notion and clarifies our thinking by replacing several long statements with a short one.

In what follows we shall see many examples of the formation of new theories by assuming definitions of new functions and new relations which are created by means of programs.

We shall mention here two characteristic forms of definitions in a formalized theory. Our considerations are based on the fact, familiar from Chapter II, that every term describes a function in a given data structure and that every formula describes a relation.

Let $T = \langle L, C, A \rangle$ be an algorithmic theory.

Suppose $\alpha(x_1, \ldots, x_n)$ is a formula in the language L with n free--variables. Let ϱ_α be a new n-ary predicate which appears neither in α nor in any formula from A.

We shall call the formula

(1) $\qquad \varrho_\alpha(x_1, \ldots, x_n) \equiv \alpha(x_1, \ldots, x_n)$

a *definition of the predicate* ϱ_α.

In algorithmic theories formula (1) usually has the form

$$\varrho_\alpha(x_1, \ldots, x_n) \equiv K\alpha'$$

where K is a program, α' is an open formula and $\alpha = K\alpha'$.

Assume additionally that L contains a binary predicate of equality.

Suppose τ is a term in L with n free individual variables $x_1, \ldots, x_n$. Let ψ_τ be a new n-argument functor which appears neither in τ nor in any formula from the set of specific axioms of T.

We shall call the formula

(2) $\psi_\tau(x_1, \ldots, x_n) = \tau(x_1, \ldots, x_n)$

a *definition of the functor* ψ_τ.

In algorithmic theories formula (2) usually has the form

$$\psi_\tau(x_1, \ldots, x_n) = K\tau'(x_1, \ldots, x_n),$$

where $\tau = K\tau'$ and K is a program and τ' is a term.

We form an extension L' of L by adding to L a set of predicates ϱ_α and a set of functors ψ_τ for some formulas α and terms τ of language L. Let $T' = \langle L', C, A' \rangle$ be an extension of $T = \langle L, C, A \rangle$ such that A' is obtained from A by simultaneous assuming definitions of form (1) and (2) for all predicates ϱ_α and functors ψ_τ.

LEMMA 12.1. *The theory T is consistent if and only if the theory T' is consistent.*

PROOF. One implication is obvious, i.e. if T' is consistent then T is also consistent.

To prove the converse, let us assume that $\mathfrak{M}$ is a model of T. We shall construct an extension $\mathfrak{M}'$ of the data structure $\mathfrak{M}$ which will be a model of T'. The universe of $\mathfrak{M}'$ is just the one of $\mathfrak{M}$ and for every n-argument predicate ϱ and every n-argument functor ψ from the language L we put

$$\varrho_{\mathfrak{M}'} = \varrho_{\mathfrak{M}} \quad \text{and} \quad \psi_{\mathfrak{M}'} = \psi_{\mathfrak{M}}.$$

For every n-argument predicate ϱ_α and n-argument functor ψ_τ from the language L' if $\varrho_\alpha \equiv \alpha(x_1, \ldots, x_n)$ and $\psi_\tau = \tau(x_1, \ldots, x_n)$ are specific axioms from the set $A' - A$, then for every $j_1, \ldots, j_n$ from $\mathfrak{M}$

$$\varrho_{\alpha\mathfrak{M}'}(j_1, \ldots, j_n) = \alpha_{\mathfrak{M}}(v),$$

$$\psi_{\tau\mathfrak{M}'}(j_1, \ldots, j_n) = \tau_{\mathfrak{M}}(v),$$

where $v(x_i) = j_i$ for $i \leqslant n$.

It follows from the above definition that $\mathfrak{M}'$ is a model of T'. Hence, by the Model Existence Theorem, if the theory T is consistent then T' is consistent.   □

THEOREM 12.2 (on inessentiality of definitions). *The theory* $T' = \langle L', C, A' \rangle$ *obtained from* $T = \langle L, C, A \rangle$, *by assuming definitions* (1), (2) *is an inessential extension of* T, *i.e. for every* α *of the language* L, $A \vdash \alpha$ *iff* $A' \vdash \alpha$.

PROOF. Let us note that every theorem of T is a theorem of T' since $A \subset A'$ (cf. Chapter II, § 5).

If a formula β of the language L is not a theorem of T, then non $A \vdash \beta$. The latter implies that there exists a model $\mathfrak{M}$ of A which is not a model of β. Hence, by the previous lemma, there exists an extension $\mathfrak{M}'$ of a model $\mathfrak{M}$ such that $\mathfrak{M}' \models A$ and non $\mathfrak{M}' \models \beta$.

This implies by the Completeness Theorem that non $A' \vdash \beta$. $\square$

Theorem 12.2 states that by admitting definitions of new predicates or new functors we cannot prove anything new about the predicates and functors of the old language.

BIBLIOGRAPHIC REMARKS

The Completeness Theorem for algorithmic logic was first proved by Mirkowska (1971). The proof is based on the lemma on the existence of Q-filters (cf. Rasiowa and Sikorski, 1968, p. 89). Another variant of the Completeness Theorem with axioms for classical quantifiers can be found in Banachowski (1977). That the Completeness Theorem implies the definability of operational semantics by means of axioms of algorithmic logic was observed by Salwicki (1980). The Gentzen-style axiomatization for algorithmic logic was proposed by Mirkowska (1971) and modified by Kreczmar (1974). The theorem on the normal form of programs has a long history (cf. Harel, 1980); for algorithmic logic it appeared in Mirkowska (1971) and Kreczmar (1974). Algorithmic logic with partial functions was proposed by Petermann (1983); the approach presented here is different.

CHAPTER IV

ALGORITHMIC PROPERTIES OF DATA STRUCTURES

1. DATA STRUCTURES IN PROGRAMMING

It is generally recognized in computer science that data structures are of vital importance in programming. The number of papers devoted to data structures is rapidly increasing. Nevertheless, no consensus of opinion has been reached. In programming practice, data structures are not treated in the right way. The languages currently in use have no tools for dealing with data structures. Among theoreticians there have been many attempts to define the semantics of programming constructions such as program connectives, procedures, coroutines, parallel processes and other constructs. There are numerous program logics. Almost all of them assume that there exists a predefined first--order theory of the data structure in question (cf. the theorems on relative completeness in Floyd–Hoare logic (cf. Cook, 1978), and the arithmetical completeness of dynamic logic (cf. Harel, 1979). In this way the problem of providing a logical theory for reasoning concerning data structures and the program properties has been overlooked. There are other theories which allow to identify (or specify) a data structure; as a rule they lack the tools for proving program properties. The same observation applies to theories presenting the constructions used in implementing data structures.

Here we propose a point of view which involves:

(1) conceiving data structures as heterogeneous algebraic systems,

(2) developing theories of data structures based on algorithmic logic and

(3) studying not only algorithmic theories in themselves but also the connections between them. We propose namely, to study interpretations as the formal counterpart of the software notion of implementation.

Many authors share the opinion that data structures are algebraic systems.

We shall present below the expressive power of algorithmic formulas and we shall apply these formulas in specifications of data structures. Among theorems of algorithmic theories there are statements about program properties as well as first-order sentences. The logical tools of AL allow us to deduce new properties from those asserted earlier.

In the structure of interpretations mentioned in (3) we find some interesting chains which start from "abstract" data structures and approach "real" data structures, i.e. those which have already been implemented in a computer, a virtual machine of a programming language or in the library of software. An example of such chains will be presented below where dictionaries are implemented in hash tables and hash tables are implemented in arrays and queues.

In this way our approach reflects the natural influence process which takes place when new algorithms require new data structures, and knowledge of new data structures (or new properties of structures) enables us to invent new algorithms.

One can view this connection from the point of view of a "theorist":

(a) the fact that an algorithms is correct is a new theorem of a data structure theory and

(b) a theory augmented by new facts increases our chances of improving algorithms and the proofs of their properties.

The two main problems concerning data structures are, first, what are the properties of a data structure and second, the structure, is it implemented?

The first question is concerned with verification of programs. We wish to examine program properties with respect to the axioms of a data structure, separating this goal from the implementation problems. It turns out that the first question provides a natural impetus for developing theories (more or less formalized) which need algorithmic language as an extension of first-order language, since the properties they deal with are algorithmic (e.g. termination, correctness, equivalence of programs, etc.).

It is astonishing to realize how many structure properties which cannot be expressed in the first-order language are of an algorithmic nature. To list a few: the property "y is a natural number" is expressed by the formula

$$(x := 0) \ (\textbf{while} \ x \neq y \ \textbf{do} \ x := s(x) \ \textbf{od} \ x = y)$$

similarly, "s is a stack"

$$(\textbf{while} \sim \text{empty}(s) \ \textbf{do} \ s := \text{pop}(s) \ \textbf{od true}),$$

"*pq* is a priority queue"

$$\textbf{while} \sim \text{empty}(pq) \textbf{ do } pq := \text{delete}(\text{min}(pq), pq) \textbf{ od true},$$

the axiom of Archimedes

$$(\forall x, y)(x > 0 \wedge y > 0) \Rightarrow (z := y)(\textbf{while } z < x \textbf{ do } z := z+y$$
$$\textbf{od true}),$$

the axiom of fields of characteristic zero

$$\sim(x := 1)(\textbf{while } x \neq 0 \textbf{ do } x := x+1 \textbf{ od true}),$$

the axiom of torsion groups

$$(\forall x)(z := x)(\textbf{while } z \neq 1 \textbf{ do } z := z \cdot x \textbf{ od true}),$$

the axiom of cyclic groups

$$(\exists y)(\forall x)(z := y)(\textbf{while } x \neq z \textbf{ do } z := z \cdot y \textbf{ od true}).$$

The second question can be approached in the following way. Suppose we are considering two data structures $\mathfrak{A}$ and $\mathfrak{B}$ and their algorithmic theories $\mathscr{T}_{\mathfrak{A}}$ and $\mathscr{T}_{\mathfrak{B}}$. We shall say that a data structure $\mathfrak{A}$ is *implemented* in a data structure $\mathfrak{B}$ whenever there is an interpretation of the algorithmic theory $\mathscr{T}_{\mathfrak{A}}$ in the algorithmic theory $\mathscr{T}_{\mathfrak{B}}$. This in turn requires an answer to the question "what is an interpretation relation among algorithmic theories?" We shall not develop a theory of interpretation. Instead we shall relate the examples of interpretations given below to software units called *classes*. Examples of software are written in LOGLAN. An acquaintance with *prefixing*, i.e. with the technique of concatenable class declarations (cf. § 12 of this chapter) is desirable. We hope the reader will see the connections. We call the reader's attention to the concatenation rule which is applied several times in the chapter to type declarations. This device was introduced in SIMULA–67 and still awaits recognition. Its properties are very interesting and worthy of study. The technique is also called *prefixing*. Making use of prefixing blocks by the names of units which introduce data structures we can profit from the distinction made earlier between programming in abstract data structures and implementations of data structures. In this way one implementation of a data structure can serve different programs. The advantages of such an approach are obvious.

Here we should mention another role of specification, namely that it allows one to check the correctness of an implementation of a data structure.

2. DICTIONARIES

A dictionary is a data structure for finite sets with the operations: insert, delete, member. Dictionaries are important, being one of the most frequently found data structures. They are used whenever we are going to:

—ask whether an element of the universe is in a given finite set,

—increase a given finite set by insertion of an element, or

—delete an element from a finite set.

There are numerous examples of applications of dictionaries, e.g. in library systems, control of contents of stores, etc. Later we shall also see other examples of structures which are extensions of dictionaries.

Dictionaries form an abstract data type since they can be implemented in various ways. Here we shall describe the algebraic structure of dictionaries. In the next section we shall develop the algorithmic formalized theory of the structure.

DEFINITION 2.1. *An algebraic structure is called a dictionary whenever its carrier consists of the two disjoint subsets E, S called sorts, and has the following operations*:

$$empty: S \to B_0,$$
$$member: E \times S \to B_0,$$
$$insert: E \times S \to S,$$
$$delete: E \times S \to S$$
$$amember: S \to E$$

where amember *is a partial operation defined iff its argument is not empty and the structure satisfies the following postulates*:

(P1) $\left(\sim empty(s) \Rightarrow member(amember(s), s) \right)$,

(P2) $empty(s)$ *iff there exists no element e such that* $member(e, s)$,

(P3) *for every s the instruction* $s := delete(amember\ (s), s)$ *can be repeated only finitely many times until s becomes empty, i.e. the following program always terminates*:

while $\sim empty(s)$ **do** $s := delete(amember(s), s)$ **od,**

(P4) *for every* $e \in E$, *for every* $s \in S$

$$member\left(e, insert(e, s)\right),$$
$$\sim member\left(e, delete(e, s)\right),$$

(P5) *for every* e, e', s

$$(e' \neq e \Rightarrow (\text{member}(e', s) \equiv \text{member}(e', \text{insert}(e, s)))),$$

(P6) *for every* e, e', s

$$(e' \neq e \Rightarrow (\text{member}(e', s) \equiv \text{member}(e', \text{delete}(e, s))))). \quad \square$$

3. THEORY OF DICTIONARIES

In this section we present and study the formalized theory of diction-aries, ATD, which is based on many-sorted algorithmic logic. In order to specify ATD theory we must define its language L and the set of specific axioms A.

L: The language of ATD

Three sets of variables are in the alphabet of the language:

V_E—the set of individual variables of the sort E,

V_S—the set of individual variables of the sort S,

V_0—the set of propositional variables.

The set of functors contains:

in—the binary functor, in: $E \times S \to S$,

del—the binary functor, del: $E \times S \to S$,

amb—the unary functor, amb: $S \to E$.

The set of predicates contains:

em—the unary predicate, em: $S \to B_0$,

mb—the binary predicate, mb: $E \times S \to B_0$.

A—the set of specific, non-logical axioms of ATD:

A1   **while** $\sim\text{em}(s)$ **do** $s := \text{del}(\text{amb}(s), s)$ **od true**,

A2 $\text{mb}(e, s) \equiv$ **begin**

$s1 := s;$ bool $:=$ **false**;

while $\sim\text{em}(s1) \wedge \sim$bool **do**

$e1 := \text{amb}(s1);$

bool $:= (e1 = e);$

$s1 := \text{del}(e1, s1)$

od

end bool,

A3 $(s := \text{in}(e, s))(\text{mb}(e, s) \wedge (e \neq e' \Rightarrow \text{mb}(e', s) \equiv \text{mb}(e', s'))),$

A4 $(s := \text{del}(e, s))(\sim\text{mb}(e, s) \wedge (e \neq e' \Rightarrow \text{mb}(e', s)$

$$\equiv \text{mb}(e', s'))),$$

A5 $(\sim\text{em}(s) \Rightarrow (e := \text{amb}(s))\textbf{true}).$

We shall prove below a few propositions in the ATD theory, they are not difficult, and the proofs are given as examples of algorithmic reasoning. The results of this section are used in the proof of the Representation Theorem in the next section.

PROPOSITION 3.1. *The program M appearing in axiom* A2 *does not loop, or more formally, the stopping formula M* **true** *is a theorem of* ATD *theory.*

PROOF. First, observe that the formula

> **while** $\sim$em(s1) **do**
> e1 := amb(s1);
>   bool := (e1 = e);
> s1 := del(e1, s1)
> **od true**

is an easy consequence of axiom A1. Next, we can apply the rule

$$\frac{(\alpha \Rightarrow \beta)}{(\textbf{while } \beta \textbf{ do } K \textbf{ od true} \Rightarrow \textbf{while } \alpha \textbf{ do } K \textbf{ od true})}$$

obtaining

> **while** $\sim$em(s1) $\wedge \sim$bool **do**
> e1 := amb(s1);
>   bool := (e1 = e);
> s1 := del (e1, s1)
> **od true.**

Now, making use of the rule $\dfrac{\alpha,\ K \textbf{ true}}{K\alpha}$ we can precede the last formula by the assignments

> s1 := s; bool := **false**;

and applying the logical axiom

> **begin** K; M **end** $\alpha \equiv K(M\alpha)$

we obtain the desired result:

> **begin**
> s1 := s; bool := **false**;
>   **while** $\sim$em(s1) $\wedge \sim$bool **do**
> e1 := amb(s1);

$$\text{bool} := (e1 = e);$$
$$s1 := \text{del}(e1, s1)$$
od
end true.   □

PROPOSITION 3.2.

$$\text{ATD} \vdash (\sim\text{em}(s) \Rightarrow (\exists e)(e = \text{amb}(s))),$$

This is an immediate consequence of axiom A5.   □

PROPOSITION 3.3.

$$\text{ATD} \vdash (\sim\text{em}(s) \Rightarrow \text{mb}(\text{amb}(s), s)).$$

PROOF. By axiom A5 and the axioms of algorithmic logic with partial functors

$$\text{ATD} \vdash (\sim\text{em}(s) \Rightarrow (\sim\text{em}(s) \wedge (e1 := \text{amb}(s))$$
$$(e1 = \text{amb}(s)))).$$

Making use of the axiom for assignment instruction

$$(s \text{ true} \Rightarrow s\gamma \equiv \overline{s\gamma}), \quad \text{where } \gamma \text{ is an open formula,}$$

we obtain

$$\text{ATD} \vdash (\sim\text{em}(s) \Rightarrow (\sim\text{em}(s) \wedge$$
$$\textbf{begin}$$
$$s1 := s; \ \text{bool} := \textbf{false};$$
$$e1 := \text{amb}(s);$$
$$\text{bool} := (e1 = \text{amb}(s));$$
$$s1 := \text{del}(e1, s1)$$
$$\textbf{end } \text{bool})).$$

From the axiom

while γ **do** M **od** $\alpha \equiv ((\sim\gamma \wedge \alpha) \vee (\gamma \wedge M\textbf{while } \gamma \textbf{ do } M \textbf{ od } \alpha))$

we have

$$\text{ATD} \vdash (\textbf{begin } s1 := s; \ \text{bool} := \textbf{false end } (\sim\text{em}(s1) \wedge$$
$$\wedge \sim\text{bool} \wedge M\text{bool}) \Rightarrow \text{mb}(\text{amb}(s), s)),$$

where M is the following program:

begin
$$e1 := \text{amb}(s);$$

$$\text{bool} := \big(e1 = \text{amb}(s)\big);$$
$$s1 := \text{del}(s1, e1)$$

end.

Thus

$$\text{ATD} \vdash \big(\sim\!\text{em}(s) \Rightarrow \text{mb}\big(\text{amb}(s), s\big)\big). \qquad\qquad \square$$

PROPOSITION 3.4.

$$\text{ATD} \vdash \big(\text{em}(s) \Rightarrow (\forall e) \sim\!\text{mb}(e, s)\big).$$

The proof is by easy verification. Observe that the precondition $\text{em}(s)$ causes the formula $\text{mb}(e, s)$ to be equivalent to

begin $s1 := s;$ bool := **false end** bool,

i.e. to **false**, independently of the choice of e. $\qquad\qquad \square$

We define below the equality relation in the set S. We shall prove the usual properties of the equality relation (reflexivity, symmetry, transitivity and extensionality) making use of this definition. Observe that the definition is algorithmic and assures us that it is possible to check the equality of s and s' mechanically. This is not always possible, cf. the Banach and Mazur theory of recursive real numbers (cf. Mazur, 1963) where we can prove that all operations in the field of recursive real numbers are effective but the equality of recursive real numbers is not a computable relation.

DEFINITION 3.1. *For arbitrary s, s'*

```
eq(s, s') ≡ begin
               s1 := s; s2 := s';
               boo := true;
               while boo ∧ ~em(s1) ∧ ~em(s2) do
                  e1 := amb(s1);
                  boo := boo ∧ mb(e1, s2);
                  if boo then
                     s1 := del (e1, s1);
                     s2 := del (e1, s2)
                  fi
               od
            end (boo ∧ em(s1) ∧ em(s2)).                    □
```

PROPOSITION 3.5. *Let* K *denote the program in the preceding definition. We then have* ATD $\vdash$ K **true**.

The proof is similar to that of Proposition 3.1.   □

The following proposition is crucial in our proof of the representation theorem for dictionaries. For this reason we give a detailed, almost formal proof.

PROPOSITION 3.6.

$$\text{ATD} \vdash \text{eq}(s, s') \equiv (\forall e)(\text{mb}(e, s) \equiv \text{mb}(e, s')).$$

PROOF. We shall prove the implication from left to right. It will suffice to prove

(1)   ATD $\vdash$ eq$(s, s') \Rightarrow (\forall e)(\text{mb}(e, s) \equiv \text{mb}(e, s'))$.

Let us assume the following abbreviations:

γ: $(\text{boo} \wedge \sim\text{em}(s1) \wedge \sim\text{em}(s2))$,
α: $(\text{boo} \wedge \text{em}(s1) \wedge \text{em}(s2))$,
M: **begin**

$\qquad$ $e1 := \text{amb}(s1)$; boo $:= \text{boo} \wedge \text{mb}(e1, s2)$;
$\qquad$ **if boo then**
$\qquad\qquad$ $s1 := \text{del}(e1, s1)$;
$\qquad\qquad$ $s2 := \text{del}(e1, s2)$
$\qquad$ **fi**
end,

I: **begin** $s1 := s$; $s2 := s'$; boo $:=$ **true end**.

With these abbreviations we can rewrite Definition 3.1 as

$$\text{eq}(s, s') \equiv I(\textbf{while } \gamma \textbf{ do } M \textbf{ od } \alpha).$$

Observe that

$$\vdash (\sim\gamma \wedge \alpha) \equiv \alpha.$$

We shall prove the following claim: for every $i \in N$

(2)   ATD $\vdash$ $(I(\textbf{if } \gamma \textbf{ then } M \textbf{ fi})^i (\sim\gamma \wedge \alpha) \Rightarrow (\forall e) \text{ mb}(e, s)$
$$\equiv \text{mb}(e, s')).$$

The implication (1) follows from claim (2) by the ω-rule.

The proof of (2) will proceed by induction on i. For $i = 0$ we have for every s,

$$\text{ATD} \vdash (\text{em}(s) \Rightarrow (\text{mb}(e, s) \equiv \textbf{false}))$$

and

$$\text{ATD} \vdash \big(I(\text{em}(s1) \wedge \text{em}(s2)) \equiv (\text{mb}(e, s) \equiv \text{mb}(e, s'))\big)$$

hence

$$\text{ATD} \vdash \big(I\alpha \Rightarrow (\forall e)(\text{mb}(e, s) \equiv \text{mb}(e, s'))\big).$$

Now assume that (2) holds for all $j < i$ and consider the formula

(3) $\quad I(\text{if } \gamma \text{ then } M \text{ fi})^{i+1}(\sim\gamma \wedge \alpha).$

By the axiom of algorithmic logic

$$\text{if } \gamma \text{ then } M \text{ fi } \beta \equiv (\gamma \wedge M\beta \vee \sim\gamma \wedge \beta)$$

it is equivalent to

$$I\big(\gamma \wedge M(\text{if } \gamma \text{ then } M \text{ fi})^i \alpha \vee \sim\gamma \wedge (\text{if } \gamma \text{ then } M \text{ fi})^i \alpha\big).$$

Applying axioms

$$K(\beta \vee \beta') \equiv (K\beta \vee K\beta') \quad \text{and} \quad K(\beta \wedge \beta') \equiv (K\beta \wedge K\beta')$$

we obtain another equivalent of (3)

$$\big(\sim\text{em}(s) \wedge \sim\text{em}(s') \wedge I(M(\text{if } \gamma \text{ then } M \text{ fi})^i \alpha) \vee$$
$$\vee (\text{em}(s) \vee \text{em}(s')) \wedge I(\text{if } \gamma \text{ then } M \text{ fi})^i \alpha).$$

Let us denote the first part of the above alternative by (4) and the second part by (5).

Observe that (4) is equivalent to the disjunction of (6) and (7)

(6) $\quad \big(\sim\text{em}(s) \wedge \sim\text{em}(s') \wedge I(\text{boo} \wedge \text{mb}(\text{amb}(s), s2) \wedge$
$$\wedge \text{begin } s1 := \text{del}(\text{amb}(s), s1) ; s2 := \text{del}(\text{amb}(s), s2) \text{ end}$$
$$(\text{if } \gamma \text{ then } M \text{ fi})^i \alpha)\big),$$

(7) $\quad \big(\sim\text{em}(s) \wedge \sim\text{em}(s') \wedge I(\sim(\text{boo} \wedge \text{mb}(\text{amb}(s), s2)) \wedge$
$$\wedge \text{begin } e1 := \text{amb}(s1); \text{boo} := \text{boo} \wedge \text{mb}(e1, s2) \text{ end } (\text{if } \gamma$$
$$\text{then } M \text{ fi})^i \alpha)\big).$$

Formula (6) can be transformed to

(8) $\quad \big(\sim\text{em}(s) \wedge \sim\text{em}(s') \wedge \text{mb}(\text{amb}(s), s') \wedge$
$$\wedge \text{begin } s1 := \text{del}(\text{amb}(s), s) ; s2 := (\text{del}(\text{amb}(s), s');$$
$$\text{boo} := \text{true end } (\text{if } \gamma \text{ then } M \text{ fi})^i \alpha).$$

Making use of the induction assumption we obtain that (8) implies the following formula:

(9) $\quad \big(\sim\text{em}(s) \wedge \sim\text{em}(s') \wedge \text{mb}(\text{amb}(s), s') \wedge$
$$\wedge (\forall e)(\text{mb}(e, \text{del}(\text{amb}(s), s)) \equiv \text{mb}(e, \text{del}(\text{amb}(s), s')))\big).$$

Let us now consider formula (7). If $i = 0$ then (7) is equivalent to
false. Assume that $i \geqslant 1$ then (7) is equivalent to the following formula:

(10) $(\sim \mathrm{em}(s) \wedge \sim \mathrm{em}(s') \wedge \sim \mathrm{mb}(\mathrm{amb}(s), s') \wedge$
$\qquad \wedge N(\gamma \wedge M(\textbf{if } \gamma \textbf{ then } M \textbf{ fi})^{i-1}\alpha \vee \sim\gamma \wedge (\textbf{if } \gamma \textbf{ then } M \textbf{ fi})^{i-1}\alpha)).$

where

$\qquad N = \textbf{begin } s1 := s; \, s2 := s'; \, \mathrm{bool} := \textbf{true}; \, e1 := \mathrm{amb}(s1);$
$\qquad\qquad\qquad\qquad\qquad \mathrm{boo} := \mathrm{boo} \wedge \mathrm{mb}(e1, s2) \textbf{ end}.$

Observe that

$\qquad \vdash N\gamma \equiv (\mathrm{mb}(\mathrm{amb}(s), s') \wedge \sim\mathrm{em}(s) \wedge \sim\mathrm{em}(s'))$

and

$\qquad \vdash (\sim\gamma \wedge (\textbf{if } \gamma \textbf{ then } M \textbf{ fi})^{i-1}\alpha) \equiv (\sim\gamma \wedge \alpha).$

From this we conclude that (10) is equivalent to

$\qquad (\sim\mathrm{mb}(\mathrm{amb}(s), s') \wedge \mathrm{em}(s) \wedge \mathrm{em}(s') \wedge \sim\mathrm{em}(s) \wedge \sim\mathrm{em}(s')),$

i.e. (10) is equivalent to **false**.

Hence we have proved that (4) implies the following formula

$\qquad (\sim\mathrm{em}(s) \wedge \sim\mathrm{em}(s') \wedge \mathrm{mb}(\mathrm{amb}(s), s') \wedge$
$\qquad\qquad \wedge (\forall e) \, \mathrm{mb}(e, \mathrm{del}(\mathrm{amb}(s), s)) \equiv \mathrm{mb}(e, \mathrm{del}(\mathrm{amb}(s), s'))).$

By axiom A4 we have

$\qquad (\forall e \neq \mathrm{amb}(s)) \mathrm{mb}(e, \mathrm{del}(\mathrm{amb}(s), s)) \equiv \mathrm{mb}(e, s) \quad \text{and}$
$\qquad (\forall e \neq \mathrm{amb}(s)) \mathrm{mb}(e, \mathrm{del}(\mathrm{amb}(s), s')) \equiv \mathrm{mb}(e, s')$

and therefore (4) implies

(11) $(\mathrm{em}(s) \wedge \sim\mathrm{em}(s') \wedge (\forall e) \mathrm{mb}(e, s) \equiv \mathrm{mb}(s, s')).$

Now consider (5), the second part of the disjunction (3). By the
inductive assumption it follows that (5) implies

(12) $((\mathrm{em}(s) \vee \mathrm{em}(s')) \wedge (\forall e) \mathrm{mb}(e, s) \equiv \mathrm{mb}(e, s')).$

Finally, from (11) and (12) we have that (3) implies

$\qquad (\forall e) \mathrm{mb}(e, s) \equiv \mathrm{mb}(e, s').$

This ends the inductive proof of claim (2). $\square$

PROPOSITION 3.7. *The following formulas are theorems of* ATD *theory*

(a) $(\forall s) \mathrm{eq}(s, s),$

(b) $(\forall s, s')(\mathrm{eq}(s, s') \Rightarrow \mathrm{eq}(s', s)),$

(c) $(\forall s, s', s'')(\mathrm{eq}(s, s') \wedge \mathrm{eq}(s, s'') \Rightarrow \mathrm{eq}(s, s'')).$ $\square$

As a consequence of Definition 3.1 we can prove following results:

PROPOSITION 3.8. *For every e, e′ ∈ E and for every s, s′ ∈ S:*

$$\text{ATD} \vdash ((e = e' \wedge \text{eq}(s, s')) \Rightarrow \text{eq}(\text{in}(e, s), \text{in}(e', s'))),$$
$$\text{ATD} \vdash (\text{eq}(s, s') \Rightarrow \text{em}(s) \equiv \text{em}(s')),$$
$$\text{ATD} \vdash (e = e' \wedge \text{eq}(s, s') \Rightarrow \text{eq}(\text{del}(e, s), \text{del}(e', s'))),$$
$$\text{ATD} \vdash \text{eq}(\text{in}(e, \text{del}(e, s)), s),$$
$$\text{ATD} \vdash (\text{em}(s) \Rightarrow \text{amb}(\text{in}(e, s)) = e),$$
$$\text{ATD} \vdash \sim\!\text{em}(\text{in}(e, s)),$$
$$\text{ATD} \vdash \text{eq}(\text{del}(e, \text{in}(e, s)), s),$$
$$\text{ATD} \vdash (\sim\!\text{eq}(s, s') \wedge \sim\!\text{mb}(e, s) \wedge \sim\!\text{mb}(e, s')$$
$$\Rightarrow \sim\!\text{eq}(\text{in}(e, s), \text{in}(e, s'))),$$
$$\text{ATD} \vdash (\text{mb}(e, s) \Rightarrow \text{eq}(s, \text{in}(e, s))). \qquad \square$$

4. REPRESENTATION THEOREM FOR MODELS OF ATD

Making use of the facts observed earlier we shall prove that every model of ATD is isomorphic with another standard, set-theoretical model. In this way we show that our choice of specific axioms of ATD was right.

DEFINITION 4.1. *We shall say that a model B of* ATD

$$B = \langle E \cup S, \text{in}_B, \text{del}_B, \text{amb}_B, \text{mb}_B, \text{em}_B, =_E \rangle$$

is an ST model (the abbreviation standing for set-theoretical or standard) iff it has the following properties:
1° *the set S consists of all finite subsets of E*

$$S = \text{Fin}(E),$$

2° *the operations in the model B are set-theoretical, i.e. for every e ∈ E, for every s ∈ S*

$$\text{in}_B(e, s) = s \cup \{e\},$$
$$\text{del}_B(e, s) = s - \{e\},$$
$$\text{mb}_B(e, s) \equiv e \in s,$$
$$\text{em}_B(s) \equiv s = \emptyset. \qquad \square$$

THEOREM 4.1. *For every model* $A = \langle E \cup S,$ in, del, amb, mb, em, $=_E \rangle$ *of* ATD, *proper for identity, there exists an* ST *model B of* ATD *with the same set E of elements. The systems A and B are isomorphic modulo* amb *operation, i.e. the reducts* A' *and* B'

$$A' = \langle E \cup S, \text{in, del, mb, em, } =_E \rangle,$$
$$B' = \langle E \cup \text{Fin}(E), \text{in}_B, \text{del}_B, \text{mb}_B, \text{em}_B, =_E \rangle$$

are isomorphic.

PROOF. We shall first construct the system B' and prove its properties. Next, we shall discuss the possibility of extension of B by a proper operation amb to a model of ATD.

With every $s \in S$ we associate the set $h(s)$

$$h(s) = \{e \in E: \text{mb}(e, s)\}.$$

The set $h(s)$ is finite by the axioms A1 and A4, since the sequence $\{e_i\}$ defined below contains all elements of $h(s)$ without repetition.

The sequence is defined by the following algorithm:

> Initialization: Put $i = 0$ and seq = empty sequence.
> WHILE the set s is not empty REPEAT the following instructions
>> PUT $e_{i+1} = \text{amb}(s)$,
>> ADJOIN the element e_{i+1} to the sequence seq,
>> REPLACE s by $\text{del}(\text{amb}(s), s)$.

The mapping

$$h: S \to \text{Fin}(E)$$

is onto, since for a given set $\{e_1, ..., e_n\}$ we can consider the element defined by the following term:

> **begin while** $\sim$em(s) **do** $s := \text{del}(\text{amb}(s), s)$ **od**;
> $s := \text{in}(e_1, s); ... ; s := \text{in}(e_n, s)$
> **end** s.

The mapping h is a one-to-one mapping. For th eproof use Proposition 3.6. Suppose $\sim$eq(s, s') then $(\exists e)\text{mb}(e, s) \land \sim \text{mb}(e, s')$ or, symmetrically, $(\exists e)(\sim \text{mb}(e, s) \land \text{mb}(e, s'))$.

It is easy to verify that

$$h(\text{in}(e, s)) = h(s) \cup \{e\} \quad \text{by axiom A3 and Proposition 3.6,}$$
$$h\text{del}(e, s) = h(s) - \{e\} \quad \text{by axiom A4 and Propositi on 3.6,}$$

$\text{mb}(e, s) \equiv e \in h(s)$ by the definition of $h(s)$,

$\text{em}(s) \equiv h(s) = \emptyset$ by Proposition 3.4.

This ends the first part of the proof. We have constructed a system B isomorphic to the reduct A.

Now, we have to extend B by an appropriate operation amb. This can be done if we accept the axiom of choice. The statement asserting the existence of a selector from the family Fin (E) is the formulation of the axiom of choice AC. $\qquad\qquad\square$

Note that in frequently occurring cases there is no need for the application of AC, e.g. in the situation where the set E is linearly ordered, or if there exists an enumeration of the elements of E.

The assumption that a model of ATD is proper for identity is important. Without it one can construct a counter example such that the set $h(s)$ is infinite.

On the other hand, it is not difficult to prove that for every model $\mathfrak{M}$ of ATD one can construct an equivalent model $\mathfrak{M}' = \mathfrak{M}/(=, \text{eq})$ proper for identity.

5. ON COMPLEXITY OF ATD

Here we shall consider some problems related to the complexity of the set of theorems of ATD and its extensions. We shall show that ATD is an undecidable theory. Later two various extensions of the theory will be presented. The theory of dictionaries over finite universes FATD can be axiomatized and we shall remark that FATD is the complement of a recursively enumerable set. The theory of dictionaries over the infinite set of natural numbers is of very high degree of undecidability, namely Π_1^1.

We begin with the criterion of undecidability of algorithmic theories. Let L be a fixed algorithmic language. For every program M of the form **while** γ **do** K **od** we define the sequence of formulas $\{\alpha_i^M\}_{i \in N}$ such that

$$\alpha_0^M = \sim\gamma,$$
$$\alpha_i^M = (\text{if } \gamma \text{ then } K \text{ fi})^{i-1}(\gamma \wedge K\sim\gamma) \quad \text{for } i > 0.$$

It is easy to observe that for every natural number $i > 0$

$$\vdash \alpha_i^M \equiv (\gamma \wedge K\gamma \wedge \ ... \ \wedge K^{i-1}\gamma \wedge K^i \sim \gamma).$$

Hence, for every data structure $\mathfrak{A}$, for every valuation v and for arbitrary natural number n, formula α_n^M is satisfied in $\mathfrak{A}$ by the valuation v iff the computation of the program M in the structure $\mathfrak{A}$ at the valuation v ends after exactly n iterations of program K. Let $\bar{x} = \{x_1, ..., x_m\}$ be the set of all variables occuring in M.

THEOREM 5.1. *Let* $T = \langle L, C, A \rangle$ *be an algorithmic theory and let M be a program of the form* **while** γ **do** K **od**. *If for every natural number n the theory* $T_n = \langle L, C, A \rangle \cup \{(\exists x)\alpha_n^M(\bar{x})\}$ *is consistent then T is undecidable.*

For the proof see Dańko (1980).   □

The above criterion can be applied to ATD theory. Let M be the program

while $\sim$ empty(s) **do** $s := $ del$(\mathrm{amb}(s), s)$ **od**.

For every set consisting of the set of axioms of ATD and of the formula $(\exists x)\alpha_n^M(\bar{x})$ there exists a model. It suffices to consider an n-element set E. Therefore the theory of dictionaries is undecidable.

Let us mention that the following formula

$$(\exists s)(\forall e)\,\mathrm{eq}\,(s, \mathrm{in}(e, s))$$

is valid in those models of ATD only for which the set E is finite. Denote by FATD the theory which have as specific axioms all axioms of ATD and the above formula.

PROPOSITION 5.2. *The theory* FATD *is the complement of a recursively enumerable set.*

PROOF. It is not difficult to observe that the set of theorems of FATD is at most Π_1^0 set. Making use of the Completeness Theorem for algorithmic logic we observe that the following conditions are equivalent:

(i) α is a theorem of FATD,

(ii) α is valid in every model of ATD which is finite.

All finite models can be enumerated and there exists a decision method for testing the validity of an algorithmic formula in a finite universe (cf. Grabowski, 1972). Hence, if α is not a theorem then in finitely many steps we shall find a counterexample. The set of theorems is an at most Π_1^0 set—the complement of a recursively enumerable set. By application of Theorem 5.1 it is an undecidable set, hence it is a $\Pi_1^0 - \Sigma_1^0$ set. $\square$

On the other hand there exists an extension of ATD which does not belong to any arithmetical class, the set of the theorems in this case lies in Π_1^1, i.e. it is an analytical set. Consider the extension of ATD which results by adding two additional non-logical functors. We admit a constant 0 (zero) of sort E and one argument functor succ (successor)

$$\text{succ}: E \to E.$$

The axioms of the extended theory NATD will be axioms of dictionaries and the following:

$$\sim\text{succ}(e) =_E 0,$$
$$\text{succ}(e) =_E \text{succ}(e') \Rightarrow e =_E e',$$
$$(e' := 0) \text{ while } \sim e -_E e' \text{ do } e' := \text{succ}(e') \text{ od true}.$$

Making use of standard techniques of recursion theory (cf. Rogers, 1967) one can prove:

THEOREM 5.3. *The set of theorems of the above-mentioned theory* NATD *is a* Π_1^1 *set*.

SKETCH OF THE PROOF. Every model of NATD is isomorphic to the standard model of arithmetic of natural numbers with operations insert, delete, member fixed as corresponding set-theoretical operations (cf. § 4 of this chapter). Any two models of NATD can differ only in the interpretation of amember operation. Let us denote a model of NATD by N_f, for N is the set of elements and amember operation f distinguishes it from other models. The following remark suffices: for every formula α, α is a theorem of NATD iff for every function f such that N_f is a model of NATD $N_f \models \alpha$.

The relation "α is valid in the structure N_f" is not in any arithmetical class. In fact it is a hyperarithmetical relation $R(f, \alpha)$, it includes the hyperarithmetical relation "α is valid in the standard model of natural numbers". Therefore α is a theorem of NATD if and only if the formula $(\forall f)R(f, \alpha)$ holds, i.e. the set of theorems of NATD is a Π_1^1 set. $\square$

6. THE THEORY OF PRIORITY QUEUES

Priority queues are similar to dictionaries. We assume additionally that elements of sort E are linearly ordered. Instead of the operator of non-deterministic choice amember for dictionaries, the structure of priority queues admits the operation min which for any priority queue gives the least element contained in it. There are many implementations of priority queues. Hence we shall think of a class of priority queues, much as one thinks of classes of groups, of rings, etc.

DEFINITION 6.1. *A data structure is called a priority queue whenever its universe consists of the two disjoint subsets*

$$E \quad and \quad S$$

called sort E and sort S, and has the following operations:

insert: $E \times S \to S$,

delete: $E \times S \to S$,

min: $S \to E$,

member: $E \times S \to B_0$,

empty: $S \to B_0$,

$\leqslant : E \times E \to B_0$,

and is such that the following axioms are valid in the structure:

PQ1 *the set E is linearly ordered by the relation $\leqslant$,*

PQ2 **while** $\sim$empty(s) **do** $s := $ delete$(\min(s), s)$ **od true**,

PQ3 $(\sim$empty$(s) \Rightarrow ((\forall e)$ member$(e, s) \Rightarrow \min(s) \leqslant e))$,

PQ4 member$(e,$ insert$(e, s))$,

PQ5 $(e \neq e' \Rightarrow ($member$(e', s) \equiv $ member$(e',$ insert$(e, s))))$,

PQ6 $\sim$member$(e,$ delete$(e, s))$,

PQ7 $(e \neq e' \Rightarrow ($member$(e', s) \equiv $ member$(e',$ delete$(e, s))))$,

PQ8 member $(e, s) \equiv$ **begin** $s1 := s;$ bool $:= $ **false**;

         while $\sim$empty$(s1) \wedge \sim$bool **do**

           $e1 := \min(s1)$;

           bool $:= (e1 = e)$;

           $s1 := $ delete$(e1, s1)$;

        od

       end bool.

We assume also the usual axioms of identity $=$.   □

Repeating the arguments of the preceding sections with the necessary alterations we can prove the following theorem:

THEOREM 6.1 (Representation Theorem). *Every model $\mathfrak{M}$ of the algorithmic theory of priority queues proper for identity is isomorphic to a standard one, that is*

$$\langle E \cup \mathrm{Fin}(E), f_1, f_2, f_3, r_1, r_2, =, \leqslant \rangle$$

where $\mathrm{Fin}(E)$ *is the family of all finite subsets of E.*

$$f_1(e, s) = s \cup \{e\}, \quad r_1(e, s) \equiv e \in s,$$
$$f_2(e, s) = s - \{e\}, \quad r_2(s) \equiv s = \varnothing,$$
$$f_3(s) = \textit{the least element of } s.$$

The proof is a mutation of the proof of the Representation Theorem for Dictionaries. As we remarked before, the proof does not make use of the axiom of choice due to the assumption that the set E is linearly ordered.   ☐

7. THE THEORY OF NATURAL NUMBERS

The structure

$$\mathfrak{N} = \langle N, 0, s, = \rangle$$

of natural numbers with 0—a zero-argument operation, s—a one--argument operation and identity is axiomatized by the following axioms AxAr:

$$(\forall x) \sim s(x) = 0,$$
$$(\forall x, y)\big(s(x) = s(y) \Rightarrow x = y\big),$$
$$(\forall y)(x := 0)\big(\textbf{while } \sim x = y \textbf{ do } x := s(x) \textbf{ od } (x = y)\big).$$

THEOREM 7.1. *Every model $\mathfrak{M}$ of AxAr is isomorphic with the standard model of Peano axioms, i.e. the algorithmic theory of natural numbers is categorical (cf. Chapter II, Theorem 4.2).*   ☐

We are now going to prove that every instance of the scheme of induction is a theorem of an algorithmic theory of natural numbers.

First, let us remark that classical quantifiers can be replaced by formulas with programs and iteration quantifiers

$$(\forall x)\alpha(x) \equiv (x := 0)\cap(x := s(x))\alpha(x),$$
$$(\exists x)\alpha(x) \equiv (x := 0)\cup(x := s(x))\alpha(x),$$

assuming that x is free in α and never occurs on the left-hand side of an assignment in α.

In the case where α is an open formula we can prove that

$$(\exists x)\alpha(x) \equiv (x := 0)(\textbf{while } \alpha(x) \textbf{ do } x := s(x) \textbf{ od true}).$$

All three equivalences can be proved formally from AxAr axioms. Indeed, all the equivalences are valid in the standard model of AxAr axioms. By categoricity they are valid in every model of AxAr hence they are provable from AxAr (by completeness of AL).

Now, let us recall that every formula of the following scheme

$$((\beta \wedge \cap K(\beta \Rightarrow K\beta)) \Rightarrow \cap K\beta)$$

is a theorem of algorithmic logic (cf. Chapter II, § 7).

By the rule

$$\frac{\alpha, K \textbf{ true}}{K\alpha}$$

we have

$$(x := 0)\left((\alpha(x)\wedge\cap(x := s(x))(\alpha(x) \Rightarrow (x := s(x))\alpha(x)))\right)$$
$$\Rightarrow \cap(x := s(x))\alpha(x).$$

Distributing the assignment $x := 0$ over implication and conjunction we obtain

$$((x := 0)\alpha(x)\wedge(x := 0)\cap(x := s(x))(\alpha(x)$$
$$\Rightarrow (x := s(x))\alpha(x)) \Rightarrow (x := 0)\cap(x := s(x))\alpha(x))$$

which is equivalent to the scheme of induction

$$(\alpha(x/0)\wedge(\forall x)(\alpha(x) \Rightarrow \alpha(x/s(x)))) \Rightarrow (\forall x)\alpha(x)).$$

Hence we have proved the following proposition:

PROPOSITION 7.2. *Every instance of the scheme of induction is a theorem of the algorithmic theory of natural numbers.*   □

Observe that in the algorithmic theory of natural numbers the operations of addition and multiplication are definable by explicit definition. In any first-order arithmetic these operations are defined implicitly by the recursive equations:

$$x+0 = 0,$$
$$x+s(y) = s(x+y),$$
$$x \cdot 0 = 0,$$
$$x \cdot s(y) = x \cdot y + x.$$

We shall give an algorithmic definition of the $+$ operation below:

$$(\text{add}) (\forall x, y) x + y \overset{\text{df}}{=} (u := 0) ((t := x) (\textbf{while} \sim u = y$$
$$\textbf{do } u := s(u); \ t := s(t) \textbf{ od } t)).$$

We shall now prove that this definition correctly defines addition.

THEOREM 7.3. *The operation of addition is well defined by the above algorithmic definition, i.e.:*

(a) $\text{AxAr} \vdash (u := 0)((t := x)(\textbf{while} \sim u = y \textbf{ do } u := s(u); t := s(t)$
$$\textbf{od true})),$$

(b) $\text{AxAr} \cup \{\text{add}\} \vdash x+0 = x,$

(c) $\text{AxAr} \cup \{\text{add}\} \vdash x+s(y) = s(x+y).$

PROOF. The proof of (a), i.e., that the program occurring in the definition (add) always terminates is easy, and resembles the proof of Proposition 3.1.

(b) The proof of $x+0 = 0$ makes use of the logical axiom

$$\textbf{while } \gamma \textbf{ do } K \textbf{ od } \alpha \equiv (\sim\gamma \wedge \alpha) \vee (\gamma \wedge K(\textbf{while } \gamma \textbf{ do } K \textbf{ od } \alpha)),$$

hence

$$x+0 = (u := 0)((t := x)(\textbf{if} \sim u = 0 \textbf{ then } u := s(u);$$
$$t := s(t) \textbf{ fi } (\textbf{while} \sim u = 0 \textbf{ do } u := s(u); t := s(t)$$
$$\textbf{od } t))).$$

Applying the axiom

$$s(\textbf{if } \gamma \textbf{ then } K \textbf{ else } M \textbf{ fi } \alpha) \equiv s((\gamma \wedge K\alpha) \vee (\sim\gamma \wedge M\alpha))$$

and observing that

$$(u := 0)((t := x) \sim u = 0) \equiv \sim 0 = 0,$$

we obtain

$$\text{AxAr} \cup \{\text{add}\} \vdash x+0 = (u := 0)((t := x)t),$$
$$\text{AxAr} \cup \{\text{add}\} \vdash x+0 = x.$$

(c) In the proof we shall use the following lemma:

LEMMA 7.4. *Let K and M be programs written in the language of arith-metic. Let the variables y, w not appear in the program M and let*

$$AxAr \cup \{add\} \vdash K(u = 0),$$
$$AxAr \cup \{add\} \vdash (u = w) \Rightarrow M(u = s(w)),$$

i.e., the program K zeroes the variable u, and the program M increase the value of u by 1.

The following programs are then equivalent:

$$M_1 : \textbf{begin } K; \textbf{ while } \sim(u = s(y)) \textbf{ do } M \textbf{ od end}$$

and

$$M_2 : \textbf{begin } K; \textbf{ while } \sim(u = y) \textbf{ do } M \textbf{ od}; M \textbf{ end},$$

i.e. for every formula α the equivalence $M_1\alpha \equiv M_2\alpha$ holds in $\mathfrak{N}$.   □

Making use of the definition (add) we have

$$x + s(y) = \textbf{begin } t := x; \quad u := 0; \quad \textbf{while } \sim u = s(y) \quad \textbf{do}$$
$$t := s(t); \ u := s(u) \textbf{ od end } t.$$

By the Lemma 7.4, $x + s(y)$ is equal to

$$\textbf{begin } t := x; u := 0; \textbf{ while } \sim u = y \textbf{ do } t := s(t); u := s(u)$$
$$\textbf{od}; t := s(t); u := s(u) \textbf{ end } t.$$

Applying the axiom of assignment $(z := \tau)\alpha(z) \equiv \alpha(z/\tau)$ we obtain

$$x + s(y) = \textbf{begin } t := x; u := 0; \textbf{ while } \sim u = y \textbf{ do } t := s(t);$$
$$u := s(u) \textbf{ od end } s(t).$$

By the fact

$$K\varphi(\tau) = \varphi(K\tau)$$

we have

$$x + s(y) = s\big(\textbf{begin } t := x; \ u := 0; \textbf{ while } \sim u = y \textbf{ do}$$
$$t := s(t); \ u := s(u) \textbf{ od end } t\big)$$

and finally

$$x + s(y) = s(x + y).$$

Similarly, one proves that the multiplication operation can be defined by an algorithm in an explicit way. Hence the algorithmic arithmetic of addition and multiplication is a conservative extension of the algorithmic theory of natural numbers. All proofs of Peano arithmetic can be reproduced in this theory.

Let us conclude with two observations.

LEMMA 7.5. *The sets of partial recursive functions and of programmable functions are equal.* □

LEMMA 7.6. *Weak second-order arithmetic and the algorithmic theory of natural numbers are equivalent, i.e. there are translations enabling one to replace every formula of one theory by an equivalent from the other.* □

8. STACKS

The universe of a data structure of stacks consists of the two disjoint sets E and S. Elements of S will be called *stacks*, while elements of the set E will simply be called *elements*. The primary relations and operations of a system of stacks are as follows:

$=$ identity in E,

empty, a distinguished subset of S, empty $\subset S$,

push: $E \times S \rightarrow S$,

pop: $S -$ empty $\rightarrow S$,

top: $S -$ empty $\rightarrow E$.

Any relational system with a similar signature will be called a *data system of stacks* provided that it satisfies the following postulates:

(P1) For every stack s there exists an iteration of pop operation such that the result is empty

$$(\forall s \in S)(\exists i \in N) \text{ empty}(\text{pop}^i(s)).$$

(P2) For every non-empty stack s

s is equal to push(top(s), pop(s)),

for every element e and for every stack s.

(P3) $e = \text{top}(\text{push}(e, s))$.

(P4) s is equal to pop(push(e, s)).

(P5) $\sim \text{empty}(\text{push}(e, s))$.

Below we shall present a formalized theory of relational systems of stacks. In order to express properties (P1)–(P5) we shall use the language of algorithmic logic, and the phrase the stack s is equal to the stack s' will be replaced by $s = s'$, making use of its algorithmic definition.

Let us note that the postulate (P1) may be informally stated:

(P1') $(\forall s) \left(s =_s \varnothing \vee (\exists e)s =_s \text{push}(e, \varnothing) \vee (\exists e, e')s =_s \text{push}(e', \right.$
$$\left. \text{push}(e, \varnothing)) \dots \right),$$

where $\varnothing$ denotes a stack such that empty$(\varnothing)$.

Let E be an arbitrary set. By the *standard system* $\mathfrak{S}$ *of stacks* over E we shall mean the system

$$\langle E \cup F\,\text{Seq}(E), \text{ precede, delete-first, first, } =, \varnothing \rangle$$

in which stacks are the finite sequences of elements of the set E. The operation precede(e, s) gives as a result the one-element sequence $\{e\}$ concatenated with the sequence s. The delete-first operation and first operation are self-explaining. A stack s is empty iff s is the empty sequence.

We shall prove below that every system of stacks $\mathfrak{S}$ is isomorphic to the standard system of stacks over the set E of elements of the system $\mathfrak{S}$.

The complexity of the ATS theory is not less than the complexity of algorithmic arithmetics of natural numbers since the latter theory may be interpreted in ATS.

9. THE THEORY OF STACKS

In the algorithmic theory of stacks, ATS, the properties of operations on stacks are considered from an axiomatic point of view. We assume certain axioms about push, pop and top operations knowing nothing about the elements to be placed in the stacks or the implementation of operations. Accordingly, the ATS theory has many different models. Two examples will illustrate the difference in approach. For a mathematician, stacks are nothing more than finite sequences of elements; operations on them are always performed at one end of the sequence in question, say on the left. For a computer scientist a stack denotes the chain of objects depicted in Figure 8.1.

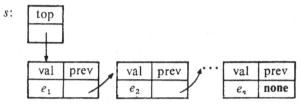

Fig. 8.1

An execution of the $s' := \text{push}(e, s)$ instruction leads to a new configuration of objects, as shown in Figure 8.2.

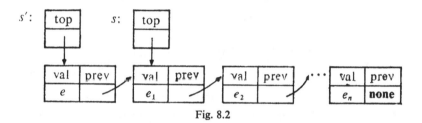

Fig. 8.2

The alphabet of the theory of stacks contains:

(a) variables:

the set V_E of individual variables of type E,

the set V_S of individual variables of type S,

the set of propositional variables V_0;

(b) predicates:

empty   one-argument predicate of type (S),

$=_E$ two-argument predicate of type $(E \times E)$,

$=_S$ two-argument predicate of type $(S \times S)$;

(c) functors:

push   two-argument functor of type $(E \times S \rightarrow S)$,

pop   one-argument functor of type $(S \rightarrow S)$,

top   one-argument functor of type $(S \rightarrow E)$;

(d) logical and program connectives and auxiliary signs.

Variables of the set V_E will be denoted by e, e', e_1, etc.

Variables of the set V_S will be denoted by s, s', s_1, etc.

The set of well-formed expressions consists of terms, open formulas programs, generalized terms and formulas (cf. Chapter III, § 10).

The specific axioms of ATS are:

A1 **while** $\sim$empty(s) **do** $s := $ pop(s) **od true**,

A2 $(\sim$empty(s) $\Rightarrow s =_S$ push$($top(s), pop(s)$))$,

A3 $e =_E$ top$($push$(e, s))$,

A4 $s =_S$ pop$($push$(e, s))$,

A5 $\sim$empty$($push$(e, s))$,

A6 $s =_S s' \equiv$ **begin** $s_1 := s$; $s_2 := s'$; bool := **true**;

                 while bool $\wedge \sim$empty(s_1) $\wedge \sim$empty(s_2) **do**

                    bool := bool $\wedge ($top(s_1) $=_E$ top(s_2)$)$;

                    $s_1 := $ pop(s_1); $s_2 := $ pop(s_2);

            od

           end $($bool $\wedge$ empty(s_1) $\wedge$ empty(s_2)$)$,

A7, A8, A9 axioms of reflexivity, symmetry and transitivity of $=_E$.

LEMMA 9.1. *The program in axiom A6 always halts, i.e. the relation* $=_S$ *is strongly programmable in terms of the remaining relations and notions.*   □

LEMMA 9.2. *For every* $s, s', s'' \in S$ *and for every* $e, e' \in E$

(a) $s =_S s$,

(b) $(s =_S s' \Rightarrow s' =_S s)$,

(c) $(s =_S s' \wedge s' =_S s'' \Rightarrow s =_S s'')$,

(d) $(e =_E e' \wedge s =_S s') \equiv$ push(e, s) $=_S$ push(e', s')

(e) $(s =_S s' \wedge \sim$empty(s)$) \equiv ($pop(s) $=_S$ pop(s')$\wedge$ top(s) $=_E$ top(s')$)$,

(f) $(s =_S s' \Rightarrow ($empty(s) $\equiv$ empty(s')$))$.

(g) $($empty(s) $\wedge$ empty(s') $\Rightarrow s =_S s'$).

PROOF.

(a) The reflexivity of $=_S$ follows immediately from A6.

(d) The formula

$$\text{push}(e, s) =_S \text{push}(e', s')$$

is equivalent to

        begin $s_1 := $ push(e, s); $s_2 := $ push(e', s'); bool := **true**,

           while bool $\wedge \sim$empty(s_1) $\wedge \sim$empty(s_2) **do**

$$\text{bool} := \text{bool} \wedge \text{top}(s_1) =_E \text{top}(s_2);$$
$$s_1 := \text{pop}(s_1); \ s_2 := \text{pop}(s_2)$$
od
end $\left(\text{bool} \wedge \text{empty}(s_1) \wedge \text{empty}(s_2)\right);$

Next, we obtain by Ax23 from Chapter II, § 5, another equivalent

begin $s_1 := \text{push}(e, s); \ s_2 := \text{push}(e', s'); \ \text{bool} := \textbf{true};$
    if $\text{bool} \wedge \sim\text{empty}(s_1) \wedge \sim\text{empty}(s_2)$ **then**
        begin $\text{bool} := \text{bool} \wedge \text{top}(s_1) =_E \text{top}(s_2);$
            $s_1 := \text{pop}(s_1); \ s_2 := \text{pop}(s_2);$
            while $\text{bool} \wedge \sim\text{empty}(s_1) \wedge \sim\text{empty}(s_2)$ **do**
                $\text{bool} := \text{bool} \wedge \text{top}(s_1) =_E \text{top}(s_2);$
                  $s_1 := \text{pop}(s_1); s_2 := \text{pop}(s_2);$
            od
        end
    fi
end $\left(\text{bool} \wedge \text{empty}(s_1) \wedge \text{empty}(s_2)\right);$

Now by A5 and Ax22 from Chapter II, § 5, the last formula is equivalent to:

begin $s_1 := \text{push}(e, s); \ s_2 := \text{push}(e', s'); \ \text{bool} := \textbf{true};$
    $\text{bool} := \text{bool} \wedge \text{top}(s_1) =_E \text{top}(s_2);$
    $s_1 := \text{pop}(s_1); \ s_2 := \text{pop}(s_2);$
    while $\text{bool} \wedge \sim\text{empty}(s_1) \wedge \sim\text{empty}(s_2)$ **do**
        $\text{bool} := \text{bool} \wedge \text{top}(s_1) =_E \text{top}(s_2);$
        $s_1 := \text{pop}(s_1); \ s_2 := \text{pop}(s_2)$
    od
end $\left(\text{bool} \wedge \text{empty}(s_1) \wedge \text{empty}(s_2)\right).$

Making use of A3, A4 and the simple facts from the semigroup of assignment instructions, we transform the last formula into

begin $\text{bool} := (e =_E e');$
    $s_1 := s; \ s_2 := s';$
    while $\text{bool} \wedge \sim\text{empty}(s_1) \wedge \sim\text{empty}(s_2)$ **do**
        $\text{bool} := \text{top}(s_1) =_E \text{top}(s_2) \wedge \text{bool};$
        $s_1 := \text{pop}(s_1); \ s_2 := \text{pop}(s_2);$
    od
end $\left(\text{bool} \wedge \text{empty}(s_1) \wedge \text{empty}(s_2)\right).$

This is equivalent to

$$e =_E e' \wedge s =_S s'.$$

(e, f) This easily follows from (d) by A2, A3, A4.

(g) Obvious.

(b) The symmetry of the $=_S$ relation follows from the fact that the instructions

$$s_1 := s; \quad s_2 := s'$$

in A6 can be permuted, and that the same can be done with

$$s_1 := pop(s_1); \quad s_2 := pop(s_2),$$

and also from the commutativity of $\vee$ and the symetry of $=_E$.

(c) Proof is by induction with respect to the depth of stack s'. We proceed in an informal way, passing to an extension of ATS by arithmetic. Define the mapping depth: $S \rightarrow N$ by

$$depth(s) \overset{df}{=} (i := 0) \,(\textbf{while} \sim empty(s) \,\textbf{do}$$
$$i := i+1; \; s := pop(s) \,\textbf{od} \; i).$$

Observe that $depth(s) = 0 \equiv empty(s)$. From this and (f) we have the base of induction. Assume for all s' of depth not greater than n that statement (c) is true. Consider a stack s' of depth $(n+1)$. Stack s' may be presented in the form

$$s' =_S push\,(top(s'),\; pop(s')).$$

From $s =_S s' \wedge s' =_S s''$ we have

$$top(s) = top(s') \wedge pop(s) = pop(s') \wedge top(s') = top(s'') \wedge$$
$$\wedge pop(s') = pop(s'').$$

Making use of the inductive assumption for stacks of depth not greater than n together with the transitivity of $=_E$, we obtain

$$top(s) =_E top(s'') \wedge pop(s) =_S pop(s'')$$

and by (e)

$$s =_S s''. \qquad\qquad\qquad \square$$

10. THE REPRESENTATION THEOREM FOR STACKS

As a simple corollary of Lemma 9.2 we observe that in every model $\mathfrak{M}$ of ATS the relation denoted by $=_S$ is a congruence and, consequently we have the following theorem:

THEOREM 10.1. *If a system* $\mathfrak{M}$ *is a model of* ATS, *then the quotient system* $\mathfrak{M}' = \mathfrak{M}/(=_E, =_S)$ *is a model of* ATS *proper for identity.* $\square$

THEOREM 10.2 (Representation Theorem). *Every model* $\mathfrak{M}$ *of* ATS *which is proper for identity is isomorphic with a standard model of it*:

$$\mathfrak{S} = \langle E \cup FSeq(E), \text{ precede, delete-first, first, } =_E, \varnothing \rangle.$$

PROOF. For every natural number i we define a partial mapping

ith from top: $S \to E$,

ith from top$(s) = ($**if** $\sim$empty(s) **then** $s := \text{pop}(s)$ **fi**$)^i$ top(s)

and another mapping

card: $S \to N$,

card(s) = the least natural number i such that empty$\big(\text{pop}^i(s)\big)$.

There is exactly one element s such that the formula empty(s) holds (by Lemma 9.2 and the assumption of the theorem). We shall denote this element by s_{empty}.

With every stack $s \in S$ we associate the finite sequence seq(s)

seq: $S \to FSeq(E)$,

seq$(s) = \{e_0, e_1, \ldots, e_{n-1}\}$,   where $n = $ card(s) and
$\qquad\qquad\qquad\qquad\qquad\quad e_i = $ ith from top(s) for
$\qquad\qquad\qquad\qquad\qquad\quad 0 \leqslant i < n$,

seq$(s_{\text{empty}}) = \varnothing$.

It is easy to observe that for every finite sequence $e_1, \ldots, e_n$, the following equality holds;

$$\text{seq}\big(\text{push}\,(e_1, \text{push}(e_2, \ldots, \text{push}\,(e_n, s_{\text{empty}}) \ldots)))$$
$$= \{e_1, e_2, \ldots, e_n\},$$

hence the mapping seq is onto $FSeq(E)$.

Let s and s' be two different stacks, $s \neq s'$. From A6 we see that either after the execution of the program in A6 the formula (bool$\wedge$ $\wedge \sim$(empty$(s) \wedge$ empty$(s'))$) holds and then card$(s) \neq$ card(s') or there exists a natural number i such that

ith from top$(s) \neq$ ith from top(s')

/

when after the execution of the program $\sim$ bool holds. In both cases $\mathrm{seq}(s) \neq \mathrm{seq}(s')$, i.e. the mapping seq is one-to-one.

It is easy to verify that

$$\mathrm{seq}\,(\mathrm{pop}(s)) = \mathrm{delete\text{-}first}\,(\mathrm{seq}(s)),$$
$$\mathrm{top}(s) = \mathrm{first}\,(\mathrm{seq}(s)),$$
$$\mathrm{seq}\,(\mathrm{push}(e, s)) = \mathrm{precede}\,(e, \mathrm{seq}(s)),$$
$$\mathrm{empty}(s) \equiv \mathrm{seq}(s) = \emptyset.$$

Hence seq is an isomorphism.   □

11. IMPLEMENTATIONS OF ARITHMETIC AND DICTIONARIES

Arithmetic of natural numbers

If we extend the language by a constant e_0 of sort E then putting

$$s =_A s' \overset{\mathrm{df}}{=} \textbf{begin } s1 := s;\ s2 := s';$$
$$\textbf{while } \sim\mathrm{empty}(s1) \wedge \sim\mathrm{empty}(s2)\ \textbf{do}$$
$$s1 := \mathrm{pop}(s1);\ s2 := \mathrm{pop}(s2)$$
$$\textbf{od}$$
$$\textbf{end }\ (\mathrm{empty}(s1) \wedge \mathrm{empty}(s2)),$$

$\mathrm{succ}(s) \overset{\mathrm{df}}{=} \mathrm{push}(e_0, s)$ where e_0 denotes a fixed element of E,

$0 \overset{\mathrm{df}}{=} \textbf{while } \sim\mathrm{empty}(s)\ \textbf{do}\ s := \mathrm{pop}(s)\ \textbf{od}\ s,$

we can prove the axioms of natural numbers:

$$\sim \mathrm{succ}\, x =_A 0,$$
$$(\mathrm{succ}(x) =_A \mathrm{succ}(y) \Rightarrow x =_A y),$$
$$(x := 0)\ (\textbf{while } x =_A y\ \textbf{do}\ x := \mathrm{succ}(x)\ \textbf{od true}),$$

which shows that ATS contains all theorems of the algorithmic arithmetic of natural numbers.

Dictionaries

In this case we implement the following "vocabulary" of notions:

$$\mathrm{amember}(s) \overset{\mathrm{df}}{=} \mathrm{top}(s),$$
$$\mathrm{insert}(e, s) \overset{\mathrm{df}}{=} \textbf{if } \sim \mathrm{member}(e, s)\ \textbf{then}\ s := \mathrm{push}(e, s)\ \textbf{fi}\ s,$$

$\text{delete}(e, s) \stackrel{\text{df}}{=}$ **begin**

        $s1 := s;$

        while $\sim\text{empty}(s2)$ **do** $s2 := \text{pop}(s2)$ **od**;

        while $\sim\text{empty}(s1)$ **do**

          if $\sim e =_E \text{top}(s1)$

          then $s2 := \text{push}(\text{top}(s1), s2);$

          fi;

          $s1 := \text{pop}(s1)$

        od

      end $s2,$

$\text{member}(e, s) \stackrel{\text{df}}{=}$ **begin** $s1 := s;$ bool $:= $ **false**;

          while $\sim\text{bool} \wedge \sim\text{empty}(s1)$ **do**

            if $e =_E \text{top}(s1)$ **then** bool $:= $ **true**

            else $s1 := \text{pop}(s1)$

            fi

          od

        end bool.

Observe that the axioms of the algorithmic theory of dictionaries may be proved from axioms of ATS and the above definitions showing that stacks can be used in order to implement insert, delete and member instructions. If this is not done in practice, the reason is to be found in the high cost of such implementation.

12. THEORY OF LINKS AND STACKS—ATSL

The aim of this section is to construct a bridge between such an abstract theory like ATS and the computer implementation of stacks to be found in § 13. We shall do this by (i) formalization of operations on attributes of stacks and links of stacks in ATSL theory, (ii) construction of a model for ATSL, (iii) interpretation of ATS theory within ATSL theory.

In ATS theory we have studied the properties of operations on stacks knowing nothing about how to perform them. Now we shall try to construct a model for ATS out of objects that can be handled by a computer. We assume that the objects of a set E of elements are computable, i.e. that there is an effective method of constructing them.

How do we construct stacks and how do we perform operations on them? An auxiliary set of links will take us nearer to a solution.

Any link object has two attributes:

> prev—pointing to a previous link—object in a stack,
> elem—pointing to an element of a set E.

The operations allowed on attributes are those of programming:

> read—to be denoted by $l.\langle$name of attribute$\rangle$,
> assign—to be denoted by $l.\langle$name of attribute$\rangle :=$.

In this way we reach the point where all operations are either from programming language (cf. axioms As1, As2, As3, As7, As8) below, or are defined explicitly (e.g. As9, As10).

The crucial fact about stacks of links is that walking along 'prev' path we shall always reach 'none', i.e. the bottom of the stack. This will be stated as axiom As4.

It will be observed that total freedom in assigning new values to the 'prev' attribute would eventually destroy property As4, and our theory could turn out to be inconsistent. In order to solve this problem we introduce the predicate ap, a guard of operation prevap, checking whether assigning a new value to the 'prev' attribute is safe.

Here we are assuming some properties of objects of classes (notion used in SIMULA, LOGLAN). We do not pretend that our understanding of their properties is complete. For example, we are not explaining the difficult question of identification of objects nor differences between copies of the same object. These questions will be studied more systematically elsewhere.

The alphabet of the language of ATSL contains the individual variables, predicates, functors, and other signs.

The set of individual variables is split into three disjoint subsets:

V_E—set of variables of type E,
V_L—set of variables of type L,
V_S—set of variables of type S.

In the following description of sets of predicates and functors we shall use the letters E, L and S to denote the sorts of arguments and results.

The predicates of the language are:

> $=_E\colon E \times E \to B_0$ where B_0 is the two-element Boolean algebra,

isnone: $L \to B_0$,

ap: $L \times L \to B_0$.

The functors of the language are:

tops: $S \to L$,   topsa: $S \times L \to S$,

new $S \in S$,   **none** $\in L$,

new $L : E \to L$,

elem: $L \to E$,   elema: $L \times E \to L$,

prev: $L \to L$,   prevap: ap $\to L$.

The notation elema: $L \times E \to L$ should be read: the first argument of elema is of the sort L, the second argument is of the sort E, the result of elema operation is of the sort L.

The sets of formulas and of programs are constructed as usual.

Notation

1. We shall use a postfix notation for tops, elem, and prev functors, i.e. instead of prev(l) we shall write l.prev.

2. Without loss of generality we can assume that functors topsa, elema and prevap will appear in the following context only:

$s := \text{topsa}(s, \tau_L)$,   where L is the type of τ_L,

$l := \text{elema}(l, \tau_E)$,   where E is the type of τ_E,

$l := \text{prevap}(l, \tau_L)$.

This allows us to use the following shortened forms below:

s.tops $:= \tau_L$,

l.elem $:= \tau_E$,

l.prev $:= \tau_L$.

Axioms

As1   isnone(**none**),

As2   isnone$\big($**new** $L(e)$.prev$\big)$,

As3 $e =_E$ **new** $L(e)$.elem,

As4   **while** $\sim$isnone(l) **do** $l := l$.prev **od** true,

As5 $(s$.tops $:= l)(s$.tops $=_L l)$,

As6   ap$(l, l') \equiv (\exists l'') l'' =_L$ prevap(l, l'),

As7 $\big($ap$(l, l') \wedge e =_E l$.elem$\big)$
        $\Rightarrow (l$.prev $:= l')\big(e =_E l$.elem $\wedge l$.prev $=_L l' \wedge \sim$isnone(l)$\big)$,

As8 $(l' =_L l.\text{prev} \Rightarrow (l.\text{elem} := e))(e =_E l.\text{elem} \land l' =_L l.\text{prev})$,

As9 $(l =_L l' \equiv$ **begin** $l1 := l;\ l2 := l';$ **bool** $:=$ **true**;

$\qquad\qquad\qquad$ **while** bool $\land \sim$isnone($l1$) $\land \sim$isnone($l2$) **do**

$\qquad\qquad\qquad\qquad$ bool $:=$ bool $\land$ ($l1.\text{elem} =_E l2.\text{elem}$);

$\qquad\qquad\qquad\qquad$ $l1 := l1.\text{prev};\ l2 := l2.\text{prev};$

$\qquad\qquad\qquad$ **od**

$\qquad\qquad$ **end** $(\text{bool} \land \text{isnone}(l1) \land \text{isnone}(l2))$,

As10 $\text{ap}(l, l') \equiv$ **begin** $l1 := l';$ **bool** $:=$ **true**;

$\qquad\qquad\qquad$ **while** bool $\land \sim$isnone($l1$) **do**

$\qquad\qquad\qquad\qquad$ **if** $l1 =_L l$ **then** bool $:=$ **false**; **else**

$\qquad\qquad\qquad\qquad\qquad\qquad\qquad\qquad$ $l1 := l1.\text{prev}$ **fi**

$\qquad\qquad\qquad$ **od**

$\qquad\qquad$ **end** bool.

It is not obvious that ATSL is a consistent theory. In order to prove this we shall construct a model of ATSL starting from system $\langle E, =_E \rangle$.

An object l from the set L will have the structure of a valuation of elem and prev variables shown in Figure 12.1

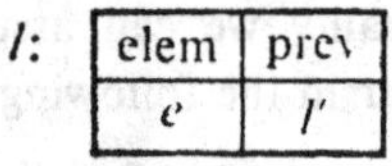

$l:$

elem	prev
e	l'

Fig. 12.1

and a similar, even simpler structure will have objects from S, as shown in Figure 12.2.

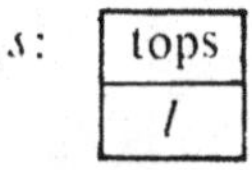

$s:$

tops
l

Fig. 12.2

In order to draw the model SL shown in Figure 12.3 we shall limit ourselves to the case where $E = \{e_1, e_1, e_3\}$; the reader will see that this limitation is inessential.

The tree SL contains diagrams of the operations: tops, elem, prev, **new** S, **new** L and none. For the remaining operations we assume the following definitions:

elema(l, e)—for a given l find its brother l' such that $l'.\text{elem} =_E e$, this l' will be the value of elema(l, e),

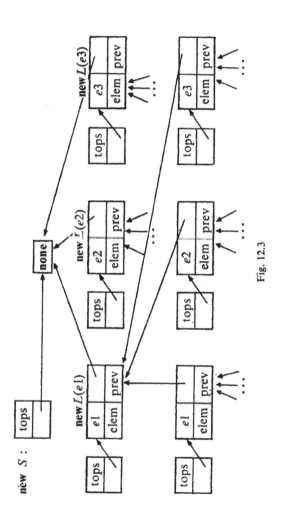

Fig. 12.3

topsa(s, l)—find s' such that s'.tops $=_L l$,

ap$(l, l') \equiv$ there is no path from l' to l,

prevap(l, l')—among the sons of l' find l'' such that l''.elem $=_E l$.elem
(remember that prevap(l, l') is defined only when ap(l, l') holds),

isnone $= \{\textbf{none}\}$.

It is not difficult then to prove the following theorem:

THEOREM 12.1. *The* SL *tree described above is a model for* ATSL.   □

We can now show that ATS theory is interpretable within ATSL.
Let us assume the following definitions:

$$\text{empty}(s) \overset{\text{df}}{=} \text{isnone}(s, \text{tops}).$$

$$\text{push}(e, s) \overset{\text{df}}{=} \textbf{begin } s1 := \textbf{new} S; \ l1 := \textbf{new} L(e);$$
$$l1.\text{prev} := s.\text{tops}; s1.\text{tops} := l1$$
$$\textbf{end } s1.$$

$$\text{pop}(s) \overset{\text{df}}{=} \textbf{begin if } \text{empty}(s) \textbf{ then ERROR fi};$$
$$s1 := \textbf{new} S; \ s1.\text{tops} := s.\text{tops.prev}$$
$$\textbf{end } s1.$$

$$\text{top(s)} \overset{\text{df}}{=} s.\text{tops.elem}.$$

$$s =_s s' \overset{\text{df}}{=} s.\text{tops} =_L s'.\text{tops}.$$

THEOREM 12.2. *The theory* ATS *is interpretable within* ATSL *theory,
i.e. axioms* A1–A6 *of* ATS *theory are theorems in* ATSL *theory augmented
by the above definitions.*

PROOF.

Ad A1 From T we have

$$(s := s1)(s =_S s1).$$

From As5 we have

$$(s1.\text{tops} := s.\text{tops.prev})(s1.\text{tops} = s.\text{tops.prev});$$

Combining these facts we have

$$(s.\text{tops} \equiv_L l \Rightarrow (\textbf{begin } s1 := \textbf{new} S; \ s1.\text{tops} := s.\text{tops.prev};$$
$$s := s1 \textbf{ end})(s.\text{tops} = l.\text{prev})).$$

Making use of definitions of empty and pop, and As4 we obtain

$$\textbf{while } \sim \text{empty}(s) \textbf{ do } s := \text{pop}(s) \textbf{ od true}.$$

Ad A2 Proof follows directly from the definitions of push and top.

Ad A3 By the definitions and As5 and As3.

Ad A4 By the definitions.

Ad A5 From As7 and the definitions.

Ad A6 Compare with As9.   □

Let $\mathfrak{M}_1$ and $\mathfrak{M}_2$ be two models of ATSL theory.

$$\mathfrak{M}_1 = \langle E \cup L_1 \cup S_1, \text{ tops, topsa, elem, elema, new } S, \text{ new } L,$$
$$\textbf{none}, \text{ ap}, \ldots \rangle,$$
$$\mathfrak{M}_2 = \langle E \cup L_2 \cup S_2, \text{ tops, topsa, elem, elema, new } S, \text{ new } L,$$
$$\textbf{none}, \text{ ap}, \ldots \rangle$$

with the same set of elements E. With this we have the following result.

THEOREM 12.3. *Models $\mathfrak{M}_1$ and $\mathfrak{M}_2$ are isomorphic.*   □

Consider the system described by the tree SL and observe the following.

THEOREM 12.4. *The least subsystem of* SL *containing* $E \cup \{\textbf{none}\} \cup \{\textbf{new } S\}$ *and closed with respect to the operations* push, pop *and* top *is the system* SL *itself, i.e. the* SL *is generated from* $E \cup \{\textbf{none}\} \cup \{\textbf{new } S\}$ *by the* push, pop *and* top *operations.*

The proof is straightforward. It is easy to see that every element in the SL tree can be obtained by a finite number of push operations, either explicitly if it is an S-element or implicitly if it belongs to L.   □

The meaning of the last theorem may be explained and utilized in the following way. It is possible to implement stacks in terms of ATSL in such a way that the operations of ATSL are internal and hidden, but the operations of ATS are external—the only ones accessible to the user.

13. IMPLEMENTATION OF STACKS IN LOGLAN PROGRAMMING LANGUAGE

The results of the previous section justify the introduction of the following program constituent. Its orthography is taken from the LOGLAN programming language designed at the University of Warsaw.

```
unit STACKS: class
begin virtual: function eq(a, b: element): Boolean;
    {we assume that eq is an equivalence relation}
    hidden protected link;
    unit element: class begin end element;
    unit link: class (elem: element); begin
        variable prev: link end link;
    unit stack: class begin variable tops: link end stack;
    function empty(s: stack): Boolean:
        begin result := (tops = none) end empty;
    function top(s: stack): element:
        begin result := s.tops.elem; end top;
    function push (e: element, s: stack): stack:
    variable l1: link, s1: stack;
    begin
        l1 := new link(e); s1 := new S;
        l1.prev := s.tops; s1.tops := l1;
        result := s1
    end push;
    function pop(s: stack): stack:
    variable s1: stack;
    begin
        if empty(s) then ERROR fi;
        s1 := new S; s1.tops := s.tops.prev;
        result := s1
    end pop;
    function eqs(s1, s2: stack): Boolean:
    variable l1, l2: link, bool: Boolean;
    begin
        l1 := s1.tops; l2 := s2.tops; bool := true;
        while bool l1 ≠ none ∧ l2 ≠ none do
            bool := bool ∧ eq(l1.elem, l2.elem);
            l1 := l1.prev; l2 := l2.prev
        od
        result := (bool ∧ l1 = none ∧ l2 = none)
    end eqs:
end STACKS.
```

STACKS may be viewed as an algebraic system of three sorts elements, links and stacks with three predicates—eq, empty and eqs, and three operations—top, push, pop. Let us denote the set of all objects that belong to a type t by $|t|$.

$$\text{STACKS} = \langle|\text{element}|\cup|\text{link}|\cup|\text{stack}|, \text{ empty, eq, eqs, top,}$$
$$\text{pop, push}\rangle.$$

About element-objects we assume nothing except that there is a binary predicate eq.

The structure of a link-object agrees with the earlier picture (see Figure 13.1) where $e \in|\text{element}|$, $l \in|\text{link}|$ and **none** is also a link-object.

elem	prev
e	l

Fig. 13.1

The structure of a stack-object is as shown in Figure 13.2.

tops
l

Fig. 13.2

From Theorem 12.4 we know that if we limit ourselves only to those objects which are generated by the push operation, then the resulting subsystem will be a model for **ATS** (neglecting links since they play only an auxiliary role). The line

hidden protected link;

serves the purpose of showing that the link is accesible only in functions declared in the STACKS type.

We should now like to show that the STACKS declaration serves as a definition of a family of similar algebraic systems.

Let a set E possess a definition in the form of type declaration

unit E: **class**...**end** E;

and let eq be a Boolean function determining the equality of two given elements of the set E:

function (eq e, e': E): Boolean:...result := ...

We are able to form a definition of the system of stacks over E concatenating the previous definition of STACKS with the one above. This is done by prefixing (a notion familiar from SIMULA 67 and LOGLAN).

> **unit** STACKS OVER E: STACKS **class**
>> **unit** E: element **class** ... **end** E;
>> **function** eq(e, e': E): Boolean: ... result := ...
> **end** STACKS OVER E.

Since STACKS prefixes STACKS OVER E and element prefixes E, every object prefixed by such a type behaves as if it possessed all the attributes of the prefixing type.

This last definition may be used as a prefix in front of a program written in the language of the defined system.

> **pref** STACKS OVER E **block**
>> **variable** e, $e1$, $e2$: E, l, l': link, s, $s1$, s': stack;
>> ...
>> {Objects of the types E, link, stack may be created only by **new** stack, **new** link, **new** E, pop, push operations. No change of attributes of link objects is possible. The program written here can use top, pop, push, empty, eqs, eq operations on stacks}
>> ...
> **end**.

14. QUEUES

We are now going to interpret dictionaries within queues, so we must introduce the algorithmic theory of queues ATQ. ATQ is a two-sorted theory. Let E and Q denote its two sorts.

Variables of sort E will be denoted by e, e', etc.; variables of sort Q will be denoted by q, q', $q1$, etc. The specific signs of the theory are listed below;

> em: $Q \rightarrow B_0$,
> put: $E \times Q \rightarrow Q$,
> out: $Q \rightarrow Q$,
> fr: $Q \rightarrow E$,
> $=_E$: $E \times E \rightarrow B_0$,
> $=_Q$: $Q \times Q \rightarrow B_0$.

Axioms of queues.

Aq1 **while** $\sim$em(q) **do** $q := $ out(q) **od true**,

Aq2 $\big(\text{em}(q) \Rightarrow (q =_Q \text{out}(\text{put}(e, q)))\big)$,

Aq3 $\big(\sim\text{em}(q) \Rightarrow \text{put}(e, \text{out}(q)) =_Q \text{out}(\text{put}(e, q))\big)$,

Aq4 $\big(\text{em}(q) \Rightarrow (e =_E \text{fr}(\text{put}(e, q)))\big)$,

Aq5 $\big(\sim\text{em}(q) \Rightarrow \text{fr}(\text{put}(e, q)) =_E \text{fr}(q)\big)$,

Aq6 $\sim$em$(\text{put}(e, q))$,

Aq7 $q =_Q q' \equiv$ **begin** $q1 := q$; $q2 := q'$; bool := **true**;

              while $\sim$em$(q1) \wedge \sim$em$(q2) \wedge$ bool **do**

                 if fr$(q1) \neq$ fr$(q2)$ **then** bool := **false fi**;

                 $q1 := $ out$(q1)$; $q2 := $ out$(q2)$;

              od

        end $\big(\text{bool} \wedge \text{em}(q1) \wedge \text{em}(q2)\big)$.

THEOREM 14.1 (Representation Theorem for ATQ Theory). *Every model of* **ATQ** *is isomorphic to the structure of finite sequences over the set E of elements of the given model with obvious operations on the sequences*

put(e, s),   *adjoin the element e to the sequence s at its end,*

fr(s),   *first element of the sequence s, if it is not empty,*

out(s),   *delete the first element of s,*

em(s),   *the sequence s is empty.*   □

After this brief presentation of the theory of queues we shall define an interpretation of dictionaries in the algorithmic theory of queues. It will consist of four definitions, which can be conceived as an extension of the theory ATQ introducing new primitive notions and four axioms.

The following vocabulary defines an interpretation of ATD theory in ATQ theory

DEFINITION 14.1.

     mb$(e, q) = $ **begin** $q1 := q$; bool := **false**;

                while $\sim$em$(q1) \wedge \sim$bool **do**

                  $e1 := $ fr$(q1)$;

                  if $e =_E e1$ **then** bool := **true fi**;

                  $q1 := $ out$(q1)$

                od

           end bool.   □

DEFINITION 14.2.

$$\text{in}(e, q) =_Q \textbf{begin } q1 := q;$$
$$\textbf{if} \sim \text{mb}(e, q1) \textbf{ then } q1 := \text{put}(e, q1) \textbf{ fi}$$
$$\textbf{end } q1.$$ □

DEFINITION 14.3.

$$\text{del}(e, q) =_Q \textbf{begin } q1 := q;$$
$$\textbf{if } \text{mb}(e, q1) \textbf{ then}$$
$$\textbf{while} \sim \text{em}(q2) \textbf{ do } q2 := \text{out}(q2) \textbf{ od};$$
$$\textbf{while} \sim \text{em}(q1) \textbf{ do}$$
$$e1 := \text{fr}(q1);$$
$$\textbf{if } e \neq e1 \textbf{ then } q2 := \text{put}(e1, q2) \textbf{ fi};$$
$$q1 := \text{out}(q1)$$
$$\textbf{od};$$
$$q1 := q2$$
$$\textbf{fi}$$
$$\textbf{end } q1.$$ □

DEFINITION 14.4.

$$\text{amb}(q) = \text{fr}(q)$$ □

We need not redefine the predicate em.

In order to prove that the vocabulary presented above is the correct implementation of dictionaries in queues we need to prove that formulas A1–A6 from Chapter 3, § 3, are theorems in the extension of the theory ATQ obtained by adding definitions 14.1–14.4 as extra axioms.

We shall limit ourselves to the proof of

A1   **while** $\sim \text{em}(q)$ **do** $q := \text{del}(\text{amb}(q), q)$ **od true.**

In the proof we shall use the Representation Theorem for ATQ Theory. $\text{amb}(q)$ is the first element in the sequence q, and $\text{del}(e, q)$ denotes the sequence obtained from the sequence q by deleting all occurrences of the element e. Hence, formula A1 is valid in every model of ATQ, and by the Completeness Theorem it is a theorem of extended ATQ.

In this way we have defined an implementation of dictionaries and proved its correctness. One can define other implementations, e.g. in arrays or in arrays of queues, i.e. hashtables.

15. BINARY TREES

Let A be a set whose elements will be called *atoms*. We shall give a specification of the structure of binary trees with atoms associated to leaves.

The structure has two sorts:

A—the sort of atoms,

T—the sort of trees.

The sorts A and T are not disjoint; we assume $A \subset T$.

The operations of the structure are as follows:

$$c: T \times T \to T,$$
$$e: T \to B_0,$$
$$a: T \to B_0,$$
$$l: T \to T,$$
$$r: T \to T,$$

l, r are partial operations, not defined if the argument is an atom.

The axioms of binary trees are:

TR1 $(\forall t \in T)(a(t) \vee e(t) \vee t = c(l(t), r(t)))$;

TR2 $(\forall t_1, t_2 \in T) l(c(t_1, t_2)) = t_1$;

TR3 $(\forall t_1, t_2 \in T) r(c(t_1, t_2)) = t_2$;

TR4 $(\forall t_1, t_2 \in T) \sim e(c(t_1, t_2)) \wedge \sim a(c(t_1, t_2))$;

TR5 $(\forall t \in T)$ **while** $\sim e(t) \wedge \sim a(t)$ **do**

        if $e(l(t)) \vee a(l(t))$ **then** $t := r(t)$

           else $t := c(l(l(t)), c(r(l(t)), r(t)))$

        fi

    od true;

TR6 $(\forall t_1 t_2 \in T)((e(t_1) \wedge e(t_2)) \Rightarrow t_1 = t_2)$.

A standard model for these axioms is the set of S-expressions. S-expressions constitute the semantic basis for "pure" LISP programming language.

DEFINITION 15.1. *The set of S-expressions over the set A is the least set of expressions such that*:

1° *it contains the set $A \cup \{nil\}$;*

2° *for every two S-expressions τ_1 and τ_2 the expression $(\tau_1 \cdot \tau_2)$ is also in the set of S-expressions.* □

THEOREM 15.1. *Every model of the axioms listed above proper for identity is isomorphic with a model in the set of S-expressions.*

PROOF. It may easily be observed that the set of S-expressions with an obvious interpretation of the functors c, l, r and predicates a and e is a model of axioms TR1—TR6. Observe that the axiom TR5 excludes elements outside S. It is obvious that TR5 rejects infinite trees. Let us note that axiom TR5 excludes elements like that shown in Figure 15.1.

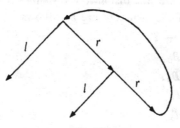

Fig. 15.1

The program in TR5 will not terminate on this input. Our axioms do accept directed acyclic graphs (dags). One can say that dags appear when we identify the subtrees of a given tree which have the same structure. In this way we have touched on the problem of identification of objects in a data structure. Our axiomatic theory deals with an abstraction of the notion of binary tree. For a more realistic treatment, the notion of reference should be included. This allows us to explain why two objects of the same structure are treated as though they are different although they are in fact basically the same.   □

These problems will be studied later (cf. Chapter VII).

Let us recall another specification of trees (cf. Kuratowski and Mostowski, 1967). A data structure $\mathfrak{A}$ of the signature

$$\mathfrak{A} = \langle A, f, a_0 \rangle$$

where $a_0 \in A$, $f \colon A \to A$, is called a *tree* iff it satisfies the axiom:

$$(\forall a \in A) \text{ while } a \neq a_0 \text{ do } a := f(a) \text{ od true.}$$

The axiom given above rejects dags and other graphs. One does not meet this specification on its own very frequently in computer science literature. In practical applications it should be combined with the previous definition of binary trees.

16. BINARY SEARCH TREES

Let E be a set linearly ordered by the relation $\leqslant$. A *binary search tree* is a labelled binary tree in which each vertex w is labelled by an element $e(w) \in E$ and where:

(a) for every vertex q in the left subtree of w: $e(q) < e(w)$,

(b) for every vertex q in the right subtree of w: $e(w) < e(q)$.

Binary search trees are usually implemented with the help of the following declaration of type:

unit N: **class** $(v\!:\!E)$; **variable** $l, r : N$; **end** N;

which is related to the following signature:

$$\langle E \cup N, v, l, r, \mathbf{new}N, \text{ul}, \text{ur}, \text{isnone}, =_E, \leqslant_E \rangle,$$

where

$\mathbf{new}N\!: E \rightarrow N,$

$v\!: N \rightarrow E, \quad l\!: N \rightarrow N, \quad r\!: N \rightarrow N,$

$\text{ul}\!: N \times N \rightarrow N, \quad \text{ur}\!: N \times N \rightarrow N,$

$\text{isnone}\!: N \rightarrow B_0,$

$=_E$ and $\leqslant_E$ are relations of identity and of linear order in E.

For programming languages the type declaration of N is to be interpreted as a description of a class of objects of the structure shown in Figure 16.1.

n:	v	e
	l	n_1
	r	n_2

Fig. 16.1

The class N also contains an empty object denoted **none**.

The operations listed above have the obvious meaning $v(n)$—read the value of v in the object n, $l(n)$, $r(n)$—and indicate the objects associated with n as the roots of its left and right subtrees respectively. The operations ul and ur update the values of l and r. In a programming language the instructions $n := \text{ul}(n', n)$ and $n := \text{ur}(n', n)$ are written $n.l := n'$ and $n.r := n'$, and we shall keep to the same convention here. We shall also write $n.v$ instead of $v(n)$, and similarly $n.l$ and $n.r$ instead of $l(n)$ and $r(n)$.

An algebraic structure of the above signature will be called a *binary search tree* if it satisfies the following axioms B1–B9:

B1 $\text{new}N(e).v = e$

{the value of the attribute v of the newly created object is e}.

B2   isnone $(\text{new}N(e).l)$,

B3   isnone $(\text{new}N(e).r)$

{in a newly created object the attributes l and r have the value **none**, hence any object obtained by $\text{new}N(e)$ should be interpreted as a leaf}.

The following definition will be used in axioms B4 and B5:

$$mb(e, n) = \textbf{begin } n1 := n; \text{ bool} := \textbf{false};$$
$$\textbf{while } \sim \text{isnone}(n1) \wedge \sim \text{bool } \textbf{do}$$
$$\textbf{if } n1 \cdot v = e \textbf{ then bool} := \textbf{true else}$$
$$\textbf{if } e < n1 \cdot v \textbf{ then } n1 := n1.l \textbf{ else}$$
$$n1 := n1.r \textbf{ fi}$$
$$\textbf{fi}$$
$$\textbf{od}$$
$$\textbf{end } \text{bool}$$

{the mb relation defined above is the relation of membership}.

B4 $(mb(e, n.l) \Rightarrow e < n.v)$.

B5 $(mb(e, n.l) \Rightarrow n.v < e)$

{for every non-empty tree with a root n, every member of its left subtree is less than the value associated with the root n and every member of its right subtree is greater than the value associated with the root n}.

B6 $(\text{isnone}(n) \vee$
$$\textbf{begin}$$
$$n' := n;$$
$$\textbf{while } \sim \text{isnone}(n') \textbf{ do}$$
$$\textbf{if } \text{isnone}(n'.l) \textbf{ then } n1 := n'.r \textbf{ else}$$
$$n1 := \text{new}N(n'.l.v);$$
$$n1.l := n' l.l ; \ n2 := \text{new}N(n'.v);$$
$$n2.l := n'.l.r; \ n2.r := n'.r;$$
$$n1.r := n2$$
$$\textbf{fi};$$
$$n' := n1$$
$$\textbf{od}$$
$$\textbf{end true})$$

{for every element n, n is the root of the finite binary tree (cf. Definition 15.1)}.

B7 $((n.r = n'' \wedge n.v = e \wedge$
          (begin
              $n2 := n'$;
              while $\sim$ isone($n2.r$) do $n2 := n2.r$ od;
              if $n2.v < n.v$ then bool := true else
                                                          bool := false fi
          end bool) $\vee$ isone(n'))
          $\Rightarrow (n.l := n')(n.r = n'' \wedge n.v = e \wedge n.l = n'))$

{if the greatest element in the tree n' is less than $n.v$ or isone (n') then the assignment associating n' as the left son of n is well defined and the remaining attributes of n are untouched}.

B8 $((n.l = n'' \wedge n.v = e \wedge$
          begin
              $n2 := n'$; while $\sim$ isone($n.2l$) do $n2 := n2.l$ od;
              if $n2.v > n.v$ then bool := true else bool := false fi
          end bool $\vee$ isone(n'))
          $\Rightarrow (n.r := n')(n.l = n'' \wedge n.v = e \wedge n.r = n'))$

{if the least element in the tree n is greater than $n.v$ or isone(n') then the assignment associating n' as the left son of n is well defined and the remaining attributes of n are untouched}

B9 The set E is linearly ordered by the relation $\leqslant$.

The set of axioms B1–B9 is consistent due to the following theorem:

THEOREM 16.1. *The algorithmic theory of binary search trees* ATBST *has a model.*

PROOF. Let us consider the set S of expressions over the set E which includes the expression ( ) representing **none** and where for every $e \in E$

$1°$ the expression $(( ) e ( ))$ is in S;

$2°$ if two expressions v and τ are in S and if for every element f occurring in v, $f < e$, for every element f occurring in τ, $f > e$, then the expression $(v e \tau)$ is in S;

$3°$ S is the least set of expressions closed with respect to $1°$ and $2°$. The interpretation of functors is as follows

$$\text{isone}(v) \stackrel{\text{df}}{\equiv} v = ( ),$$
$$\textbf{new}\,N(e) \stackrel{\text{df}}{=} (( )e( )),$$

and for every $v' \neq (\ )$, if it is of the form $(v e \tau)$, we put

$$v(v') = e, \quad l(v') = v, \quad r(v') = \tau,$$

The operations ul and ur are partial operations defined in the following way. Let n denote an expression $(v e \tau)$ and let n' be another expression. The operation $ul(n, n')$ is defined iff all elements of E occurring in n' are less than e and $ul(n, n') = (n' e \tau)$, i.e. we replace the left subtree of n by n' provided that tree n' contains only elements less than e.

The definition of ur is dual.

It is easy to verify that all axioms are valid in this structure.   □

We shall now formulate the following theorem:

THEOREM 16.2. *Every model of axioms* B1–B9 *is isomorphic with a standard model defined above.*

The axiom B6 ensures that every tree can be traversed in a finite time. The problem of directed acyclic graphs will not appear since axioms B4 and B5 assures that no two subtrees of a tree have the same structure.   □

17. AN INTERPRETATION OF THE THEORY OF PRIORITY QUEUES

We aim to prove that there exist an interpretation of the theory of priority queues in the theory of binary search trees ATBST. The interpretation retains the structure of the universe and extends the set of operations. The definitions of member, insert, delete, min operations are algorithmic. One can prove the axioms of priority queues in the theory, which results from joining new axioms to ATBST. In this way we approach our goal of verification of implementation. Implementation in this case consists of a set of definitions. This is a correct implementation since one can prove.

Let us consider the following definitions:

DEFINITION 17.1.

$$\min(n) \overset{df}{=} \text{if isnone}(n) \text{ then ALARM else } n1 := n \text{ fi}$$

$$(\textbf{while} \sim \text{isnone}(n1.l) \textbf{ do } n1 := n1.l \textbf{ od } n1.v).   □$$

LEMMA 17.1. *For every $n \neq$ **none** the value of* $\min(n)$ *is defined.*

For the proof it suffices to observe that every computation of the instruction **while** $\sim$isnone $(n1.l)$ **do** $n1 := n1.l$ **od** is finite. This follows from the Representation Theorem 16.2.   □

DEFINITION 17.2.

$$\text{member}(e, n) \overset{\text{df}}{\equiv} \textbf{begin } n1 := n; \text{ result} := \textbf{false};$$

$$\textbf{while } \sim \text{result} \wedge \sim \text{isnone}(n1) \textbf{ do}$$

$$\textbf{if } e = n1.v \textbf{ then } \text{result} := \textbf{true else}$$

$$\textbf{if } e < n1.v \textbf{ then } n1 := n1.l$$

$$\textbf{else } n1 := n1.r \textbf{ fi}$$

$$\textbf{fi}$$

$$\textbf{od}$$

$$\textbf{end } \text{result}. \qquad □$$

LEMMA 17.2. *The program in Definition* 17.2 *always terminates.*   □

LEMMA 17.3. *If the value of* $\min(n)$ *is defined then*

$$(\forall e) \ \big(\text{member}(e, n) \Rightarrow \min(n) \leqslant e\big). \qquad □$$

DEFINITION 17.3.

$$\text{insert}(e, n) \overset{\text{df}}{=} \textbf{begin } n1 := n; \text{ bool} := \textbf{false}; n3 := n1;$$

$$\textbf{while } \sim \text{isnone}(n1) \wedge \sim \text{bool } \textbf{do } n2 := n1;$$

$$\textbf{if } e = n1.v \textbf{ then } \text{bool} := \textbf{true else}$$

$$\textbf{if } \ e < n1.v \ \textbf{ then } \ n1 := n1.l$$

$$\textbf{else } n1 := n1.r \textbf{ fi}$$

$$\textbf{fi}$$

$$\textbf{od};$$

$$\textbf{if } \sim \text{bool } \textbf{then}$$

$$\textbf{if } \text{isnone}(n2) \textbf{ then } n3 := \textbf{new}N(e) \textbf{ else}$$

$$\textbf{if } e < n2.v \textbf{ then } n2.l := \textbf{new}N(e)$$

$$\textbf{else } n2.r := \textbf{new}N(e) \textbf{ fi}$$

$$\textbf{fi}$$

$$\textbf{fi}$$

$$\textbf{end } n3. \qquad □$$

LEMMA 17.4. *Let* M *denote the program in Definition* 17.3. *For every* $e \in E$, *for every* $n \in N$:

(i) M member$(e, n3)$,

(ii) *for every* $e' \neq e$,

member$(e, n1) \equiv M$ member$(e, n3)$.   □

In order to save space we shall informally indicate the structure of a deleting procedure. The reader will find it, in a modified version in the following section.

DEFINITION 17.4.

delete$(e, n) \overset{\text{df}}{=}$ **begin**

{search n}

{suppose e is found at $n1 \wedge$ father of $n1 = n2$}

{if $n1$ is a leaf—delete $n1$}

{if $n1$ has exactly one son—make the father $n2$ of $n1$ father of the son}

{if $n1$ has two sons—find the least element min$(n1.l)$ in the right subtree of $n1$. Delete min$(n1.r)$ from the tree with the root$(n1.r)$. Make this element the root of the subtree $n1$}

$n3 :=$ tree constructed above.

end $n3$.   □

LEMMA 17.5. *Let K denote the program sketched above. For every $e \in E$ and for every $n \in N$:*

(i) $K \sim$ member$(e, n3)$,

(ii) *for every $e' \neq e$,* member$(e', n) \equiv K$ member$(e', n3)$.   □

Making use of Lemmas 17.1–17.5 we can formulate the next theorem.

THEOREM 17.6 (on the interpretation of the theory of priority queues). *All axioms of priority queues are provable from the axioms of binary search trees and definitions of the operations* insert, delete, member min, empty.   □

This means that given a model of the theory of binary search trees we can define a model of the theory of priority queues. Moreover, since all definitions are algorithmic we can construct such a model in an effective way. The theorem on the interpretation of the theory of priority queues in the theory of binary search trees justifies the implementation of priority queues given below.

18. AN IMPLEMENTATION OF PRIORITY QUEUES

In this section we shall give a declaration of an encapsulated data type priority queue in binary search trees.

```
unit BST: class (type E; function less(e, e:E): Boolean);
    unit node: class (v:E);
        variable l, r:node;
    end node;
    unit min: function (n:node):E;
    begin
        while n.1 ≠ none do n := n l od;
        result := n.v
    end min;
    unit member: function (e:E, n:node): Boolean;
        variable n1:node, bool: Boolean;
    begin
        n1 := n; bool := false;
        while none ≠ n1 ∧ ~bool do
            if n1.v = e then bool := true else
                if e < n1.v then n1 := n1.l else
                                n1:= n1.r fi
            fi
        od;
        result := bool
    end member;
    unit empty: function (n:node): Boolean;
    begin
        if n = none then result := true else
                                result := false fi
    end empty;
    unit insert: function (e:E, n:node): node;
        variable n1, n2, n3:node, bool: Boolean;
    begin
```

```
    n1 := n; n3 := n; bool := false;
    while ~none = n1 ∧ ~bool do
        n2 := n1;
        if e = n1.v then bool := true else
            if e ⩽ n1.v then n1 := n1.l else
                                    n1 := n1.r fi
        fi
    od;
    if ~bool then
        if none = n3 then n3 := newN(e) else
            if e < n2.v then n2.l := newN(e) else
                            n2.r := newN(e) fi
        fi
    fi;
    result := n3
end insert;
unit delete: function (e:E, n:node): node;
    variable n1, n2, n3, n4, n5: node, bool, leftson:
                                            Boolean;
begin
    n1 := n; n3 := n; bool := false;
    while ~none = n1 ∧ ~bool do
        n2 := n1;
        if e = n1.v then bool := true else
            if e < n1.v then n1 := n1.l else n1:= n1.r fi
        fi
    od;
    if bool then {e found in n1 and n2 is the father
        of n1}
        if e < n2.v then leftson := true else
            leftson := false
        fi; {leftson iff n1 is the leftson of n2}
        if n1.l = none ∧ n1.r = none then {n1 is a leaf}
            if leftson then n2.l := none else
                n2.r := none fi
            else {n1 is not a leaf}
                if n1.l = none then {n1 has no leftson}
```

```
                    if n1 = n then n3 := n1.r else
                       if leftson then n2.l := n1.r else
                                               n2.r:= n1.r fi
                    fi
                    else
                    if n1.r = none then  {n1 has no right son}
                       if n1 = n then n3 := n1.l else
                          if leftson then  n2.l := n1.l else
                                               n2.r := n1.r fi
                       fi
                    else  {n1 has two sons}
                       n4 := n1.r;
                       while n4.l ≠ none do n5 := n4;
                                               n4 := n4.l od;
                       n5.l := n4.r;
                       n1.v := n4.v
                       fi
                    fi
                 fi
              fi {if bool};
              result := n3
         end delete
    end BST.
```

There exists another possibility where one can avoid making E
a formal parameter of type BST. In order to do this, we apply a concatenation of type declaration and virtual procedure

```
         unit BST' : class;
           unit E : class; end E;
           unit less : virtual function (e, e : E): Boolean; end less;
           unit node : class(v : E);
              variable l, r : node
           end node;
           unit min ...
           unit member ...
           unit insert ...
           unit empty ...
           unit delete ...
         end BST'.
```

Units BST and BST' are two different implementations of a problem-oriented language. Different environments are required in order to apply BST and BST'. LOGLAN allows parametrized-type declarations like BST. Notice that concatenation of type declarations is another solution of generic-type declarations. BST' can be conceived of as a description of a whole family of data structures. It represents a pattern which is to be completed by a user.

That is to say, the declaration

> **unit** myBST : BST' **class**;
> > **unit** Elem : E **class** ... **end** Elem;
> > > **unit** less : **function** $(e, e'$: Elem): Boolean ... **end** less;
> > **end** myBST

represents an extension of BST' by one concrete set Elem and the corresponding relation less.

In order to apply such a problem-oriented language we write

> **pref** myBST **block**
> > {declarations e.g. n, n: node, e, e': Elem}
> **begin**
> > {instructions e.g. $n := $ delete (e, n)}
> **end**.

19. ARRAYS

This frequently used structure allows us to treat finite sequences of elements of a given sort E together with the operations: access i-th component of a sequence and update i-th component of a sequence. The idea seems simple but there are some hidden traps, however. The literature on arrays quotes the instructions and various interpretations of their meaning (cf. van Emde Boas and Janssen, 1977).

By a *data structure of one-dimensional arrays* we shall understand any system

$$\langle E \cup Ar \cup N, \text{ put, det, lower, upper, newar, succ, emptyc,}$$
$$\text{emptyar, } 0, =, \leqslant \rangle,$$

where E, Ar, N are sets of data structure. N is the set of natural numbers, E—a non-empty set of elements, AR—a non-empty set of arrays. The operations of the data structure are as follows:

$$\text{put}: Ar \times N \times E \rightarrow Ar, \quad \text{get}: Ar \times N \rightarrow E,$$
$$\text{lower}: Ar \rightarrow N, \quad \text{upper}: Ar \rightarrow N,$$
$$\text{newar}: N \times N \rightarrow Ar, \quad \text{empty} \in E,$$
$$\text{emptyar} \in Ar, \quad \text{succ}: N \rightarrow N,$$
$$0 \in N.$$

$=$ is the identity relation, $\leqslant$ is the ordering in the set of natural numbers. Instead of $\text{succ}(x)$ we shall write $x+1$.

Variables of sort E will be denoted by e, e', e_1, etc, variables of sort Ar will be denoted by a, a' etc., variables of sort N will be denoted by i, j, l, u.

Specific axioms of arrays.

(1) $\quad (\sim a = \text{emptyar} \Rightarrow \text{lower}(a) \leqslant \text{upper}(a));$

(2) $\quad ((\sim a = \text{emptyar} \wedge \text{lower}(a) \leqslant i \leqslant \text{upper}(a))$
$$\Rightarrow \text{get}(\text{put}(a, i, e), i) = e);$$

(3) $\quad ((\sim a = \text{emptyar} \wedge \text{lower}(a) \leqslant i \leqslant \text{upper}(a)) \Rightarrow$
$$(\text{lower}(\text{put}(a, i, e)) = \text{lower}(a) \wedge$$
$$\wedge \text{upper}(\text{put}(a, i, e)) = \text{upper}(a)));$$

(4) $\quad (l \leqslant u \Rightarrow (\text{lower}(\text{newar}(l, u) = l \wedge$
$$\wedge \text{upper}(\text{newar}(l, u)) = u));$$

(5) $\quad (l < u \Rightarrow \textbf{begin } a := \text{newar}(l, u); \; l := l; \; \text{bool} := \textbf{true};$
$$\textbf{while } l \leqslant u \wedge \text{bool } \textbf{do}$$
$$\text{bool} := (\text{get}(a, i) = \text{emptyar}); i := i+1$$
$$\textbf{od}$$
$$\textbf{end } \text{bool});$$

(6) $\quad ((\sim a = \text{emptyar} \wedge \text{lower}(a) \leqslant i \leqslant \text{upper}(a)) \Rightarrow$
$$\textbf{begin } a := \text{put}(a, i, e); \; j := \text{lower}(a); \; \text{bool} := \textbf{true};$$
$$\textbf{while } j \leqslant \text{upper}(a) \wedge \text{bool } \textbf{do}$$
$$\textbf{if } i \neq j \textbf{ then } \text{bool} := (\text{get}(a, j) = \text{get}(a', j)); j := j+1 \textbf{ fi}$$
$$\textbf{od}$$
$$\textbf{end } \text{bool}).$$

To the above axioms we add axioms of natural numbers (cf. § 7) and axioms asserting that operations are undefined in certain circumstances. We shall give one example of an axiom of this type

(7) $\quad ((a = \text{emptyar} \vee i < \text{lower}(a) \vee i > \text{upper}(a)) \Rightarrow$
$$\text{get}(a, i) = \text{ERROR } a'),$$

where ERROR denotes the never-terminating program **while true do od**.

One can verify that this set of axioms is consistent since there exists a standard model of it. In the model arrays are conceived as triples, each triple consisting of a finite sequence of elements of sort E and a pair of natural numbers l, u. The length of the sequence is equal to $(u - l + 1)$.

The appropriateness of the specification given above is verified by the following theorem:

THEOREM 19.1 (Representation Theorem). *Every model of the theory of arrays is isomorphic with a standard model.*   □

In this example we can already observe the modularity of our approach. The theory of arrays includes the theory of natural numbers. The specifications are joined in order to define the more complicated objects of arrays. In the following section we shall see another example of this technique. The specification given above is sufficient to explain the computational phenomena of arrays if the programming language satisfies certain assumptions: 1° every array is identified by only one name—variable of array type, 2° every array is created at declaration time and is accessible as long as its name is accessible, 3° the only operations admissible are those of the indexed variables: read or update a value of an indexed variable.

There is a class of programming languages which admits arrays richer in operating possibilities, 4° an array can possess more than one name, i.e., many variables can point to the same array, 5° it is possible to make an assignment on an array variable and compare their values, 6° arrays are created (and deleted) dynamically during computations of programs, there is no syntactic guarantee that the value of a variable points to an array (cf. 2° above), 7° it is possible to read the lower and upper bounds of an array.

For languages like LOGLAN and others, our theory of arrays is not sufficient, and the notion of reference must therefore be introduced.

In what follows we shall use a notational convention close to that of programming languages:

$$get(a, i) \qquad \text{will be replaced by} \quad a[i],$$
$$a := put(a, i, e) \quad \text{will be replaced by} \quad a[i] := e.$$

20. HASHTABLES

The reader has no doubt seen a few examples of interpretation—implementation where an implementation of a data structure retained sorts and simply introduced new operations. Hashtables are a good example of a different kind of situation. A concise definition would read: a *hashtable* is an array of queues. Two modules of queues and of arrays are needed in order to implement hashtables. Moreover, a sort E of elements is mentioned in the definition of hashtables below. The specification of this sort is almost void; we assume only that there exists a function $h:E \to N$ enumerating the elements of the sort E. It is assumed additionally that the image $h(E)$ is a finite set. In fact our definition of hashtables will be generic for the whole family of similar data structures. They differ in sorts E and functions h.

The *data structure of hashtables* consists of five sorts:

$$N, E, Q, Ar, HT.$$

The language of our theory is the union of the languages of queues and of arrays. Additionally, we have a functor $h:E \to N$. We shall consider queues of elements from set E and arrays of the queues.

To the axioms of queues of elements (cf. § 14) and of arrays of queues we add axioms defining operations on dictionaries:

$$insert(e, s) \stackrel{df}{=} \textbf{begin } i := h(e);$$
$$s[i] := in(e, s[i])$$
$$\textbf{end } s;$$
$$delete(e, s) \stackrel{df}{=} \big(s[h(e)] := del(e, s[h(e)])\big)s;$$
$$member(e, s) \stackrel{df}{=} mb(e, s[h(e)]);$$

amember(s)—in order to find a member of s it is satisfactory to find a non-empty queue among $s[l], ..., s[u]$ and an element in it.

The proof of correctnes of the implementation given above is easy (cf. § 14). Again we can make use of the Representation Theorem for arrays in order to convince ourselves that the definitions above induce a model of dictionaries.

21. RATIONAL NUMBERS

In this section we shall present some results concerning programmability in the field of rational numbers Ω (cf. A. Kreczmar, 1977). First, we shall prove that the stopping property of Euclid's algorithm:

$$E: \textbf{while } x \neq y \textbf{ do}$$
$$\textbf{if } x > y \textbf{ then } x := x-y \textbf{ else } y := y-x \textbf{ fi}$$
$$\textbf{od}$$

characterizes the field of rational numbers up to isomorphism.

THEOREM 21.1. *For every ordered field $\mathfrak{F}$, if $\mathfrak{F}$ is a model for the formula*

(Euc) $(\forall x, y)((x > 0 \wedge y > 0) \Rightarrow E \textbf{ true})$

then F is isomorphic to Ω.

PROOF. Every ordered field contains a subalgebra isomorphic with the field of rational numbers Ω. Hence it is sufficient to prove that every element e of the Euclidean ordered field $\mathfrak{F}$ is of the form k/m where k and m are integers. Consider two arbitrary positive elements x_0, y_0 of $\mathfrak{F}$. By the (Euc) axiom we know that the computation of Euclid's algorithm is finite. The sequence of consecutive values of the variables x, y is finite. Let us denote it by

$$(x_0, y_0), \ldots, (x_n, y_n).$$

All values x_i, y_i are positive and $x_n - y_n$. There exist positive integers k_n and m such that $x_0 = k \cdot x_n$ and $y_0 = m \cdot x_n$. Hence $x \cdot y^{-1} = (k \cdot 1) \cdot (m \cdot 1)^{-1}$, i.e. every element of the field $\mathfrak{F}$ can be represented as a rational number.

THEOREM 21.2. *For every ordered field $\mathfrak{F}$, if $\mathfrak{F}$ is Euclidean then $\mathfrak{F}$ is Archimedean (cf. § 1 of this chapter).*

PROOF. Suppose that $\mathfrak{F}$ is not Archimedean. There then exist two elements x_0, y_0 such that for every natural number n, $(n \cdot x_0) < y_0$. This implies that for these x_0 and y_0 Euclid's algorithm does not terminate since for every n,

$$y_0 - n \cdot x_0 > x_0.$$ □

THEOREM 21.3. *A total function* $f: \mathfrak{Q} \to \mathfrak{Q}$ *is programmable in* $\mathfrak{Q}$ *iff there exist three total recursive functions* g, h, j *such that*

$$f\big((n-k)/m\big) = \big(g(n)-h(k)\big)/j(m)$$

for all natural numbers n, k, m.   □

22. COMPLEX NUMBERS

We shall prove that the algorithmic theory of complex numbers is hyperarithmetical (cf. Grabowski, 1978). On the other hand the set of Engeler's algorithmic properties, i.e. Boolean combinations of formulas $K\alpha$ (where α is an open formula and K is a program), is axiomatizable and Π_2^0-complete.

We shall study the properties of the field of complex numbers

$$\mathfrak{C} = \langle C, \, +, \, -, \, *, \, /, \, 0, \, 1, \, = \rangle.$$

Observe that the class $\mathcal{K}$ of algebraically closed fields of characteristic zero with an infinite degree of transcendency is axiomatizable by algorithmic formulas. Indeed, in Section 1 we have seen the axiom χ of fields of characteristic zero

(χ) $\qquad \sim (z := 1)(\textbf{while } z \neq 0 \textbf{ do } z := z+1 \textbf{ od true}).$

Let A' denote the set of axioms of algebraically closed fields of characteristic zero. Let $\{P_i(x, y_1, \dots, y_n)\}$ denote a sequence of all n-th degree polynomials with rational coefficients and indeterminates x, $y_1, \dots, y_n$. There exists a program K which for given data $(x, y_1, \dots, y_n, i)$ computes the value of the i-th polynomial in the sequence $\{P_i(x, y_1, \dots \dots, y_n)\}$ and assigns it to the variable z. Consider the following algorithmic formula $\varphi_n(x, y_1, \dots, y_n)$:

> **begin**
> $\qquad z := 1; \ i := 0;$
> $\qquad$ **while** $z \neq 0$ **do** $i := i+1; \ K(i, z)$ **od**
> **end true**.

It defines the property: "x is algebraic with respect to $y_1, \dots, y_n$". To the set A' we add formulas

$$(\forall y_1, \dots, y_n)(\exists x) \sim \varphi_n(x, y_1, \dots, y_n) \qquad \text{for every natural } n$$

The resulting set will be denoted by A. It is easily seen that the class $\mathcal{K}$ is characterized by the set A of axioms and that A is a recursive set.

THEOREM 22.1. (Vaught, 1973). *Let $\mathfrak{F}_1$, $\mathfrak{F}_2$ be two fields of the class $\mathcal{K}$. The algorithmic theories of $\mathfrak{F}_1$ and $\mathfrak{F}_2$ are equal, i.e. fields $\mathfrak{F}_1$ and $\mathfrak{F}_2$ are algorithmically equivalent.*

PROOF. Suppose the contrary; then there is a sentence α such that

$$\mathfrak{F}_1 \models \alpha \quad \text{and} \quad \mathfrak{F}_2 \models {\sim}\alpha.$$

Consequently $A \cup \{\alpha\}$ and $A \cup \{{\sim}\alpha\}$ are consistent sets. By the Downward Skolem–Löwenheim Theorem (Theorem 3.3, Chapter III) there exist enumerable fields $\mathfrak{F}_1^e$ and $\mathfrak{F}_2^e$ which are models for $A \cup \{\alpha\}$ and $A \cup \{{\sim}\alpha\}$ respectively. $\mathfrak{F}_1^e$ and $\mathfrak{F}_2^e$ are fields of characteristic zero, algebraically closed and with an infinite degree of transcendency. By Steinitz's Theorem (Vaught, 1973) they are isomorphic. This is a contradiction.   □

From Theorem 22.1 above we see that the set of theorems of the algorithmic theory of the field of complex numbers, i.e., theory $\text{Th}(\mathbb{C})$ forms an analytical set. In fact, by the Completeness Theorem $A \models \alpha$ iff $A \vdash \alpha$. Hence, $A \vdash \alpha$ iff for every enumerable set D such that $A \cup Ax \subseteq D$ (here Ax denotes the set of logical axioms of AL) and D is closed under the inference rules, the formula α is in D.

The theory $\text{Th}(\mathbb{C})$ of complex numbers is not arithmetical since for every arithmetical property p it is possible to construct an appropriate formula in the language of the field $\mathbb{C}$ which defines p. Natural numbers are definable in $\mathbb{C}$, hence we can relativize each occurrence of an individual bounded variable to natural numbers. This transformation in effective, hence we have proved that the set of first order sentences valid in the standard model of arithmetics is recursively reducible to $\text{Th}(\mathbb{C})$.

To estimate the location of $\text{Th}(\mathbb{C})$ in the analytical hierarchy we first observe that it is either hyperarithmetical or Π_1^1.

LEMMA 22.2. *The field of complex recursive numbers belongs to the class $\mathcal{K}$.*

PROOF. The complex numbers whose real and imaginary parts have effective decimal representation form the algebraically closed field of characteristic zero (cf. Rice, 1954; Mazur, 1963).

It remains to be proved that its degree of transcendency is not finite. Let us suppose the contrary, i.e. that there exists a finite set of recursive complex numbers $a_1, \ldots, a_n$ such that for every recursive number x

there exists a polynomial $f \in Q[x, x_1, \ldots, x_n]$ such that $f(x, a_1, a_2, \ldots, a_n)$ = 0. By a diagonalization argument we shall prove that there exist recursive numbers x such that for every polynomial f, $f(x, a_1, \ldots, a_n) \neq 0$.

We begin with an effective enumeration of all polynomials from $Q[x, x_1, \ldots, x_n]$. A polynomial $f(x, a_1, \ldots, a_n)$ can be treated as a polynomial of single variable x with coefficients determined by $a_1, \ldots, a_n$. These coefficients are effectively enumerable. Each coefficient is a recursive number and a limit of a recursive sequence which is recursively convergent. By Rice's theorem at least one complex number which is the root of a polynomial with recursive coefficients is the recursive limit of an effectively given recursive sequence. In order to obtain other roots of the polynomial in a uniform way we uniformly and effectively generate the coefficients of the quotient polynomial. Now we can effectively enumerate all these numbers which are roots of polynomials from the sequence defined above. Let us denote this sequence of recursive complex numbers by $c_1, c_2, \ldots$ There is a uniform algorithm of the generation of the subsequent approximation of the i-th number so defined. The construction of the necessary recursive real number x is easy. Ensure that the i-th decimal digit of the real part of c_i differs from the i-th decimal digit of x. We compute c_i with accuracy 10^{-i+1}. If the two last digits of the real part of this approximation are not 00 or 99, then we define the i-th decimal digit of x simply to be different from the i-th digit of c_i. If these two digits are 9 or 0 then we put the i-th decimal digit of x equal to 5.   □

LEMMA 22.3. *The field of recursive complex numbers is definable in the algorithmic theory of natural numbers.*   □

THEOREM 22.4. *The algorithmic theory of the field of complex numbers is hyperarithmetical.*

PROOF. Let $\mathfrak{F}$ be a field isomorphic to the field of complex recursive numbers definable in the system of natural numbers $\mathfrak{N}$. By Lemma 22.2, $\mathfrak{F}$ belongs to the class $\mathcal{K}$ and, by Theorem 22.3, $\text{Th}(\mathbb{C}) = \text{Th}(\mathfrak{F})$. For every formula α in the language of arithmetic of complex numbers we can effectively construct a first-order formula α' such that

$$\mathbb{C} \models \alpha \quad \text{iff} \quad \mathfrak{N} \models \alpha'.$$ □

Consider now some simpler algorithmic formulas.

THEOREM 22.5. *An enumerable set of open (i.e. quantifier-free and program-free) formulas is satisfiable in $\mathfrak{C}$ iff its every finite subset is satisfiable in $\mathfrak{C}$.*   □

The proof makes use of two facts and the following definition:

A field $\mathfrak{F}$ satisfies the *finite covering condition* iff for every algebraic variety A and every enumerable set $\{B_i\}_{i\in\omega}$ of algebraic varieties over $\mathfrak{F}$, if $A \subset \bigcup_{i\in\omega} B_i$, there exists a finite subset $I \subset \omega$ such that $A \subset \bigcup_{i\in I} B_i$. (By an *algebraic variety* we mean the set of zeros of a finite set of polynomials).

THEOREM 22.6. *Let $\{A_i\}$, $\{B_i\}$ be two enumerable sets of algebraic varieties over a field $\mathfrak{F}$ which satisfies the "finite covering condition". If*

$$\bigcup_{i\in\omega} (B_i - A_i) = \mathfrak{F}^n$$

then there exists a finite subset $I \subset \omega$ such that

$$\bigcup_{i\in I} (B_i - A_i) = \mathfrak{F}^n.$$

Proof of this theorem can be found in Kreczmar (1977).   □

THEOREM 22.7 (T. Mostowski). *The fields of complex numbers and real numbers satisfy the finite covering condition.*

The proof is to be found in Kreczmar (1977).   □

Theorem 22.5 is an easy corollary of Theorems 22.6 and 22.7.

Another interesting corollary of the results quoted above is the following theorem:

THEOREM 22.8 (Kfoury, 1972). *There is an effective method of transforming every total program K in $\mathfrak{C}$ (i.e. such that $\mathfrak{C} \models K\mathbf{true}$) into a loop-free program M equivalent to K.*

The proof follows from Theorem 22.5 and the observation that the halting formula of the program K is equivalent to an infinite disjunction of open formulas.   □

Theorem 22.8 asserts the algorithmic triviality of the field of complex numbers.

DEFINITION 22.1. *Any Boolean combination of formulas of the form $K\gamma$, where γ is an open formula and K is a program, will be called an algorithmic property.* $\square$

THEOREM 22.9 (Kreczmar, 1977). *The set of algorithmic properties valid in field $\mathbb{C}$ is a Π_2^0-complete set.* $\square$

The following theorem gives an axiomatization for the algorithmic properties valid in the field $\mathbb{C}$. Recall that χ denotes the axiom of fields of characteristic zero.

THEOREM 22.10. *For every algorithmic property β*

$$\mathbb{C} \models \beta \quad \text{iff} \quad \chi \vdash \beta.$$

PROOF. It is obvious that $\mathbb{C}$ is of characteristic zero. Now suppose that for a field F of characteristic zero β is not valid. Without loss of generality we can consider β to be of the form

$$\Big(\bigvee_{i \in \omega} \alpha_i(x_1, \ldots, x_n) \Rightarrow \bigvee_{j \in \omega} \beta_j(x_1, \ldots, x_n) \Big).$$

If β is not valid in F then

$$F \models (\exists x_1, \ldots, x_n) \bigvee_{i \in \omega} \alpha_i(x_1, \ldots, x_n) \wedge \bigwedge_{i \in \omega} \sim \beta_j(x_1, \ldots, x_n).$$

The same formula is valid in the algebraic closure F' of F since it is in existential form. Thus there exists $k \in \omega$ such that a set $\{\alpha_k, \sim \beta_i, i \in \omega\}$ is satisfiable in F'. Hence its every finite subset is satisfiable in F', and therefore it is also satisfiable in $\mathbb{C}$.

If every finite subset of some enumerable set of open formulas is satisfiable in $\mathbb{C}$ then by Theorem 22.5, the set $\{\alpha_k, \sim \beta_i, i \in \omega\}$ is satisfiable in $\mathbb{C}$. This proves that $\mathbb{C} \models \sim \beta$. $\square$

There are numerous applications of this fact (cf. Kreczmar, 1977). We shall end this section with an example showing that certain functions are not programmable in $\mathbb{C}$.

Consider the predicate $r(z)$—the number z is real. If it were strongly programmable over $\mathbb{C}$ then there would exist a program $K(z, x)$, such that the formula $K(x = 0)$ would define the subset of real numbers in the set of complex numbers. By Theorem 22.8 we can assume K to be a loop-free program. Hence the formula $K(x = 0)$ would be equivalent to an open formula $\alpha(z)$ of one variable. But this means that the set defined by α is finite or cofinite. The straight line of reals is neither finite nor cofinite in the field $\mathfrak{C}$. Hence the relation $r(z)$ is not programmable in $\mathfrak{C}$.

23. REAL NUMBERS

In this section we shall study a few theories of real numbers. The languages used may be classified as follows:

$\mathscr{L}_E$—we admit only Boolean combinations of formulas $K\alpha$, no classical quantifiers;

$\mathscr{L}_A$—iteration quantifiers admitted, no classical quantifiers;

$\mathscr{L}_F$—no restrictions.

Let $\mathfrak{R} = \langle R, +, -, \cdot, /, 0, 1, = \rangle$ be the field of real numbers. By $\mathfrak{RO}$ we shall denote the ordered field of reals. Observe that in $\mathscr{L}_F$ there exists a formula defining the ordering relation

$$x \leqslant y \overset{\text{df}}{=} (\exists z)(x+z^2 = y).$$

For $\mathscr{L}_E$ and $\mathscr{L}_A$ the cases of $\mathfrak{R}$ and $\mathfrak{RO}$ should be discussed separately.

In a manner similar to that of the preceding section we can prove the following:

THEOREM 23.1 (Kreczmar, 1977). *An enumerable set of open formulas is satisfiable in $\mathfrak{R}$ iff its every finite subset is satisfiable in $\mathfrak{R}$.* □

THEOREM 23.2 (Kreczmar, 1977). *The set of algorithmic properties valid in field $\mathfrak{R}$ is a Π_2^0-complete set.* □

DEFINITION 23.1. *A field $\mathfrak{F}$ is called formally real iff for every natural number n*

$$x_1^2 + \ldots + x_n^2 \neq -1.$$ □

It is easy to observe that formally real fields are of characteristic zero.

Let us denote by θ the axioms of formally real fields, i.e. the axioms of fields and the scheme of axioms

$$(\forall x_1, \ldots, x_n)(x_1^2 + \ldots + x_n^2 \neq -1), \quad n \geqslant 1.$$

THEOREM 23.3 (Kreczmar, 1977). *For every algorithmic property β,*

$$\mathfrak{R} \models \beta \quad \text{iff} \quad \theta \vdash \beta.$$

The proof is similar to that of Theorem 22.10 and is omitted.   □

The above theorems do not hold in the ordered field of reals $\mathfrak{RO}$. Making use of the fact that every Archimedean ordered field is embeddable in $\mathfrak{RO}$, together with the observation that the Archimedean axiom is a universally quantified formula, we obtain the next theorem.

THEOREM 23.4 (Engeler, 1967). *For every Boolean combination of formulas Kα*

$$\mathfrak{RO} \models \beta \quad iff \quad \Omega \vdash \beta$$

where Ω denotes the axioms of Archimedean ordered fields.   □

In contrast with the field $\mathfrak{C}$ of complex numbers, we have the following result.

THEOREM 23.5 (Grabowski and Kreczmar, 1978). *The set of Boolean combinations of formulas Kα valid in $\mathfrak{RO}$ is a Π_1^1-complete set.*

PROOF. By Kleene's normal form theorem (cf. Rogers, 1967) it is sufficient to prove that every set A definable by the formula

$$(\forall f)(\exists w)\, r\,(f(w), x),$$

where r is a recursive relation and f is a function, is definable in $\mathscr{L}_E(\mathfrak{RO})$—the theory of Boolean combinations of formulas $K\alpha$ valid in $\mathfrak{RO}$. We shall use the well-known fact that every real number x can be represented in a unique way as a continued fraction

$$x = a_0 + \cfrac{1}{a_1 + \cfrac{1}{a_2 + \ldots}}$$

where a_0 is an integer and a_i are natural numbers. Every continued fraction obviously represents a real number.

We shall construct a program $K(x, j, a)$ such that for any real x and any natural number j the value of the output variable a is equal to a_j—the j-th denominator in the expansion of x into a continued fraction. The integral part of a real number x, i.e., entier(x) is the programmable function in $\mathfrak{RO}$. Hence, our program K takes the following form:

```
begin
    z := x;  i := 0;
    while i ≠ j do
        a := entier(z);
        if z ≠ a then z := 1/(z−a) else z := 0 fi;
        i := i+1
    od
end.
```

Now, observe that every recursive relation is programmable in $\mathfrak{RO}$. Let us assume that a program $T(z, y, x)$ and an open formula α are so defined that $T\alpha$ computes the Kleene predicate $T^*_{1,\,1}(z, y, x)$ (cf. Rogers, 1967). The formula;

$$(\forall x) \ \textbf{begin} \ w := 0; \ p := \textbf{true}; \ \textbf{while} \ p \ \textbf{do} \ w := w+1;$$
$$K(x, w, a); \ T(n, a, n); \ p := \alpha \ \textbf{od} \ \textbf{end} \ \textbf{true},$$

where p is a propositional variable and n denotes the constant $(1 + ... + 1)$, n-times, defines then in $\mathfrak{RO}$ a Π^1_1-complete set E^1 (Rogers, 1967).   □

THEOREM 23.6 (Grabowski and Kreczmar, 1978). *Every analytical set is definable in* $\mathscr{L}_F(\mathfrak{RO})$.   □

COROLLARY (Grabowski and Kreczmar, 1978). *The algorithmic theory of the field of real numbers is not an analytical set.*

The algorithmic theory of the ordered field of real numbers is not an analytical set.   □

24. CONCLUDING REMARKS

The map of data structures shown in Figure 24.1 summarizes the discussion of this chapter. Implementability relations are represented by arrows.

Obviously, many interesting and important structures have been left out of the map. Moreover, the map can be enriched by the information about the costs of implementation treated as quasi distances associated with arrows.

Observe that the formulas

$$(\textbf{if} \ \alpha \ \textbf{then} \ K \ \textbf{fi})^i \ \alpha \quad ((\textbf{if} \ \alpha \ \textbf{then} \ K \ \textbf{fi})^i \sim \alpha)$$

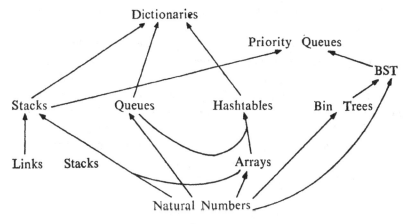

Fig. 24.1

express the properties: computations of the program **while** α **do** K **od** include at least (at most) i iterations of the program K, respectively. This shows that AL can be used in considerations concerning computational complexity. Moreover, the complexity of algorithms (and of interpretations) can be derived from specific axioms of data structures. It is not difficult to see that algorithms interpreting priority queues, say, in stacks, have a cost proportional to the depth of the stack in question.

The theory of interpretations mentioned here is the counterpart of the theories considered in classical logic by Szczerba (1977).

An extension of first order logic which admits the quantifiers "there exists a finite set such that ...", "for all finite sets..." is called a *weak second-order logic*. It can be proved that for every algorithmic formula there exists an equivalent weak second-order formula.

By the representation theorems for theories of stacks, dictionaries etc., one arrives at the following observation: in a model of the theory of stacks (dictionaries...) every weak second-order formula is equivalent to an algorithmic formula (cf. Grabowski, 1981). This means that models of stacks (dictionaries...) are expressive. They are also arithmetic in the sense that every partial recursive function is programmable in the model.

A data structure is called *constructive* iff there exists an enumeration α of its elements such that:

(i) every programmable relation is recursively enumerable with respect to α, and

(ii) every recursively enumerable relation is programmable.

Among many other data structures that can be axiomatized let us mention the constructive systems of Malcev, Markov, and Turing (cf. Malcev, 1965). Programs of some special form correspond to the normal algorithms of Markov, Turing machines, etc. In this way we can uniformly approach the Church thesis (cf. Malcev, 1965). By exhaustion of known definitions of the notion of an algorithm we verify that in every constructive system notions of programmability and of effective computability coincide.

BIBLIOGRAPHIC REMARKS

There are many approaches to the problems of data structures. Three of the most basic are domain identification (cf. Scott, 1976), algebraic specification (cf. Goguen, 1977, Guttag, 1977) and construction of domains (cf. Constable, 1982). The approach presented here goes back to the early papers of Engeler (1967).

It has been pointed out in a paper by Hoare (cf. Hoare, 1972) that the job of programming should be subdivided into two stages:

(i) specification and subsequent implementation of data structure,

(ii) design and verification of the abstract program in the data structure.

Among programming languages, SIMULA–67 was the first succesful realization of this principle, because it allows concatenation of type declarations, but it was learned only a few years later. LOGLAN brings a full solution to problems connected with the concatenation of type declarations; it removes all the limitations imposed on concatenation of types which occur in SIMULA.

Logic-based theories of data structures have been studied either from theoretical point of view (cf. for example Engeler, 1973; Kfoury, 1972; Kreczmar, 1977; Grabowski, 1978; Urzyczyn, 1981, 1982), with the aim of describing structures occurring in program languages (Salwicki, 1980, 1981, 1982; Mirkowska, 1981) or from the point of view of implementation (cf. Oktaba, 1981; Bartol, 1981).

Obviously other structures appear in programming practice which are either of a geometrical nature (as in computer graphics), or are used for data processing in management (e.g. banking or real-time applications). In every case one can conceive of an algorithmic theory of the data structure in question. Its axiomatization serves various purposes: identification of the domain, verification of correctness statements about programs, testing of the appropriateness of an implementation and, most important, a proper insight into data structure problems.

The objection can be made that the theories presented here are static or abstract since they do not reflect important phenomena related to problems of identification of objects. This aspect of data structures has been successfully studied by Oktaba (1981). The algorithmic theory of references which enables dynamization of objects will be presented in Chapter VII.

CHAPTER V

PROPOSITIONAL ALGORITHMIC LOGIC

The aim of propositional algorithmic logic PAL is to investigate the properties of program connectives:

> **begin...end,**
> **if...then...else...fi,**
> **while...do...od,**
> **either...or...ro** (the connective of non-deterministic choice).

In this it resembles the program of classical propositional calculus, where we study the properties of the propositional connectives *and, or* (disjunction), and *not*. Classical propositional logic provides us with useful inference rules for proofs and this is also true of PAL, which provides us with the inference rules necessary for proving the properties of program schemes. We are also interested in tautologies, i.e. expressions which are true by virtue of their syntactic composition, independently of the various interpretations which may be associated with the signs occurring in them.

We intend here to study PAL, in which:

(1) schemes of programs are constructed from program variables and propositional formulas by means of program connectives,

(2) formulas are either propositional variables or are composed of simpler formulas by means of logical connectives or are composed of program schemes and shorter formulas by means of the modalities possible, $\Diamond$, and necessary, $\square$.

The semantics of PAL is based on the notion of a semantic structure—directed graph of states in which edges are labelled by program variables; a valuation of propositional variables is associated with every state of a semantic structure. This enables us to introduce the next important notion, that of computation. Given a program M, a semantic structure $\mathfrak{M}$ and a state s determine a tree of acceptable computations of M in $\mathfrak{M}$ starting from the initial state s. Among the various properties of the tree we shall mention: strong termination (all branches of the tree are finite), looping (the existence of an infinite computation);

correctness of M with respect to an input condition α and output condition β, etc.

The meaning of the formulas of a PAL language depends on the meaning of their components in the usual way. Here we shall mention the modal expressions $\Diamond M\alpha$, $\square M\alpha$:

$\Diamond M\alpha$—after a finite computation of the program M its resulting state satisfies the formula α,

$\square M\alpha$—all computations of the program M are finite and all final states satisfy the formula α.

It is clear that with the help of these modal phrases we are able to construct formulas expressing the important properties of programs like termination, looping, correctness, and partial correctness. For example, the last property, program M being partially correct with respect to a postcondition α, can be expressed by $\sim\Diamond M\sim\alpha$.

In PAL we study the properties of semantics: we are looking for axioms and inference rules. These are discovered by studying the properties of the semantic consequence operation. Here we encounter several difficulties. First, we observe that the logic does not have the compactness property. This has already been observed in the case of the first-order deterministic AL. Thus in order to assure completeness we are forced to introduce the infinitary rule of inference

$$\frac{\{(\square(\text{if } \gamma \text{ then } M \text{ fi})^i(\sim\gamma\wedge\alpha) \Rightarrow \beta)\}_{i\in N}}{(\square\text{while } \gamma \text{ do } M \text{ od } \alpha \Rightarrow \beta)} ,$$

where γ, α, β are formulas and M is a program scheme. This rule is sound if the following equivalence holds:

$$(*) \qquad \square \text{ while } \gamma \text{ do } M \text{ od } \alpha \equiv \underset{i\in N}{\text{l.u.b.}} \ \square(\text{if } \gamma \text{ then } M \text{ fi})^i(\sim\gamma\wedge\alpha).$$

However, we shall see that $(*)$ is not always valid, it holds in certain cases and not in others. This is the source of our greatest difficulties. The following question remains unanswered: what are the nesessary any sufficient conditions for equivalence $(*)$?

We introduce an assumption of a finite degree of non-determinism of the interpretation of program variables. Under this assumption, $(*)$ holds and we can prove the soundness of the infinitary rule of inference. However, this property of a finite degree of non-determinism is not expressible in PAL, although for every n we can express that the degree of non-determinism is at most n.

In view of the lack of a general axiom we consider the family of systems PAL_n, where n denotes the degree of non-determinism of semantic structures. Thus we consider separately the case of deterministic interpretations of program variables and the case of bounded non-deterministic interpretations of program variables. For all these systems we shall prove the Completeness Theorem. The case of deterministic interpretations of program variables is treated in a way similar to the proofs in earlier considerations on AL. The other cases are much more difficult. We propose a method of proof of the Completeness Theorem which is a combination of the algebraic method of Rasiowa and Sikorski (cf. Rasiowa and Sikorski, 1968) for classical logic with the Kripke method for modal logic (cf. Kripke, 1963).

Propositional algorithmic logic is surprisingly powerful. One would naturally expect that program scheme properties such as termination, looping, partial correctness, correctness, etc., might be expressed by PAL formulas. It turns out, that in addition, we can define data structures by means of axioms written in the language of PAL. It is possible to study propositional theories of stacks, natural numbers, etc. These theories are cathegorical in the sense that all normalized models of a theory are isomorphic. Another propositional theory of natural numbers can be constructed which not only describes the sequence of natural numbers but also allows us to program every recursive function. The propositional approach is also recommended for providing a theory of control for a given concurrent program.

Propositional logic of programs is closely related to modal logics. The properties of relations like transitivity, reflexivity and associativity can be expressed by formulas of PAL. One can construct algorithmic theories which are the counterparts of systems studied in modal logic.

1. SYNTAX AND SEMANTICS

We shall consider a formalized language L_0, an extension of a propositional language in which there are propositional variables and program variables, and apart from the usual propositional connectives there are program connectives.

Let V_0 denote an at most enumerable set of propositional variables and V_p an at most enumerable set of *program variables*. Let F_0 be the set of all classical propositional formulas composed in the usual way by means of the propositional connectives: disjunction $\vee$, conjuction $\wedge$, negation $\sim$ and implication $\Rightarrow$, and the two logical constants, **true** and **false**.

The set of all well-formed expressions in the language L_0 will be augmented by schemes of programs, hence let us first define what a program scheme is.

By the *scheme of a program* we understand any element of the set of expressions Π which is the least set containing the program variables V_p and a program constant Id, and is closed under the following rules:

—If M, N are schemes of programs then the expressions of the forms **begin** M; N **end**, **either** M **or** N **ro** are schemes of programs,

If γ is a classical formula $\gamma \in F_0$ and M, N are schemes of programs, then the following expressions are schemes of program:

$$\textbf{while } \gamma \textbf{ do } M \textbf{ od}, \quad \textbf{if } \gamma \textbf{ then } M \textbf{ else } N \textbf{ fi}.$$

Now, we can define the set of all formulas F as the least set containing F_0 and such that;

—if α is a formula and M is a scheme of a program, then $\Box M\alpha$, $\Diamond M\alpha$ are formulas,

—if α, β are formulas, then $\sim \alpha$, $(\alpha \vee \beta)$, $(\alpha \wedge \beta)$, $(\alpha \Rightarrow \beta)$ are formulas.

The semantics of the language L_0 is based on the notions of interpretation and valuation. A *valuation* is a function which assigns an element of the two-element Boolean algebra B_0 to every propositional variable. An *interpretation* assigns to every program variable a binary relation in a non-empty set S. The elements of S will be called *states*. Every state will be understood to be an abstraction of a concrete situation on which the behaviour of the program and the value of any formula depends. Every state carries information about the valuationes-of propositional variables. Depending on the choice of the set of states and the kind of relation which is assigned to program variables, we can obtain various semantics for a given algorithmic language.

DEFINITION 1.1. *By a semantic structure we shall mean a system*

$$\langle S, \mathscr{I}, w \rangle$$

where S is a non-empty set of states—the universe of the structure, $\mathscr{I}$ is an interpretation of the program variables

$$\mathscr{I}:V_p \rightarrow 2^{S \times S} \quad and \quad \mathscr{I}(\mathrm{Id}) = \{(s, s):s \in S\}$$

*and w is a function which assigns to every state a valuation of proposi-
tional variables*

$$w:S \rightarrow B_0^{V_0}.$$

For a given structure $\mathfrak{M} = \langle S, \mathscr{I}, w \rangle$ and a given state $s \in S$ the Boolean value of the formula α is denoted by $\alpha_{\mathfrak{M}}(s)$ and is defined for classical connectives as follows:

$$\mathbf{false}_{\mathfrak{M}}(s) = 0, \quad \mathbf{true}_{\mathfrak{M}}(s) = 1.$$
$$p_{\mathfrak{M}}(s) = w(s)(p), \quad p \in V_0,$$
$$(\alpha \vee \beta)_{\mathfrak{M}}(s) = \alpha_{\mathfrak{M}}(s) \cup \beta_{\mathfrak{M}}(s),$$
$$(\alpha \Rightarrow \beta)_{\mathfrak{M}}(s) = \alpha_{\mathfrak{M}}(s) \rightarrow \beta_{\mathfrak{M}}(s),$$
$$(\alpha \wedge B)_{\mathfrak{M}}(s) = \alpha_{\mathfrak{M}}(s) \cap \beta_{\mathfrak{M}}(s), \quad (\sim \alpha)_{\mathfrak{M}}(s) = -\alpha_{\mathfrak{M}}(s).$$

In this way any formula determines a one-argument relation in S. The meaning of the formula $\Box M\alpha$ or $\Diamond M\alpha$ will be defined after some preliminary definitions.

Let us denote by $K_{\mathfrak{M}}$ a relation which is assigned to a program variable K by interpretation $\mathscr{I}$ in the semantic structure $\mathfrak{M} = \langle S, \mathscr{I}, w \rangle$. For a given state s, $K_{\mathfrak{M}}(s)$ is the set of all states s' such that $(s, s') \in K_{\mathfrak{M}}$.

By a *configuration* (cf. Chapter II, §2) we shall understand an ordered pair $\langle s, M_1, ..., M_n \rangle$, where s is a state in the structure $\mathfrak{M}$ and $M_1, ..., M_n$ is a sequence of schemes of programs (which may be empty).

For a given interpretation $\mathscr{I}$ of the program variables let $\mapsto$ denote a binary relation of successorship in the set of all configurations such that:

$\langle s;M_1, ..., M_n \rangle \mapsto \langle s';M_2, ..., M_n \rangle$ where M_1 is an atomic program, i.e. $M_1 \in V_p$ and $(s, s') \in \mathscr{I}(M_1)$,

$\langle s; \textbf{either } M_1 \textbf{ or } M_2 \textbf{ ro}, M_3, ..., M_n \rangle$
$\mapsto \langle s; M_1, M_3, ..., M_n \rangle,$

$\langle s; \textbf{either } M_1 \textbf{ or } M_2 \textbf{ ro}, M_3, ..., M_n \rangle$
$\mapsto \langle s; M_2, M_3, ..., M_n \rangle,$

$\langle s; \textbf{if } \gamma \textbf{ then } M_1 \textbf{ else } M_2 \textbf{ fi}, ..., M_n \rangle$
$$\mapsto \begin{cases} \langle s; M_1, M_3, ..., M_n \rangle & \text{if } \gamma_{\mathfrak{M}}(s) = 1, \\ \langle s; M_2, M_3, ..., M_n \rangle & \text{if } \gamma_{\mathfrak{M}}(s) = 0, \end{cases}$$

$\langle s;\ \mathbf{begin}\, M_1, M_2\ \mathbf{end},\ M_3, ..., M_n \rangle$

$\mapsto \langle s;\ M_1, M_2, M_3, ..., M_n \rangle,$

$\langle s;\ \mathbf{while}\ \gamma\ \mathbf{do}\ M_1\ \mathbf{od},\ M_2, ..., M_n \rangle$

$$\mapsto \begin{cases} \langle s;\ M_1, \mathbf{while}\ \gamma\ \mathbf{do}\ M_1\ \mathbf{od},\ M_2, ..., M_n \rangle & \text{if } \gamma_{\mathfrak{M}}(s) = 1, \\ \langle s;\ M_2, ..., M_n \rangle & \text{otherwise.} \end{cases}$$

Let N_0 be an initial segment of the set of natural numbers.

A sequence $\{c_i\}_{i \in N_0}$ of configurations will be called a *computation of the program scheme M in the structure $\mathfrak{M}$ at the initial state s* iff $c_1 = \langle s; M \rangle$ and for all i, $c_i \mapsto c_{i+1}$, and the sequence $\{c_i\}_{i \in N_0}$ is maximal in the sense of relation $\mapsto$.

If the computation is a finite sequence $c_1, ..., c_n$ and the last configuration c_n is of the form $\langle s'; \rangle$, i.e. the second part of the configuration c_n is the empty sequence, then the computation will be called *successful*. The state s' in a successful computation will be called the *result of the computation of the program M in the structure $\mathfrak{M}$*.

The set of all results of the program M in the structure $\mathfrak{M}$ at the initial state s will be denoted by $M_{\mathfrak{M}}(s)$.

Hence, for a given structure $\mathfrak{M}$, to every program scheme M we can assign a binary relation $M_{\mathfrak{M}}$ such that

$$sM_{\mathfrak{M}}s' \quad \text{iff} \quad s' \in M_{\mathfrak{M}}(s).$$

EXAMPLE. Consider the program scheme M of the form

$$\mathbf{while} \sim (a1 \lor a2 \lor a3 \lor a4 \lor a0)\ \mathbf{do}\ K\ \mathbf{od};$$

Let $\mathfrak{M} = \langle S, \mathscr{I}, w \rangle$ be a semantic structure such that

$$S = \{0, 1, 2, ...\}, \quad \mathscr{I}(K) = \{(i+5, i):\ i = 0, 1, ...\},$$

$$w(i) = v_i \quad \text{where } v_i(aj) = 1 \text{ iff } j = i.$$

The program scheme M describes in $\mathfrak{M}$ a binary relation such that

$$(i, j) \in M_{\mathfrak{M}} \text{ iff } i(\bmod\ 5) = j. \qquad \square$$

Now we are ready to define the value of the formulas $\square M\alpha$ and $\lozenge M\alpha$ in a given structure $\mathfrak{M}$ at a given initial state s.

$(\lozenge M\alpha)_{\mathfrak{M}}(s) = 1$ iff there exists a successful computation of the program M at the initial state s in $\mathfrak{M}$ such that its resulting state satisfies α.

$(\Box M\alpha)_{\mathfrak{M}}(s) = 1$ iff all computations of the program M at the state s in the structure $\mathfrak{M}$ are successful and all the results satisfy the formula α.

We shall say that the formula α is *valid in the semantic structure* $\mathfrak{M} = \langle S, \mathscr{I}, w \rangle$ (or $\mathfrak{M}$ *is a model of* α), iff α is satisfied by every state $s \in S$, i.e. $\alpha_{\mathfrak{M}}(s) = 1$ for all s. In symbols, $\mathfrak{M} \models \alpha$.

If α is valid in every semantic structure $\mathfrak{M}$, then α is called a *tautology*, in symbols $\models \alpha$.

In what follows we shall write $\mathfrak{M}, s \models \alpha$ to denote that $\alpha_{\mathfrak{M}}(s) = 1$.

EXAMPLE. For every program scheme M and for every formula α the following formulas are tautologies:

$$(\Box\, M\alpha \Rightarrow \Diamond M\alpha), \quad (\Box\, M \sim \alpha \Rightarrow \sim \Diamond M\alpha),$$
$$(\Box\, M\alpha \equiv (\Box\, M\,\text{true} \wedge \sim \Diamond M \sim \alpha)),$$
$$(\Diamond M\alpha \equiv (\Diamond M\,\text{true} \wedge \sim \Box\, M \sim \alpha)),$$
$$(\sim \Box\, M\alpha \vee \sim \Box\, M \sim \alpha). \qquad\qquad \Box$$

2. SEMANTIC PROPERTIES OF PROGRAM SCHEMES

In this section we shall discuss the basic properties of program schemes. The **while**-scheme will be our main interest.

Note that there is a strict correspondence between the set of all computations of the program **while** γ **do** M **od** and the set of all computations of the program (**if** γ **then** M **fi**)^i, where i is a natural number.

Consider an arbitrary successful computation $\mathcal{O}$ of the program scheme **while** γ **do** M **od**. Let $j_1, \ldots, j_n$ be numerals of those configurations in which the list of programs begins with **while** γ **do** M **od**. We shall construct another sequence of configurations such that:

—every configuration j_k of the form

$$\langle s'; \textbf{while } \gamma \textbf{ do } M \textbf{ od}, M_2, \ldots \rangle$$

will be replaced by the two configurations

$$\langle s'; (\textbf{if } \gamma \textbf{ then } M \textbf{ fi})^{n-k+1}, M_2, \ldots \rangle,$$
$$\langle s'; \textbf{if } \gamma \textbf{ then } M \textbf{ fi}, (\textbf{if } \gamma \textbf{ then } M \textbf{ fi})^{n-k}, M_2, \ldots \rangle, \quad k = 1, \ldots, n,$$

—in all configurations between j_k and j_{k+1} we put (**if** γ **then** M **fi**)$^{n-k}$ instead of **while** γ **do** M **od**. The resulting sequence of configurations $\mathcal{O}'$ is a successful computation of the program (**if** γ **then** M **fi**)^n.

Conversely, if we have a successful computation of the program (if γ then M fi)^n whose result satisfies the formula $\sim\gamma$, then, we can similarly construct a computation of the program while γ do M od.

The following facts are immediate consequences of the above observations.

FACT 1. If there exists a successful computation of the program while γ do M od, then there exists an i such that the program (if γ then M fi)^i has a successful computation with the same result.

FACT 2. If there exists a successful computation of the program (if γ then M fi)^i for a certain i such that its result satisfies the formula $\sim\gamma$, then there is a successful computation of the program while γ do M od with the same result.

FACT 3. If the program (if γ then M fi)^i has an unsuccessful computation (or infinite computation), then for every $j > i$ the programs (if γ then M fi)^j and while γ do M od have an unsuccessful (or infinite) computation.

For a given program scheme and a given semantic structure $\mathfrak{M}$ the set of all results of the program can be characterized as follows:

LEMMA 2.1. *For every formula γ, programs M, N and every state $s \in S$ in the structure $\mathfrak{M}$ the following hold*:

$$\text{(if } \gamma \text{ then } M \text{ else } N \text{ fi)}_{\mathfrak{M}}(s) = \begin{cases} M_{\mathfrak{M}}(s) & \text{if} \quad \mathfrak{M}, s \models \gamma, \\ N_{\mathfrak{M}}(s) & \text{if} \quad \mathfrak{M}, s \models \sim\gamma, \end{cases}$$

$$\text{(either } M \text{ or } N \text{ ro)}_{\mathfrak{M}}(s) = M_{\mathfrak{M}}(s) \cup N_{\mathfrak{M}}(s),$$

$$\text{(begin } M; N \text{ end)}_{\mathfrak{M}}(s) = \bigcup_{s' \in M_{\mathfrak{M}}(s)} N_{\mathfrak{M}}(s'),$$

$$\text{(while } \gamma \text{ do } M \text{ od)}_{\mathfrak{M}}(s) = \bigcup_{i \in N} \{s' \in \text{(if } \gamma \text{ then } M \text{ fi)}_{\mathfrak{M}}^i(s): \mathfrak{M}, s' \models \sim\gamma\}. \qquad \square$$

In connection with the last equality in the above lemma let us consider the following example.

EXAMPLE 2.1. Let

$\qquad M: $ if p then K' else K fi,

$\qquad M_1: $ while q do M od,

where q, p are propositional variables and K', K are program variables.

Consider the semantic structure

$\qquad \mathfrak{M} = \langle S, \mathscr{I}, w \rangle$

such that

$$S = \{(i,j): i,j = 0, 1, 2, \ldots\},$$
$$\mathscr{I}(K') = \{((0,0), (i,i)): i = 1, 2, 3, \ldots\},$$
$$\mathscr{I}(K) = \{((i,j+1), (i,j)): i,j = 0, 1, 2, \ldots\},$$
$$w(i,j)(p) = 0 \quad \text{iff} \quad j = 1,$$
$$w(i,j)(q) = 1 \quad \text{iff} \quad j = 0.$$

The tree of all possible computations of the program M_1 in the semantic structure $\mathfrak{M}$ at the initial state $(0, 0)$ is described below (Figure 2.1).

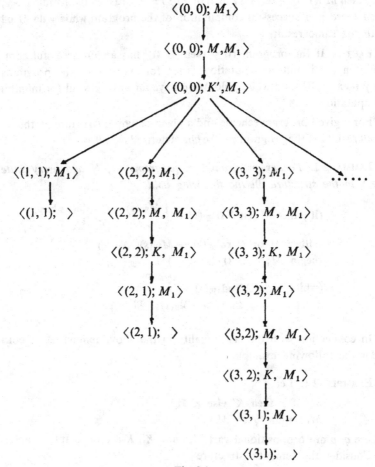

Fig. 2.1

It is easy to see that each computation of the program M is finite but there is no common upper bound on the length of the computations. However, we shall see that if we consider only those interpretations for which $K'_{\mathfrak{M}}(s)$ is a finite set for all states s, then there exists an i_0 such that

$$\textbf{(while } q \textbf{ do } M \textbf{ od)}_{\mathfrak{M}}(s) = \textbf{(if } q \textbf{ then } M \textbf{ fi)}^{i_0}_{\mathfrak{M}}(s). \qquad \square$$

The next lemmas will be of great importance in our further discussion.

LEMMA 2.2. *For any state s in the structure* $\mathfrak{M}$ *and for arbitrary* $\gamma \in F_0$, $\alpha \in F$, $M \in \Pi$ *the following equality holds*:

$$(\Diamond\textbf{while } \gamma \textbf{ do } M \textbf{ od } \alpha)_{\mathfrak{M}}(s)$$
$$= \text{l.u.b } \underset{i \in N}{\Diamond}(\textbf{if } \gamma \textbf{ then } M \textbf{ fi})^i \, (\sim\gamma \wedge \alpha)_{\mathfrak{M}}(s).$$

PROOF. Suppose that

(1) $(\Diamond\textbf{while } \gamma \textbf{ do } M \textbf{ od } \alpha)_{\mathfrak{M}}(s) = \textbf{1},$

(2) $\underset{i \in N}{\text{l.u.b. }} \Diamond(\textbf{if } \gamma \textbf{ then } M \textbf{ fi})^i(\sim\gamma \wedge \alpha)_{\mathfrak{M}}(s) = \textbf{0}.$

By (2), for every natural number i, $i \in N$

(3) $\Diamond(\textbf{if } \gamma \textbf{ then } M \textbf{ fi})^i \, (\sim\gamma \wedge \alpha)_{\mathfrak{M}}(s) = \textbf{0}.$

Hence,

(4)   for every i, either $(\textbf{if } \gamma \textbf{ then } M \textbf{ fi})^i_{\mathfrak{M}}(s)$ is empty or for all $s' \in (\textbf{if } \gamma \textbf{ then } M \textbf{ fi})^i_{\mathfrak{M}}(s)$, $\mathfrak{M}, s' \models \gamma$ or $\mathfrak{M}, s' \models \sim\alpha$.

If, for a certain i, $(\textbf{if } \gamma \textbf{ then } M \textbf{ fi})^i_{\mathfrak{M}}(s) = \varnothing$, then all computations of the program $\textbf{while } \gamma \textbf{ do } M \textbf{ od}$ are unsuccessful and consequently $\mathfrak{M}, s \models \sim\Diamond\textbf{while } \gamma \textbf{ do } M \textbf{ od } \alpha$, which contradicts (1).

Let us assume that for every i, $(\textbf{if } \gamma \textbf{ then } M \textbf{ fi})^i_{\mathfrak{M}}(s) \neq \varnothing$. We then have

(5) $(\forall i)(\forall s' \in (\textbf{if } \gamma \textbf{ then } M \textbf{ fi})^i_{\mathfrak{M}}(s))$ $\mathfrak{M}, s' \models \gamma$ or
$$\mathfrak{M}, s' \models (\sim\gamma \wedge \sim\alpha)$$

Let i be an arbitrary natural number. If we have a computation of $(\textbf{if } \gamma \textbf{ then } M \textbf{ fi})^i$ whose result satisfies $(\sim\gamma \wedge \sim\alpha)$ then we can construct the computation of $\textbf{while } \gamma \textbf{ do } M \textbf{ od}$ (see Fact 1), whose result does not satisfy α. Let us skip all such computations. All the remaining computations of $(\textbf{if } \gamma \textbf{ then } M \textbf{ fi})^i$ have results which satisfy the formula γ.

There are two possible cases: either (a) the computation of the program $(\textbf{if } \gamma \textbf{ then } M \textbf{ fi})^i$ can be extended to a computation of a program (if γ then M fi)^j, $j > i$, such that its result satisfies $\sim\gamma$, or (b) there is no such extension.

In case (a), by (5), the result of the extended computation does satisfy $\sim\alpha$. In case (b), we can construct an infinite computation of the program scheme **while** γ **do** M **od**. Thus

$$(\Diamond\text{while } \gamma \text{ do } M \text{ od } \alpha)_{\mathfrak{M}}(s) = 0,$$

which contradicts (1).

Conversely, suppose that

$$\underset{i \in N}{\text{l.u.b }} \Diamond(\text{if } \gamma \text{ then } M \text{ fi})^i(\sim\gamma \wedge \alpha)_{\mathfrak{M}}(s) = 1.$$

Hence, for a certain i_0, $\mathfrak{M}, s \models \Diamond(\text{if } \gamma \text{ then } M \text{ fi})^{i_0}(\sim\gamma \wedge \alpha)$. Consider a successful computation of the program (if γ then M fi)^{i_0} with a result satisfying $(\sim\gamma \wedge \alpha)$. After a simple transformation we shall obtain a successful computation of **while** γ **do** M **od** such that its result satisfies α. Thus $\mathfrak{M}, s \models \Diamond\text{while } \gamma \text{ do } M \text{ od } \alpha$. $\square$

From Lemma 2.2 we conclude that every formula of the form

$$\Diamond\text{while } \gamma \text{ do } M \text{ od } \alpha$$

defines an infinite operation. For the formula $\square$ **while** γ **do** M **od** α the problem is more complicated. It is a simple consequence of Fact 1 and Fact 2 that:

$$\mathfrak{M}, s \models \square(\text{if } \gamma \text{ then } M \text{ fi})^i(\sim\gamma \wedge \alpha)$$

for a certain i implies

$$\mathfrak{M}, s \models \square \text{ while } \gamma \text{ do } M \text{ od } \alpha.$$

But the converse is not true in general, as was shown in Lemma 2.1. However, if we consider only special interpretations, then a lemma analogous to Lemma 2.2 can be obtained. These special interpretations will have the so called *finite degree of non-determinism property*, referred to for short as the FDN property.

DEFINITION 2.1. *We shall say that a structure* $\mathfrak{M} = \langle S, \mathscr{I}, w \rangle$ *has the property of finite degree of non-determinism (FDN property) iff for every program variable K and every state $s \in S$, the set $K_{\mathfrak{M}}(s)$ is finite.* $\square$

The structure with FDN property we shall call simply FDN structure.

LEMMA 2.3. *For every structure* $\mathfrak{M} = \langle S, \mathscr{I}, w \rangle$ *with the* FDN *property, and for every state* $s \in S$ *and* $\gamma \in F_0$, $\alpha \in F$, $M \in \Pi$ *the following equality holds*:

$$\square(\textbf{while } \gamma \textbf{ do } M \textbf{ od } \alpha)_{\mathfrak{M}}(s)$$
$$= \text{l.u.b.} \underset{i \in N}{\square}(\textbf{if } \gamma \textbf{ then } M \textbf{ fi})^i \ (\sim\gamma \wedge \alpha)_{\mathfrak{M}}(s).$$

PROOF. Suppose that

(6) $\mathfrak{M}, s \models \square \textbf{ while } \gamma \textbf{ do } M \textbf{ od } \alpha$

and

(7)   non $\mathfrak{M}, s \models \square(\textbf{if } \gamma \textbf{ then } M \textbf{ fi})^i(\sim\gamma \wedge \alpha)$
            for every natural number i.

By (7), for every i there are three possible situations:

A. There exists an unsuccessful computation of (**if** γ **then** M **fi**)^i at the initial state s.

B. There exists an $s' \in (\textbf{if } \gamma \textbf{ then } M \textbf{ fi})^i_{\mathfrak{M}}(s)$ such that

$$\mathfrak{M}, s' \not\models (\sim\gamma \wedge \alpha).$$

C. There exists an $s' \in (\textbf{if } \gamma \textbf{ then } M \textbf{ fi})^i_{\mathfrak{M}}(s)$ such that

$$\mathfrak{M}, s' \models \gamma.$$

However, if for a certain i case A holds, then we can construct an unsuccessful computation of the program **while** γ **do** M **od**, contrary to (6).

Analogously, if for a certain i case B holds, then we can construct a computation of **while** γ **do** M **od** which has a result not satisfying α. This contradicts (6).

Suppose that case C holds for every natural number i. Thus for all i there exists a computation of **while** γ **do** M **od** in which program M is executed i times. Since the degree of any vertex in the tree of all possible computations of the program **while** γ **do** M **od** is finite, then by König's Lemma (cf. Kuratowski, 1967) there exists an infinite computation of **while** γ **do** M **od**. This contradicts (6).

Thus

$$\mathfrak{M}, s \models \square \textbf{ while } \gamma \textbf{ do } M \textbf{ od } \alpha$$

implies

$$\text{l.u.b.} \underset{i \in N}{\square}(\textbf{if } \gamma \textbf{ then } M \textbf{ fi})^i(\sim\gamma \wedge \alpha)_{\mathfrak{M}}(s) = \textbf{1}.$$

The converse implication is obvious.   □

It appears that for structures with the FDN property the above results can be made even stronger. To simplify future considerations let us first prove an auxiliary lemma.

LEMMA 2.4. *Let* α, α_i, $i \in N$, *be formulas and* $\mathfrak{M}$ *be a semantic structure with the* FDN *property. If*:

(i) *for every state s in* $\mathfrak{M}$, $\alpha_{\mathfrak{M}}(s) = \text{l.u.b.} \ \alpha_{i\mathfrak{M}}(s)$ *and*
$$\hspace{6cm} {\scriptstyle i \in N}$$

(ii) *for every i and* $j \leqslant i$, $\mathfrak{M} \models (\alpha_j \Rightarrow \alpha_i)$

then for every program variable $K \in V_p$ *we have*:

(iii) *for every state s*,

$$(\Box K\alpha)_{\mathfrak{M}}(s) = \text{l.u.b.}_{i \in N}(\Box K\alpha_i)_{\mathfrak{M}}(s),$$

$$(\lozenge K\alpha)_{\mathfrak{M}}(s) = \text{l.u.b.}_{i \in N}(\lozenge K\alpha_i)_{\mathfrak{M}}(s),$$

(iv) *for every i and* $j \leqslant i$,

$$\mathfrak{M} \models (\Box K\alpha_j \Rightarrow \Box K\alpha_i), \quad \mathfrak{M} \models (\lozenge K\alpha_j \Rightarrow \lozenge K\alpha_i).$$

PROOF. To prove (iii) let us suppose that

(8) $(\Box K\alpha)_{\mathfrak{M}}(s) = 1$

and

(9) $\text{l.u.b.}_{i \in N}(\Box K\alpha_i)_{\mathfrak{M}}(s) = 0.$

By (9), for every natural number i we have

(10) $(\Box K\alpha_i)_{\mathfrak{M}}(s) = 0.$

There are two possible situations: $K_{\mathfrak{M}}(s) = \varnothing$ and $K_{\mathfrak{M}}(s) \neq \varnothing$. $K_{\mathfrak{M}}(s) = \varnothing$ implies that $(\Box K\alpha)_{\mathfrak{M}}(s) = 0$, contrary to (8). $K_{\mathfrak{M}}(s) \neq \varnothing$ implies by (10) that for every i there exists an $s' \in K_{\mathfrak{M}}(s)$ such that $\alpha_{i\mathfrak{M}}(s') = 0.$

Since by assumption $K_{\mathfrak{M}}(s)$ is finite, then for at least one $s' \in K_{\mathfrak{M}}(s)$ there are infinitely many formulas α_i which are not satisfied by s'. By (ii) non $s' \models \alpha_i$ for all $i \in N$. Thus $\text{l.u.b.}_{i \in N} \alpha_{i\mathfrak{M}}(s') = 0$. By (i) $\alpha_{\mathfrak{M}}(s') = 0$ and therefore $(\Box K\alpha)_{\mathfrak{M}}(s) = 0$, which contradicts (8). Conversely, let us assume that $\text{l.u.b.}_{i \in N}(\Box K\alpha_i)_{\mathfrak{M}}(s) = 1$. Hence there exists an i such that $(\Box K\alpha_i)_{\mathfrak{M}}(s) = 1$, i.e. $K_{\mathfrak{M}}(s) \neq \varnothing$ and for every $s' \in K_{\mathfrak{M}}(s)$, $\alpha_{i\mathfrak{M}}(s') = 1$. Thus, for every $s' \in K_{\mathfrak{M}}(s)$ there exists an i such that

$$\mathfrak{M}, s' \models \alpha_i.$$

In consequence, l.u.b. $\alpha_{i\mathfrak{M}}(s') = 1$ for every $s' \in K_{\mathfrak{M}}(s)$. By the second
$\quad\quad\quad\quad\quad{i\in N}$
assumption (ii)

$$\mathfrak{M}, s' \models \alpha \quad \text{for every } s' \in K_{\mathfrak{M}}(s),$$

i.e. $(\Box K\alpha)_{\mathfrak{M}}(s) = 1$.

Thus, the first part of (iii) is proved. The other equality will be
proved analogously. Suppose

(11) $\mathfrak{M}, s \models \Diamond K\alpha \quad$ and $\quad$ l.u.b.$(\Diamond K\alpha_i)_{\mathfrak{M}}(s) = 0$.
$\quad\quad\quad\quad\quad\quad\quad\quad\quad\quad\quad{i\in N}$

Thus, for every $i \in N$, $(\Diamond K\alpha_i)_{\mathfrak{M}}(s) = 0$.

This means by the definition of the value of the formula that either
$K_{\mathfrak{M}}(s) = \varnothing$ or $K_{\mathfrak{M}}(s) \neq \varnothing$ and for all $s' \in K_{\mathfrak{M}}(s)$, $\alpha_{i\mathfrak{M}}(s') = 0$ for
$i \in N$. If $K_{\mathfrak{M}}(s) = \varnothing$ then $(\Diamond K\alpha)_{\mathfrak{M}}(s) = 0$, which contradicts (11).

Assuming that

$$K_{\mathfrak{M}}(s) \neq \varnothing,$$

then

l.u.b. $\alpha_{i\mathfrak{M}}(s') = 0 \quad$ for all $s' \in K_{\mathfrak{M}}(s)$.
${i\in N}$

By assumption (i)

$$\alpha_{\mathfrak{M}}(s') = 0 \quad \text{for all } s' \in K_{\mathfrak{M}}(s)$$

and therefore

$$(\Diamond K\alpha)_{\mathfrak{M}}(s) = 0, \quad \text{contradiction!}$$

Conversely, suppose that

(12)   l.u.b.$(\Diamond K\alpha_i)_{\mathfrak{M}}(s) = 1$.
$\quad\quad{i\in N}$

Then there exists an $i \in N$ such that $(\Diamond K\alpha_i)_{\mathfrak{M}}(s) = 1$. This means
that for a certain $s' \in K_{\mathfrak{M}}(s)$,

l.u.b. $\alpha_{i\mathfrak{M}}(s') = 1$.
${i\in N}$

By assumption (i), $\mathfrak{M}, s' \models \alpha$ for some $s' \in K_{\mathfrak{M}}(s)$, and hence $(\Diamond K\alpha)_{\mathfrak{M}}(s)$
$= 1$, and the proof of (iii) is finished.

To prove (iv), assume that $\mathfrak{M} \models (\alpha_j \Rightarrow \alpha_i)$ and suppose that for
some state s

$$\mathfrak{M}, s \models \Box K\alpha_j.$$

Then

$$K_{\mathfrak{M}}(s) \neq \varnothing \quad \text{and for all } s' \in K_{\mathfrak{M}}(s), \quad \mathfrak{M}, s' \models \alpha_j.$$

By assumption, for all $s' \in K_{\mathfrak{M}}(s)$, $\mathfrak{M}, s' \models \alpha_i$ and therefore

$$\mathfrak{M}, s \models \Box K\alpha_i.$$

Thus $\mathfrak{M} \models (\Box K\alpha_j \Rightarrow \Box K\alpha_i)$. Analogously, we can prove that $\mathfrak{M} \models (\Diamond K\alpha_j \Rightarrow \Diamond K\alpha_i)$. $\qquad\Box$

Let pref denote a finite sequence of program variables with modality signs,

$$\text{pref} \in (\{\Box K\}_{K \in V_p} \cup \{\Diamond K\}_{K \in V_p})^*.$$

The following lemma is a generalization of Lemma 2.2.

LEMMA 2.5. *Let $\mathfrak{M}$ be a structure with the* FDN *property. In that case*

$$\text{pref} \,(\bigcirc \textbf{ while } \gamma \textbf{ do } M \textbf{ od } \alpha)_{\mathfrak{M}}(s)$$
$$= \text{l.u.b. } \underset{i \in N}{\text{pref}} \bigcirc (\textbf{if } \gamma \textbf{ then } M \textbf{ fi})^i (\sim\!\gamma \wedge \alpha)_{\mathfrak{M}}(s),$$

for every $\gamma \in F_0, \alpha \in F$, $M \in \Pi$ and every state s. $\bigcirc$ denotes either $\Box$ everywhere or $\Diamond$.

PROOF. The proof is by induction on the length of pref. The basic step has been proved in Lemmas 2.2 and 2.3. To apply Lemma 2.4 it is sufficient to prove that for every $i \in N$ and $j < i$ the formula

$$(\bigcirc(\textbf{if } \gamma \textbf{ then } M \textbf{ fi})^j(\sim\!\gamma \wedge \alpha) \Rightarrow \bigcirc(\textbf{if } \gamma \textbf{ then } M \textbf{ fi})^i(\sim\!\gamma \wedge \alpha))$$

is valid in the structure $\mathfrak{M}$. But this follows immediately from the fact that each computation of

$$(\textbf{if } \gamma \textbf{ then } M \textbf{ fi})^j$$

with a result satisfying $\sim\!\gamma$ determines a successful computation of

$$(\textbf{if } \gamma \textbf{ then } M \textbf{ fi})^i$$

with the same result.

Thus, using Lemma 2.4, we shall obtain the required equalities. $\qquad\Box$

As a simple consequence of the above lemmas we have:

$$\text{l.u.b. } \underset{i \in N}{\text{pref}} \Box(\textbf{if } \gamma \textbf{ then } M \textbf{ fi})^i(\sim\!\gamma \wedge \alpha)_{\mathfrak{M}}(s)$$
$$= (\text{pref } \Box \textbf{ while } \gamma \textbf{ do } M \textbf{ od } \alpha)_{\mathfrak{M}}(s),$$

$$\text{l.u.b. } \underset{i \in N}{\text{pref}} \Diamond(\textbf{if } \gamma \textbf{ then } M \textbf{ fi})^i(\sim\!\gamma \wedge \alpha)_{\mathfrak{M}}(s)$$
$$= (\text{pref } \Diamond \textbf{ while } \gamma \textbf{ do } M \textbf{ od } \alpha)_{\mathfrak{M}}(s).$$

The algorithmic language presented in this chapter allows us to describe certain important properties of program schemes:

$$\mathfrak{M} \models \Box M \textbf{ true},$$

i.e. all computations of the program scheme M are successful,

$$\mathfrak{M} \models (\alpha \Rightarrow \Diamond M \beta),$$

i.e. if the initial state satisfies α then it is possible to have a result of M which satisfies β,

$$\mathfrak{M} \models \Box (\textbf{if } \gamma \textbf{ then } M \textbf{ fi})^i \sim \gamma,$$

i.e. the number of iterations of M in all computations of the program scheme **while** γ **do** M **od** is less than $(i+1)$.

3. PROPERTIES OF SEMANTIC STRUCTURES

In the sequel we shall study different semantic structures. The aim of this section is to present several definitions and some of their properties.

DEFINITION 3.1. *We shall say that the semantic structure* $\mathfrak{M}$ $= \langle S, \mathscr{I}, w \rangle$ *is proper iff the set S of states is composed of valuations of propositional variables, $S \subset B_0^{V_0}$ and w is the identity function. In what follows we shall write simply* $\mathfrak{M} = \langle S, \mathscr{I} \rangle$. $\qquad \Box$

LEMMA 3.1. *Every semantic structure* $\mathfrak{M} = \langle S, \mathscr{I}, w \rangle$ *such that w is a one-to-one function is isomorphic to some proper structure.* $\qquad \Box$

DEFINITION 3.2. *We shall say that the semantic structure* $\mathfrak{M}$ $= \langle S, \mathscr{I}, w \rangle$ *is normalized iff for every two states $s, s' \in S$,*

$$s = s' \quad \textit{iff} \quad \textit{for every formula } \alpha, \, \alpha_{\mathfrak{M}}(s) = \alpha_{\mathfrak{M}}(s'). \qquad \Box$$

For every FDN semantic structure $\mathfrak{M}$ we can construct a normalized structure $\mathfrak{M}^*$ such that for every algorithmic formula $\alpha \in F$

$$\mathfrak{M} \models \alpha \quad \text{iff} \quad \mathfrak{M}^* \models \alpha.$$

Let $\mathfrak{M}$ be an FDN semantic structure $\mathfrak{M} = \langle S, \mathscr{I}, w \rangle$ and $\approx$ an equivalence relation in S such that

$$s \approx s' \quad \text{iff} \quad (\forall \alpha \in F) \alpha_{\mathfrak{M}}(s) = \alpha_{\mathfrak{M}}(s'), \quad s, s' \in S.$$

We shall construct a new structure $\mathfrak{M}^*$ as follows:

$$S^* = S/\approx, \quad |s| = \{s' : s \approx s'\},$$

$$\mathscr{I}^* : V_p \to 2^{S^* \times S^*} \text{ and for all } K \in V_p,$$

$$\mathscr{I}^*(K) = \{(|s_1|, |s_2|) : (\exists s_1' \in |s_1|)(\exists s_2' \in |s_2|)(s_1', s_2') \in \mathscr{I}(K)\},$$

$$w^*(|s|) = w(s).$$

Notice that w^* is a well-defined function

$$w^* : S^* \to B_0^{V_0}$$

since if $s', s'' \in |s|$, then $w(s') = w(s'')$.

In several proofs we shall make use of the following definition of ordering in the set of all formulas F:

DEFINITION 3.3. *We shall say that the formula α' is submitted to the formula β', $\alpha' \prec \beta'$, iff the pair (α', β') belongs to the transitive closure of the relation $\prec$, which is a set of the following pairs:*

$$(\alpha, \bigcirc K\alpha) \quad \text{for } K \in V_p,$$

$$(\alpha, \alpha \vee \beta), \quad (\alpha, \alpha \wedge \beta), \quad (\alpha, \sim\alpha),$$

$$(\bigcirc M_1(\bigcirc M_2\alpha), \bigcirc \text{begin } M_1; M_2 \text{ end } \alpha),$$

$$(\bigcirc M_1\alpha, \bigcirc \text{ either } M_1 \text{ or } M_2 \text{ ro } \alpha),$$

$$(\bigcirc M_2\alpha, \bigcirc \text{ either } M_1 \text{ or } M_2 \text{ ro } \alpha),$$

$$(\bigcirc(\text{if } \gamma \text{ then } M \text{ fi})^i(\sim\gamma \wedge \alpha), \bigcirc \text{ while } \gamma \text{ do } M \text{ od } \alpha)$$

$$\text{for } i \in N,$$

$$((\gamma \wedge \bigcirc M_1\alpha), \bigcirc \text{ if } \gamma \text{ then } M_1 \text{ else } M_2 \text{ fi } \alpha),$$

$$((\sim\gamma \wedge \bigcirc M_2\alpha), \bigcirc \text{ if } \gamma \text{ then } M_1 \text{ else } M_2 \text{ fi } \alpha),$$

where M_1, M_2 are any program schemes, γ is a classical formula, α, β are formulas and $\bigcirc$ denotes $\square$ or $\lozenge$. $\square$

LEMMA 3.2. *For every formula $\alpha \in F$ and for every state s in the structure $\mathfrak{M}$,*

$$\alpha_{\mathfrak{M}^*}(|s|) = \alpha_{\mathfrak{M}}(s).$$

PROOF. (By induction w.r.t. the ordering relation $\prec$ defined above). It is obvious that for every propositional variable $p \in V_0$ and for every state $s \in S$,

$$p_{\mathfrak{M}^*}(|s|) = p_{\mathfrak{M}}(s).$$

Assume that Lemma 3.2 holds for all formulas which are submitted to a formula $\bar{\alpha}$.

A. Consider the formula $\bar{\alpha} = \Diamond K\alpha$, $K \in V_p$.

Let $s \in S$ and $\mathfrak{M}, s \models \Diamond K\alpha$. Then $(\exists s_1 \in K_{\mathfrak{M}}(s))\,\mathfrak{M}, s_1 \models \alpha$. By the definition of the structure $\mathfrak{M}^*$ we have

$$K_{\mathfrak{M}*}(|s|) \neq \varnothing, \quad (\exists s_1 \in K_{\mathfrak{M}}(s))(|s|, |s_1|) \in K_{\mathfrak{M}*} \quad \text{and}$$
$$\mathfrak{M}, s_1 \models \alpha.$$

Hence,

$$\mathfrak{M}^*, |s| \models \Diamond K\alpha.$$

Conversely, if $\mathfrak{M}^*, |s| \models \Diamond K\alpha$ for a certain $s \in S$, then there is an element s_1 such that $|s_1| \in K_{\mathfrak{M}*}(|s|)$ and $\mathfrak{M}^*, |s_1| \models \alpha$. From the definition of $\mathfrak{M}^*$, there are $s_1' \in |s_1|$ and $s' \in |s|$ such that $(s', s_1') \in K_{\mathfrak{M}}$. By the induction hypothesis $\mathfrak{M}, s_1 \models \alpha$. Since $s_1, s_1' \in |s_1|$, we have $\mathfrak{M}, s_1' \models \alpha$ and therefore $(\Diamond K\alpha)_{\mathfrak{M}}(s') = (\Diamond K\alpha)_{\mathfrak{M}}(s) = 1$.

B. Consider the formula $\bar{\alpha} = \Box K\alpha$.

If $\mathfrak{M}, s \models \Box K\alpha$, then $K_{\mathfrak{M}}(s) \neq \varnothing$. Thus $K_{\mathfrak{M}*}(|s|) \neq \varnothing$. Let $|s_2| \in K_{\mathfrak{M}*}(|s|)$, then there are elements s_2', s' such that $s_2' \in |s_2|$, $s' \in |s|$ and $(s', s_2') \in K_{\mathfrak{M}}$. Since $\mathfrak{M}, s' \models \Box K\alpha$, we have $\mathfrak{M}, s_2' \models \alpha$. Hence $\alpha_{\mathfrak{M}}(|s_2'|) = 1 = \alpha_{\mathfrak{M}*}(|s_2|)$. Consequently, $\mathfrak{M}^*, |s| \models \Box K\alpha$.

Conversely, if, for a certain $s \in S$, $\mathfrak{M}^*, |s| \models \Box K\alpha$, then $K_{\mathfrak{M}*}(|s|) \neq \varnothing$. Thus $K_{\mathfrak{M}}(s') \neq \varnothing$ for some $s' \in |s|$. If we take $s_1 \in K_{\mathfrak{M}}(s')$, then $|s_1| \in K_{\mathfrak{M}*}(|s'|)$, i.e. $|s_1| \in K_{\mathfrak{M}*}(|s|)$, and therefore $\mathfrak{M}^*, |s_1| \models \alpha$. By the induction assumption for α we have $\mathfrak{M}, s_1 \models \alpha$. Hence $\mathfrak{M}, s' \models \Box K\alpha$ and therefore $\mathfrak{M}, s \models \Box K\alpha$.

C. Consider the formula $\bar{\alpha} = \Box$ **while** γ **do** M **od** α.

Suppose that $\mathfrak{M}, s \models \Box$ **while** γ **do** M **od** α. By Lemma 2.3, this is equivalent to the following

$$\mathop{\text{l.u.b.}}_{i \in N} \Box (\textbf{if } \gamma \textbf{ then } M \textbf{ fi})^i (\sim \gamma \wedge \alpha)_{\mathfrak{M}}(s) = 1.$$

By the induction hypothesis we have

$$\mathop{\text{l.u.b.}}_{i \in N} \Box (\textbf{if } \gamma \textbf{ then } M \textbf{ fi})^i (\sim \gamma \wedge \alpha)_{\mathfrak{M}*}(|s|) = 1,$$

and therefore by Lemma 2.3

$$\mathfrak{M}^*, |s| \models \Box \textbf{ while } \gamma \textbf{ do } M \textbf{ od } \alpha.$$

All the remaining cases can be discussed analogously. $\qquad \Box$

DEFINITION 3.4. *Two semantic structures $\mathfrak{M}$ and $\mathfrak{M}'$ are algorithmically equivalent iff, for every formula $\alpha \in F$,*

$$\mathfrak{M} \models \alpha \quad \textit{iff} \quad \mathfrak{M}' \models \alpha. \qquad \square$$

LEMMA 3.3. *Every* FDN *semantic structure is algorithmically equivalent to some normalized* FDN *structure.* $\qquad \square$

Let us now compare proper and normalized structures. Since the value of any formula is defined in a unique way by a given structure and a given state, then for all valuations v_1, v_2 in the proper structure $\mathfrak{M}$ we have

$$v_1 = v_2 \quad \text{iff} \quad (\forall \alpha) \alpha_{\mathfrak{M}}(v_1) = \alpha_{\mathfrak{M}}(v_2).$$

Thus, if $\mathfrak{M}$ is a proper structure, it is normalized.

The following lemmas describe some properties of relations which can be expressed in propositional algorithmic language.

LEMMA 3.4. *In every semantic normalized structure $\mathfrak{M}$, the following properties are satisfied*:

$$\mathfrak{M} \models \{(\Diamond K\alpha \Rightarrow \Box K\alpha)\}_{\alpha \in F} \quad \textit{iff} \quad (\forall s) \text{ card } (K_{\mathfrak{M}}(s)) \leqslant 1,$$

$$\mathfrak{M} \models \{(\Diamond K(\alpha \wedge \beta) \wedge \Diamond K(\alpha \wedge \sim \beta)) \Rightarrow \Box K\alpha\}_{\alpha, \beta \in F} \quad \textit{iff}$$
$$(\forall s) \text{ card } (K_{\mathfrak{M}}(s)) \leqslant 2,$$

$$\mathfrak{M} \models \{(\Diamond K(\alpha \wedge \beta) \wedge \Diamond K(\alpha \wedge \sim \beta) \wedge \Diamond K(\sim \alpha \wedge \beta))$$
$$\Rightarrow \Box K(\alpha \vee \beta)\}_{\alpha, \beta \in F} \quad \textit{iff} \quad (\forall s) \text{ card } (K_{\mathfrak{M}}(s)) \leqslant 3.$$

PROOF. We shall prove the first equivalence.

Obviously, if $K_{\mathfrak{M}}(s)$ is at most one-element set, then for every formula α, $(\Diamond K\alpha \Rightarrow \Box K\alpha)$ is valid in $\mathfrak{M}$.

Conversely, suppose

$$\mathfrak{M} \models \{(\Diamond K\alpha \Rightarrow \Box K\alpha)\}_{\alpha \in F}$$

and for some states s, s_1, s_2 in $\mathfrak{M}$, $s_1 \in K_{\mathfrak{M}}(s)$, $s_2 \in K_{\mathfrak{M}}(s)$, $s_1 \neq s_2$. Thus there exists a formula β such that

$$\beta_{\mathfrak{M}}(s_1) \neq \beta_{\mathfrak{M}}(s_2).$$

Hence, $\mathfrak{M}, s \models \Diamond K\beta$ and non $\mathfrak{M}, s \models \Box K\beta$, a contradiction.

The proof of the second property can be found in Lemma 10.1. The proof of the third property is left to the reader. $\qquad \square$

LEMMA 3.5. *Let $\mathfrak{M}$ be a normalized* FDN *semantic structure. The following equivalences then hold*:

(i) $\mathfrak{M} \models \{(\beta \Rightarrow \Diamond K\beta)\}_{\beta \in F}$ *iff* $K_{\mathfrak{M}}$ *is a reflexive relation*,

(ii) $\mathfrak{M} \models \{(\Diamond K(\sim \Diamond K\beta) \Rightarrow \sim \beta)\}_{\beta \in F}$ *iff* $K_{\mathfrak{M}}$ *is a symmetric relation*,

(iii) $\mathfrak{M} \models \{(\Diamond K(\Diamond K\beta) \Rightarrow \Diamond K\beta)\}_{\beta \in F}$ *iff* $K_{\mathfrak{M}}$ *is a transitive relation*,

(iv) $\mathfrak{M} \models \{(\Diamond K(\sim \Diamond M\beta) \Rightarrow \sim \beta) \wedge (\Diamond M(\sim \Diamond K\beta) \Rightarrow \sim \beta)\}_{\beta \in F}$ *iff relations* $M_{\mathfrak{M}}$ *and* $K_{\mathfrak{M}}$ *are mutually inverse*.

PROOF. All four properties have similar proofs. We shall illustrate the method of proving showing the second equivalence, as follows.

Ad (ii). Let $\mathfrak{M}$ be a normalized FDN semantic structure and

(1) $\qquad \mathfrak{M} \models (\Diamond K(\sim \Diamond K\alpha) \Rightarrow \sim \alpha)$ for every formula α.

Suppose s, s' are two fixed states such that

(2) $\qquad (s, s') \in K_{\mathfrak{M}}$

and let $\{s_1, ..., s_n\}$ be the set of all states such that $(s', s_i) \in K_{\mathfrak{M}}$ for $i \leqslant n$. Suppose $s_i \neq s$ for all $i \leqslant n$. Since $\mathfrak{M}$ is a normalized structure, for every $i \leqslant n$ there exists a formula α_i such that

$$\mathfrak{M}, s \models \alpha_i \quad \text{and} \quad \mathfrak{M}, s_i \models \sim \alpha_i.$$

Let $\alpha = (\alpha_1 \wedge \alpha_2 \wedge ... \wedge \alpha_n)$. Thus

(3) $\qquad \mathfrak{M}, s \models \alpha \quad \text{and} \quad \mathfrak{M}, s' \models \sim \Diamond K\alpha.$

By (2) $\mathfrak{M}, s \models \Diamond K(\sim \Diamond K\alpha)$ and as a consequence of (1) $\mathfrak{M}, s \models \sim \alpha$, contrary to (3). Thus (2) implies $(s', s) \in K_{\mathfrak{M}}$.

Conversely, assume that for all s, s'

(4) $\qquad$ if $\quad (s, s') \in K_{\mathfrak{M}}, \quad$ then $\quad (s', s) \in K_{\mathfrak{M}}.$

Suppose that

(5) $\qquad \mathfrak{M}, s \models \alpha \quad \text{and} \quad \mathfrak{M}, s \models \Diamond K(\sim \Diamond K\alpha).$

Thus there exists a state s', $(s, s') \in K_{\mathfrak{M}}$ such that non $\mathfrak{M}, s' \models \Diamond K\alpha$. By assumption (4) $(s', s) \in K_{\mathfrak{M}}$ and furthermore non $\mathfrak{M}, s \models \alpha$, which contradicts (5). $\qquad \square$

At the end of this section we shall present a negative result which is of great significance for further considerations. We shall prove that the FDN property is not expressible in the propositional algorithmic language.

THEOREM 3.6. *There exists no formula* α *such that for every semantic structure* $\mathfrak{M}$

$$\mathfrak{M} \models \alpha \quad \textit{iff} \quad \mathfrak{M} \textit{ has the FDN property.}$$

PROOF. Suppose, on the contrary, that there exists a formula α_0 such that for every $\mathfrak{M}$

(6) $\mathfrak{M} \models \alpha_0$ iff $\mathfrak{M}$ has FDN property.

Let us consider the family of structures $\{\mathfrak{M}_i\}_{i \in N}$ such that $\mathfrak{M}_i = \langle S_i, \mathscr{I}_i, w_i \rangle$, where $S_i \cap S_j = \emptyset$ for $i \neq j$ and $S_i = \{s_i, s_{i1}, \dots, s_{ii}\}$, $\mathscr{I}_i(K) = \{(s_i, s_{ij}): j \leqslant i\}$, $\mathscr{I}_i(K') = \emptyset$ for all program variables K' different from K, $\mathfrak{M}_i, w_i(s_i) \models q$ iff $q = p_i$ and $\mathfrak{M}_i, w_i(s_{ij}) \models q$ iff $q = q_j$.

The family $\{\mathfrak{M}_i\}_{i \in N}$ can be described more intuitively by the graphs shown in Figure 3.1.

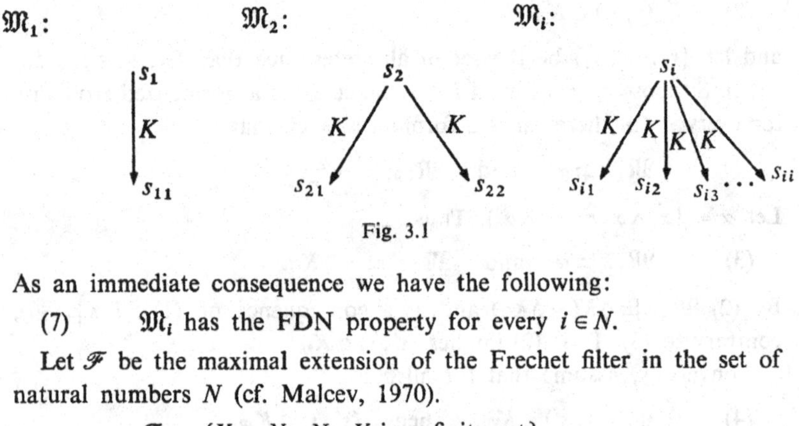

Fig. 3.1

As an immediate consequence we have the following:

(7) $\mathfrak{M}_i$ has the FDN property for every $i \in N$.

Let $\mathscr{F}$ be the maximal extension of the Frechet filter in the set of natural numbers N (cf. Malcev, 1970).

$$\mathscr{F} \supset \{X \subset N: N-X \text{ is a finite set}\}.$$

Let us denote by $\mathfrak{M}^* = \langle S^*, \mathscr{I}^*, w^* \rangle$ the product of all structures $\{\mathfrak{M}_i\}_{i \in N}$ modulo filter $\mathscr{F}$ (cf. Malcev, 1970)

$$\mathfrak{M}^* = \underset{i \in N}{\times} \mathfrak{M}_{i/\mathscr{F}}.$$

For every $u \in \underset{i \in N}{\times} \mathfrak{M}_i$ let u_i denote the i-th element of u and let $|u| = \{u': \{i: u_i = u_i'\} \in \mathscr{F}\}$.

Hence $S^* = \{|u|: u \in \underset{i \in N}{\times} \mathfrak{M}_i\}$ and

$$(|u|, |u'|) \in \mathscr{I}^*(K) \quad \text{iff} \quad \{i: (u_i, u_i') \in \mathscr{I}_i(K)\} \in \mathscr{F},$$
$$\mathfrak{M}^*, w^*(|u|) \models q \quad \text{iff} \quad \{i: \mathfrak{M}_i, w_i(u_i) \models q\} \in \mathscr{F}.$$

Let $\mathfrak{M}$ be a restriction of $\mathfrak{M}^*$ to the states s, s^j for $j \in N$, where $s = (s_1, s_2, \ldots)$ and $s^j = (s_{11}, \ldots, s_{jj}, s_{j+1,j}, s_{j+2,j}, \ldots)$. For every $j \in N$, $|s| \neq |s^j|$ since $\{i: s_i = s_i^j\} = \emptyset$.

For every $j \in N$, $(|s|, |s^j|)$ belongs to $\mathscr{I}^*(K)$ since $N = \{i: (s_i, s_i^j) \in \mathscr{I}_i(K)\} \in F$ and $(|s^k|, |s^l|) \notin K_{\mathfrak{M}}$ since $\{j: (s_j^k, s_j^l) \in \mathscr{I}_j(K)\} = \emptyset$. Moreover $|s^k| \neq |s^l|$ for $k \neq l$, since $\{i: s_i^k = s_i^l\}$ is finite and therefore does not belong to $\mathscr{F}$. Thus there are infinitely many successors of the state $|s|$.

We obtain

(8) $\mathfrak{M}$ does not have the FDN property,

as a consequence of the above considerations.

Note that the situation in which $|t|, |t'|, |t''|$ are different states in $\mathfrak{M}$ and $(|t|, |t'|) \in K_{\mathfrak{M}}$, $(|t'|, |t''|) \in K_{\mathfrak{M}}$ is impossible since in that case we would find a corresponding triple t_i, t_i', t_i'' of states in the structure $\mathfrak{M}_i$ such that $(t_i, t_i') \in K_{\mathfrak{M}_i}$, $(t_i', t_i'') \in K_{\mathfrak{M}_i}$, in contradiction to the definition of the structure $\mathfrak{M}_i$.

By induction on the length of the formula we can prove that for every $|u| \in \mathfrak{M}$ and for every formula α of PAL,

(9) $\mathfrak{M}, |u| \models \alpha$ iff $\{i: \mathfrak{M}_i, u_i \models \alpha\} \in \mathscr{F}$.

They key part of the proof is the case when formula α is of the form $\square$ **while** γ **do** M **od** β. From the previous observation we have

$$\mathfrak{M}, |u| \models \square \textbf{ while } \gamma \textbf{ do } M \textbf{ od } \beta \equiv ((\gamma \wedge \square M(\sim\gamma \wedge \beta)) \vee$$
$$\vee (\sim\gamma \wedge \beta)),$$

since there is at most one iteration of the program M during a computation of **while** γ **do** M **od** in the structure $\mathfrak{M}$. By the inductive hypothesis we shall obtain

$$\{i: \mathfrak{M}_i, u_i \models ((\gamma \wedge \square M(\sim\gamma \wedge \beta)) \vee (\sim\gamma \wedge \beta))\} \in \mathscr{F}$$

and therefore

$$\{i: \mathfrak{M}_i, u_i \models \square \textbf{ while } \gamma \textbf{ do } M \textbf{ od } \beta\} \in \mathscr{F}.$$

To complete the proof let us notice that by (6) and (7) $\mathfrak{M}_i \models \alpha_0$ for every $i \in N$. Hence for every state $|u|$ in the structure $\mathfrak{M}$

$$\{i: \mathfrak{M}_i, u_i \models \alpha_0\} = N$$

and therefore by (9), $\mathfrak{M}, |u| \models \alpha_0$. As a consequence $\mathfrak{M} \models \alpha_0$, as opposed to (8). $\square$

THEOREM 3.7. *The* FDN *property is not expressible in* PAL, *i.e., the following property does not hold: there exists a set of formulas Z such that for every semantic structure* $\mathfrak{M}$,

$$\mathfrak{M} \models Z \quad \text{iff} \quad \mathfrak{M} \text{ has the FDN property.} \qquad \square$$

REMARK. Theorem 3.6 can be strenghtened. Namely, FDN property is not expressible in PAL in the class of all normalized structures.

4. THE SEMANTIC CONSEQUENCE OPERATION IS NOT COMPACT

DEFINITION 4.1. *A semantic structure* $\mathfrak{M} = \langle S, \mathcal{I}, w \rangle$ *is a model of the set of formulas Z iff* $\mathfrak{M}$ *is a model of every formula* α *from this set.* $\square$

DEFINITION 4.2. *We shall say that* α *is a semantic consequence of the set of formulas* Z, $Z \models \alpha$ *iff every model of* Z *is a model of* α. $\square$

The semantic consequence operation $\models$ has certain classical properties (cf. Chapter II, § 4). An important difference is exhibited in the following lemma which implies non-compactness of $\models$ (see also Chapter II, Theorem 4.1).

LEMMA 4.1. *There exists a set of formulas* Z *and a formula* α *such that* $Z \models \alpha$ *and such that for every finite subset* Z_0 *of* Z *there exists a model of* Z_0 *which is not a model of* α.

PROOF. Consider the following example. Assume that

$$Z = \{\square \text{ begin } K_1 ; K_2^i \text{ end } q_0 \}_{i \in N},$$

where K_1, K_2 are program variables and q_0—a propositional variable, and $\alpha = \sim (\Diamond \text{ begin } K_1 ; \text{ while } q_0 \text{ do } K_2 \text{ od end true})$. It is easy to show that α is a semantic consequence of the set Z.

If $\langle S, \mathcal{I}, w \rangle$ is a model of Z, then for every state $s \in S$ and for each natural number i, every computation of the program **begin** $K_1 ; K_2^i$ **end** is successful and all results satisfy the formula q_0. Hence, there exists no finite computation of the program

$$\text{begin } K_1 ; \text{ while } q_0 \text{ do } K_2 \text{ od end.}$$

This implies that $\langle S, \mathcal{I}, w \rangle$ is a model for α.

Now, assume that $V_0 = \{q_i\}_{i \in N}$. For every finite subset X of N, let us construct an interpretation $\mathcal{I}$ in the following way: For every valuations v, v', v'' of the propositional variables $v \mathscr{I}(K_1) v'$ iff $v'(q_i) = \mathbf{1}$ for $i \in X$, and $v'(q_i) = \mathbf{0}$ for $i \notin X$, $v' \mathscr{I}(K_2) v''$ iff $v''(q_i) = v'(q_{i+1})$ for $i = 0, 1, \ldots$

Let $\overline{\mathfrak{M}}$ be a semantic structure $\langle W, \mathscr{I} \rangle$.

First let us observe that all computations of the program **begin** K_1; K_2^i **end** in the structure $\overline{\mathfrak{M}}$ are finite in the interpretation $\mathscr{I}$ for all valuations and for all $i \in N$. Let v be a fixed valuation. The value of the propositional variable q_i in the valuation v' obtained after execution of the program K_1 is $\mathbf{1}$ if $i \in X$. After the execution of the whole program we have a resulting valuation v'' such that $v''(q_0) = \mathbf{1}$. Thus $\langle W, \mathscr{I} \rangle$ is a model of $Z_X = \{ \square$ **begin** K_1; K_2^i **end** $q_0 \}_{i \in X}$.

$\langle W, \mathscr{I} \rangle$ is not a model for α. Let us take as i_0 the smallest natural number such that $i_0 \notin X$ and let $v \mathscr{I}(K_1) v^0$ and $v^j \mathscr{I}(K_2) v^{j+1}$ for $j < i_0$ and some valuation v. The sequence of configurations

$\langle v^0$; **while** q_0 **do** K_2 **od** $\rangle$,
$\langle v^0$; K_2; **while** q_0 **do** K_2 **od** $\rangle$,
$\langle v^1$; **while** q_0 **do** K_2 **od** $\rangle$,
$\langle v^1$; K_2; **while** q_0 **do** K_2 **od** $\rangle$,
. . . .
$\langle v^{i_0}$; **while** q_0 **do** K_2 **od** $\rangle$,
$\langle v^{i_0}$; $\rangle$

is a successful computation of the program **while** q_0 **do** K_2 **od**. Thus, $\overline{\mathfrak{M}} \models \sim \alpha$. ☐

5. THE SYNTACTIC CONSEQUENCE OPERATION

We shall now characterize the semantic consequence operation described above in syntactic terms.

Theorem 4.1 assures us that it is not possible to construct a complete and recursive axiomatization of PAL with finite rules of inference. We thus allow rules of an infinite character.

All axioms Ax1–Ax11 of algorithmic logic AL (cf. Chapter II, § 5) and the following schemes are axioms of PAL:

$$(\square M\alpha \Rightarrow \Diamond M\alpha), \quad (\Diamond K\text{true} \Rightarrow \square K\text{true}),$$
$$\bigcirc \text{ begin } M_1; \ M_2 \text{ end } \alpha \equiv \big(\bigcirc M_1(\bigcirc M_2 \alpha)\big),$$
$$\bigcirc \text{ if } \gamma \text{ then } M_1 \text{ else } M_2 \text{ fi } \alpha$$
$$\equiv \big((\gamma \wedge \bigcirc M_1 \alpha) \vee (\sim\gamma \wedge \bigcirc M_2 \alpha)\big),$$

$\bigcirc$ **while** γ **do** M_1 **od** α

$\equiv ((\sim\gamma\wedge\alpha)\vee(\gamma\wedge\bigcirc M_1(\bigcirc$ **while** γ **do** M_1 **od** $\alpha)))$,

$\Diamond$ **either** M_1 **or** M_2 **ro** $\alpha \equiv (\Diamond M_1\alpha\vee\Diamond M_2\alpha)$,

$\square$ **either** M_1 **or** M_2 **ro** $\alpha \equiv (\square M_1\alpha\wedge\square M_2\alpha)$,

$\square M(\alpha\wedge\beta) \equiv (\square M\alpha\wedge\square M\beta)$,

$\Diamond M(\alpha\vee\beta) \equiv (\Diamond M\alpha\vee\Diamond M\beta)$,

$(\square M\sim\alpha \Rightarrow \sim\Diamond M\alpha)$,

$(\square M \text{ **true** } \Rightarrow(\sim\Diamond M\alpha \Rightarrow \square\ M\sim\alpha))$,

$\sim \bigcirc M$ **false**.

$\bigcirc \text{Id}\,\alpha \equiv \alpha$.

We assume the following rules of inference:

$$\frac{\alpha, (\alpha \Rightarrow \beta)}{\beta}, \qquad \frac{(\alpha \Rightarrow \beta)}{(\bigcirc M\alpha \Rightarrow \bigcirc M\beta)},$$

$$\frac{\{(\text{pref}(\bigcirc(\text{**if** } \gamma \text{ **then** } M \text{ **fi**})^i(\alpha\wedge\sim\gamma)) \Rightarrow \beta)\}_{i\in N}}{(\text{pref }(\bigcirc \text{ **while** } \gamma \text{ **do** } M \text{ **od** } \alpha) \Rightarrow \beta)}.$$

In all the above schemes K denotes a program variable, M, M_1, M_2 denote schemes of programs, γ is a classical propositional formula and α, β are arbitrary formulas from F. All occurrences of $\bigcirc$ in a formula must be understood either as $\Diamond$ throughout or as $\square$ throughout; pref is an arbitrary prefix (see § 2 of this chapter).

The set of all axioms and inference rules defines the syntactic consequence operation C in the usual way. For any set Z of formulas, $C(Z)$ is the least set which contains Z and all axioms of PAL and is closed under the rules of inference. System $\langle L_0, C\rangle$ will be called the *propositional algorithmic logic* PAL.

A formula α is called a *theorem* of PAL iff α is an element of $C(\emptyset)$, $\vdash \alpha$ for short.

By a *formal proof of a formula α from the set of formulas Z* we shall understand a finite path tree labelled by formulas such that its root is a formula α, all leaves are axioms and every vertex is obtained from the set of predecessors by one of the inference rules (cf. Chapter II, § 5).

We shall write $Z \vdash \alpha$, $\alpha \in C(Z)$ iff α has a formal proof from the set Z.

LEMMA 5.1. *All axioms of PAL are propositional algorithmic tautologies.*

PROOF. The proof is by an easy verification. As an example we shall consider two formulas:

A. $(\Diamond K \textbf{ true} \Rightarrow \Box K \textbf{ true})$, $K \in V_p$ and

B. $(\Box M \sim \alpha \Rightarrow \sim \Diamond M \alpha)$.

Let $\mathfrak{M} = \langle S, \mathscr{I}, w \rangle$ be a fixed semantic structure, and s an arbitrary element of S.

A. Assume that $\mathfrak{M}, s \models \Diamond K \textbf{ true}$. Then $K_{\mathfrak{M}}(s) \neq \varnothing$ and all computations are one-element sequences. Thus $\mathfrak{M}, s \models \Box K \textbf{ true}$.

B. Assume that $\mathfrak{M}, s \models \Box M \sim \alpha$. Then all computations of the program M are successful and all results satisfy the formula $\sim \alpha$. Hence, there exists no finite computation which satisfies the formula α. This means that $\mathfrak{M}, s \models \sim \Diamond M \alpha$.   □

LEMMA 5.2. *The set of all formulas valid in all* FDN *semantic structures is closed under all rules of inference mentioned above.*

PROOF. Let $\mathfrak{M}$ be an FDN semantic structure. We shall prove that for any inference rule, if all premises are valid in $\mathfrak{M}$ then the conclusion is valid in $\mathfrak{M}$.

Consider the rule $\dfrac{(\alpha \Rightarrow \beta)}{(\Box M \alpha \Rightarrow \Box M \beta)}$. Assume that $\mathfrak{M} \models (\alpha \Rightarrow \beta)$ and $\mathfrak{M}, s \models \Box M \alpha$ for some state s. Thus all computations of the program M in the structure $\mathfrak{M}$ at the initial state s are successful and all results s' satisfy the formula α, i.e. $\mathfrak{M}, s' \models \alpha$. By assumption $\mathfrak{M}, s' \models \beta$ and therefore $\mathfrak{M}, s \models \Box M \beta$. As a consequence we have $\mathfrak{M}, s \models (\Box M \alpha \Rightarrow \Box M \beta)$. Hence $(\Box M \alpha \Rightarrow \Box M \beta)$ is valid in $\mathfrak{M}$.

Consider the rule $\dfrac{\{(\text{pref} \ \Box(\textbf{if } \gamma \textbf{ then } M \textbf{ fi})^i (\sim \gamma \wedge \alpha) \Rightarrow \beta)\}_{i \in N}}{(\text{pref} \ \Box \textbf{ while } \gamma \textbf{ do } M \textbf{ od } \alpha \Rightarrow \beta)}$.

Assume that for all $i \in N$ the formula

$$\left(\text{pref} \ \Box(\textbf{if } \gamma \textbf{ then } M \textbf{ fi})^i (\sim \gamma \wedge \alpha) \Rightarrow \beta\right)$$

is valid in an FDN structure $\mathfrak{M}$.

Suppose that for a fixed state s in a structure $\mathfrak{M}$

$$(\text{pref} \ \Box \textbf{ while } \gamma \textbf{ do } M \textbf{ od } \alpha)_{\mathfrak{M}}(s) = 1 \quad \text{and} \quad \beta_{\mathfrak{M}}(s) = 0.$$

This means that for all $i \in N$,

$$\left(\text{pref} \ \Box(\textbf{if } \gamma \textbf{ then } M \textbf{ fi})^i (\sim \gamma \wedge \alpha)\right)_{\mathfrak{M}}(s) = 0.$$

Thus

$$\text{l.u.b. pref } \square(\textbf{if } \gamma \textbf{ then } M \textbf{ fi})^i \, (\sim\gamma\wedge\alpha)_{\mathfrak{M}}(s) = 0.$$
$$\scriptstyle i\in N$$

Applying Lemma 2.5 we arrive at a contradiction.   □

As a natural consequence of the previous two lemmas we have the following theorem:

THEOREM 5.3. *For every formula α of the language L_0, if α is a theorem of PAL then α is valid in every semantic structure with the FDN property.*   □

LEMMA 5.4. *Propositional algorithmic calculus is consistent.*

PROOF. Suppose the contrary. There then exists a formula α such that α and $\sim\alpha$ are theorems in PAL. By the adequacy theorem for every FDN structure $\mathfrak{M}$ and state s

$$\alpha_{\mathfrak{M}}(s) = 1 \quad \text{and} \quad (\sim\alpha)_{\mathfrak{M}}(s) = 1.$$

Since the value of the formula is defined in a unique way, we shall arrive at a contradiction.   □

The question naturally arises, whether every formula valid in every FDN structure possesses a proof in PAL.

In Sections 8–10 of this chapter we shall discuss some classes of interpretations and extensions of PAL which have the completeness property.

By a *theory* based on PAL we shall understand a system $\langle L_0, C, A \rangle$ consisting of the language L_0 of propositional algorithmic logic, the syntactic consequence operation C and the set of formulas $A \subset F$, called *specific axioms*.

By a *model of the theory* $T = \langle L_0, C, A \rangle$ we shall mean any model of the set A.

We can prove the following adequacy theorem for any algorithmic theory $T = \langle L_0, C, A \rangle$:

THEOREM 5.5. *If α is a theorem of a theory T, then every FDN model of T is a model of α.*

The proof follows from Lemmas 5.1, 5.2.   □

6. EXAMPLES OF PROPOSITIONAL THEORIES

EXAMPLE 6.1. Propositional theory of arithmetic.

Let us consider a theory $Ar = \langle L_0, C, Axar \rangle$ based on PAL. We shall assume that the algorithmic language L_0 contains two program variables N, P and one propositional variable z. $Axar$ is the set of all formulas of the form:

(1)
$$\Box N \sim z, \quad (z \equiv \sim \Box P \text{ true}),$$
$$(\Diamond N\alpha \Rightarrow \Box N\alpha), \quad (\Diamond P\alpha \Rightarrow \Box P\alpha),$$
$$(\alpha \Rightarrow \Box N \Diamond P\alpha), \quad (\sim z \Rightarrow (\alpha \Rightarrow \Box P \Diamond N\alpha)),$$
$$\Box \text{ while } \sim z \text{ do } P \text{ od true},$$

where α is any formula.

This set of axioms was discovered by V. Pratt and A. Salwicki (cf. Mirkowska, 1981).

The $Axar$ theory is consistent since it posseses a model.

Consider the structure $\mathfrak{N} = \langle N, \mathscr{I}, w \rangle$, where N is the set of all natural numbers and $\mathscr{I}(N) = \{(i, i+1): i \in N\}$, $\mathscr{I}(P) = \{(i+1, i): i \in N\}$; $w(I)(z) = 1$ iff $i = 0$. By an easy verification we infer that $\mathfrak{N}$ is a model of $Axar$. We shall call this model $standard$. Let us see what the meaning of $Axar$ axioms is. If $\mathfrak{M}$ is a normalized model of $Axar$, then

$$\mathfrak{M} \models \Box N \sim z$$

says that a state obtained by $N_\mathfrak{M}$ does not satisfy z;

$$\mathfrak{M} \models \{(\Diamond N\alpha \Rightarrow \Box N\alpha)\}_{\alpha \in F},$$
$$\mathfrak{M} \models \{(\Diamond P\alpha \Rightarrow \Box P\alpha)\}_{\alpha \in F}$$

say that $N_\mathfrak{M}$ and $P_\mathfrak{M}$ are functions;

$$\mathfrak{M} \models (\sim z \equiv \Box P \text{ true})$$

says that $P_\mathfrak{M}$ is defined only for states which do not satisfy z;

$$\mathfrak{M} \models \{(\alpha \Rightarrow \Box N \Diamond P\alpha)\}_{\alpha \in F},$$
$$\mathfrak{M} \models \{(\sim z \Rightarrow (\alpha \Rightarrow \Box P \Diamond N\alpha))\}_{\alpha \in F}$$

say that $N_\mathfrak{M} = P_\mathfrak{M}^{-1}$ for all states in which $\sim z$;

$$\mathfrak{M} \models \Box \text{ while } \sim z \text{ do } P \text{ od true}$$

says that from any state we must return to the state that satisfies z after a finite iteration of P.

On the basis of the above information we can easily prove the following lemma:

LEMMA 6.1. *Every normalized model of Axar is isomorphic to the standard model* $\mathfrak{N}$.

PROOF. Suppose that $\mathfrak{M} \models Axar$ and that $\mathfrak{M}$ is a finite structure. Let $S = \{1, 2, ..., n\}$ be the set of all states in $\mathfrak{M}$. By the last of axioms (1) there exists a state j such that $\mathfrak{M}, j \models z$. Let $1, ..., k$ be all states j for which $\mathfrak{M}, j \models z$. By the first of axioms (1), for every state $i \leqslant n$, the set $N_{\mathfrak{M}}(i)$ is non-empty. Hence

$$N_{\mathfrak{M}}(i) \subset \{k+1, ..., n\}, \quad i \leqslant n.$$

It follows that there is a sequence of states $j_1, ..., j_m$ such that $j_1 = j_m$, $j_i \in \{k+1, ..., n\}$ and $(j_i, j_{i+1}) \in N_{\mathfrak{M}}$ for $i \leqslant m$. This means that there is an infinite computation of **while** $\sim z$ **do** P **od** starting from the initial state j_1, contrary to the axiom $\square$ **while** $\sim z$ **do** P **od true**.

Hence if $\mathfrak{M}$ is a model of $Axar$ then $\mathrm{card}(\mathfrak{M}) \geqslant \chi_0$. We shall prove that for every normalized model of $Axar$ there is a unique state s_0 such that

$$\mathfrak{M}, s_0 \models z.$$

Suppose, conversely, that there are two states s_1, s_2 and

(2) $\mathfrak{M}, s_1 \models z, \quad \mathfrak{M}, s_2 \models z, \quad s_1 \neq s_2.$

Thus the set $Z = \{\alpha: \alpha_{\mathfrak{M}}(s_1) \neq \alpha_{\mathfrak{M}}(s_2)\}$ is not-empty. Let $\tilde{\alpha}$ be a minimal formula in Z with respect to the ordering $\prec$ defined in Definition 3.3.

The formula $\tilde{\alpha}$ cannot take the form of pref $(\beta_1 \vee \beta_2)$, pref $(\beta_1 \wedge \beta_2)$, pref $(\beta_1 \Rightarrow \beta_2)$, pref $\sim \beta_1$ since then pref β_1 or pref β_2 would be in Z.

The formula $\tilde{\alpha}$ cannot take the form of

pref $\bigcirc$ **begin** M_1 ; M_2 **end** β,   pref $\bigcirc$ **while** γ **do** M **od** β,

pref $\bigcirc$ **if** γ **then** M_1 **else** M_2 **fi** β,   pref $\bigcirc$ **either** M_1 **or** M_2 **ro** β,

since then it would be possible to find a formula which is submitted to $\tilde{\alpha}$ and which belongs to Z.

Formulas of the form pref z remain to be considered, but in this case it is sufficient to restrict the prefix to $\square N^i$, where $i > 0$. However, from the axiom

$$\mathfrak{M} \models \square N \sim z$$

we have $\mathfrak{M}, s_1 \models \sim \square N^i z$ and $\mathfrak{M}, s_2 \models \sim \square N^i z$.

Thus the states s_1, s_2 satisfy exactly the same formulas and therefore $s_1 = s_2$, a contradiction with (2).

Hence in every normalized model $\mathfrak{M}$ of *Axar* there is exactly one state which satisfies z.

Since $N_{\mathfrak{M}}$ and $P_{\mathfrak{M}}$ are functions, the only possible situation is described by Figure 6.1.

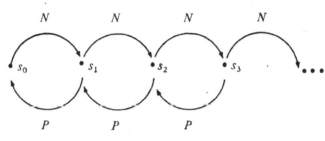

Fig. 6.1

This is obviously isomorphic to the standard model $\mathfrak{N}$ since the mapping h,

$$h(n+1) = N_{\mathfrak{M}}\big(h(n)\big),$$
$$h(0) = s_0,$$

defines a one-to-one homorphism from $\mathfrak{N}$ onto $\mathfrak{M}$. □

LEMMA 6.2. *Any two models of Axar are algorithmically equivalent.*

PROOF. This follows from Theorem 3.2 and Lemma 6.1, since any two isomorphic structures are algorithmically equivalent. □

REMARK. One can conceive of the propositional theory of arithmetic as the theory of a calculator. We are given a black box (cf. Figure 6.2) with a lamp z and two buttons N and P. The axioms (1) are all we know.

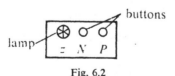

Fig. 6.2

Their interpretation is as follows: after pressing the button N the lamp z is switched off. If the lamp z is switched on the button P is blocked;

button P pressed a finite number of times causes the lamp z to light up.

From Lemma 6.1 we know that inside the black box there is a register for a natural number (don't ask us how this is implemented). Suppose we have three such modules (cf. Figure 6.3)

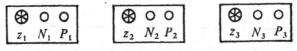

Fig. 6.3

and consider the following program

$$\text{PLUS: } \textbf{begin while } \sim z_3 \textbf{ do } P_3 \textbf{ od};$$
$$\textbf{while } \sim z_2 \textbf{ do } N_3 \text{ ; } P_2 \textbf{ od};$$
$$\textbf{while } \sim z_1 \textbf{ do } N_3 \text{ ; } P_1 \textbf{ od};$$
$$\textbf{end.}$$

We can imagine that a lid is constructed which, when put over the three modules, brings into operation a new button + which, when pressed, causes the sum of registers R_1 and R_2 to be evaluated and placed in the register R_3.   □

EXAMPLE 6.2. Propositional theory of stacks. We shall now describe a propositional version of the algorithmic theory of stacks.

Let $St = \langle L, C, Axst \rangle$ be an algorithmic propositional theory, where L is an algorithmic language as described in § 1 which contains the propositional variables e, $t1$, $t2$ and the program variables push, pop. $Axst$ is the set of specific axioms and contains all formulas of the following form:

$$\Box \text{push}((\sim e \wedge \sim t2 \wedge t1) \vee (\sim e \wedge \sim t1 \wedge t2)),$$
$$(e \equiv \sim (t1 \vee t2)),$$
$$\Box \textbf{ while } \sim e \textbf{ do} \text{ pop } \textbf{od} \text{ true},$$
$$(\alpha \Rightarrow \Box \text{ push}(\Diamond \text{pop}\alpha)),$$
$$(e \Rightarrow \Diamond \text{pop true}),$$
$$(\Box \text{ pop } \alpha \equiv \Diamond \text{ pop } \alpha),$$
$$((\Diamond \text{push}(\alpha \wedge \beta) \wedge \Diamond \text{push}(\alpha \wedge \sim \beta)) \Rightarrow \Box \text{ push } \alpha),$$

where α, β are arbitrary formulas of the language.

The *St* theory is consistent since the following structure is a model of *Axst*: $\mathfrak{M} = \langle S, \mathcal{I}, w \rangle$ where $S = \{1, 2\}^* \cup \emptyset$.

$$\mathcal{I}(\text{pop}) = \{(is, s): s \in S, i = 1, 2\},$$

$$\mathcal{I}(\text{push}) = \{(s, 1s): s \in S\} \cup \{(s, 2s): s \in S\},$$

$$w(s) = v_s \quad \text{such that} \quad v_s(t1) = \mathbf{1} \quad \text{iff } s = 1i_1 i_2 \dots,$$
$$v_s(t2) = \mathbf{1} \quad \text{iff } s = 2i_1 i_2 \dots,$$
$$v_s(e) = \mathbf{1} \quad \text{iff } s = \emptyset.$$

This model, known as the *standard model*, is illustrated in Figure 6.4.

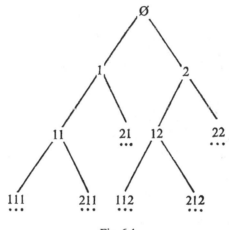

Fig. 6.4

LEMMA 6.3. *Every normalized model of St is isomorphic with the standard one.*

The proof is similar to the one given in the previous example.   □

LEMMA 6.4. *Any two models of St are algorithmically equivalent.*   □

7. LINDENBAUM ALGEBRA

We shall now describe the Lindenbaum algebra of a theory $T = \langle L, C, A \rangle$ based on PAL and some of its properties, which will be useful in further considerations.

Let T be a theory based on PAL and let $\approx$ be an equivalence relation in the set of all formulas F defined as follows:

$$\alpha \approx \beta \quad \text{iff} \quad (\alpha \Rightarrow \beta) \text{ and } (\beta \Rightarrow \alpha) \text{ are theorems in } T.$$

It is easy to verify that $\approx$ is a congruence with respect to $\vee$, $\wedge$, $\sim$, and if $\alpha \approx \beta$ then, for every program M, $\Box M\alpha \approx \Box M\beta$ and $\Diamond M\alpha \approx \Diamond M\beta$.

By $||\alpha||$ we denote the set of all formulas β such that $\alpha \approx \beta$. The following theorem characterizes the algebra $F/\approx$ (cf. Chapter III, §1).

THEOREM 7.1. *The system* $\langle F/\approx, \cup, \cap, - \rangle$ *is a Boolean algebra, where* $||\alpha|| \cup ||\beta|| = ||(\alpha \vee \beta)||$, $-||\alpha|| = ||\sim\alpha||$, $||\alpha|| \cap ||\beta|| = ||(\alpha \wedge \beta)||$ *and*:

  (i) $||\alpha|| \leqslant ||\beta||$ *iff* $(\alpha \Rightarrow \beta)$ *is theorem in* T,
  (ii) α *is a theorem in* T *iff* $||\alpha|| = \mathbf{1}$,
  (iii) $||\sim\alpha|| \neq \mathbf{0}$ *iff* α *is not a theorem in* T. $\Box$

THEOREM 7.2. *For the arbitrary formulas* $\alpha \in F$, $\gamma \in F_0$ *and for any program scheme* M, *the following equalities hold*:

$$||\,\mathrm{pref}\ \Box\ \textbf{while}\ \gamma\ \textbf{do}\ M\ \textbf{od}\ \alpha||$$
$$= \mathrm{l.u.b.}\,||\ \mathrm{pref}\ (\Box(\textbf{if}\ \gamma\ \textbf{then}\ M\ \textbf{fi})^i(\sim\gamma \wedge \alpha))||,$$
$$\;_{i \in N}$$
$$||\,\mathrm{pref}\ \Diamond\ \textbf{while}\ \gamma\ \textbf{do}\ M\ \textbf{od}\ \alpha||$$
$$= \mathrm{l.u.b.}\,||\ \mathrm{pref}\ (\Diamond(\textbf{if}\ \gamma\ \textbf{then}\ M\ \textbf{fi})^i(\sim\gamma \wedge \alpha))||,$$
$$\;_{i \in N}$$

where pref *is an arbitrary prefix*.

The proofs of Theorems 7.1, 7.2 are similar to the proofs of Lemmas 1.1–1.3 in Chapter III. $\Box$

COROLLARY. *Under the same assumption as in Theorem 7.2*

$$||\sim\mathrm{pref}\ \Box\ \textbf{while}\ \gamma\ \textbf{do}\ M\ \textbf{od}\ \alpha||$$
$$= \mathrm{g.l.b.}\,||\sim\mathrm{pref}\ \Box(\textbf{if}\ \gamma\ \textbf{then}\ M\ \textbf{fi})^i(\sim\gamma \wedge \alpha)||,$$
$$\;_{i \in N}$$
$$||\sim\mathrm{pref}\ \Diamond\ \textbf{while}\ \gamma\ \textbf{do}\ M\ \textbf{od}\ \alpha||$$
$$= \mathrm{g.l.b.}\,||\sim\mathrm{pref}\ \Diamond(\textbf{if}\ \gamma\ \textbf{then}\ M\ \textbf{fi})^i(\sim\gamma \wedge \alpha)||.$$
$$\;_{i \in N}$$
          $\Box$

By the above theorem, the Lindenbaum algebra $F/\approx$ can be considered as a Boolean algebra with an at most enumerable set of infinite operations (Q)

$$\text{l.u.b.} \underset{i \in N}{} \| \text{ pref } \square \text{ if } \gamma \text{ then } M \text{ fi } (\sim\gamma \wedge \alpha) \|,$$

(Q)

$$\text{l.u.b.} \underset{i \in N}{} \| \text{ pref } \Diamond \text{ if } \gamma \text{ then } M \text{ fi } (\sim\gamma \wedge \alpha) \|$$

for all $M \in \Pi$, $\alpha \in F$, $\gamma \in F_0$ and an arbitrary prefix pref.

Let us recall that by a *Q-filter in the Boolean algebra $F/\approx$ with the set of infinite operations Q* we understand a maximal filter that preserves all Q-operations, i.e. a maximal filter $\mathscr{F}$ such that $\text{l.u.b.} \underset{i \in N}{} \| \text{ pref } \bigcirc(\text{if } \gamma$ then $M \text{ fi})^i(\sim\gamma \wedge \alpha) \| \in \mathscr{F}$ implies that there exists an i_0 such that

$$\| \text{pref} \bigcirc (\text{if } \gamma \text{ then } M \text{ fi})^{i_0}(\sim\gamma \wedge \alpha) \| \in \mathscr{F}, \quad \text{cf. Appendix A.}$$

Making use of the Rasiowa–Sikorski Lemma (Rasiowa and Sikorski, 1968) we obtain the following:

LEMMA 7.3. *If the theory T is consistent then the Lindenbaum algebra of that theory is a non-degenerate algebra and the family of all Q-filters in $F/\approx$ is a non-empty set (cf. Appendix A).*   □

8. DETERMINISTIC AND TOTAL INTERPRETATIONS OF ATOMIC PROGRAMS

In this section we shall consider a special kind of semantic structures called *functional semantic structures*.

By $\mathfrak{M}_f$ we denote a semantic structure $\langle S, \mathscr{I}, w \rangle$ which assigns a total function in S to every program variable. We shall say that $\mathfrak{M}_f$ is a *functional semantic structure*.

Let us extend the set of axioms defined in § 5 of this chapter by the axioms of the following two schemes:

$\square K \text{ true,}$

$(\Diamond K\alpha \Rightarrow \square K\alpha), \quad \text{for } K \in V_p$

and all formulas $\alpha \in F$. Denote the new consequence operation by C_f and corresponding propositional calculus by PAL_f.

We shall say that α is *functionally valid* if it is valid in every functional structure.

LEMMA 8.1. *If α is a theorem of PAL_f then α is valid in every functional structure $\mathfrak{M}_f$.*

PROOF. To prove this lemma it is sufficient to discuss the axioms

$$\Box K \textbf{ true} \quad \text{and} \quad (\Diamond K\alpha \Rightarrow \Box K\alpha), \quad \text{where } K \in V_p.$$

The validity of these formulas follows immediately, since for every state s, the set $K_{\mathfrak{M}_f}(s)$ has exactly one element (see also Lemma 3.4). $\Box$

LEMMA 8.2. *The propositional algorithmic logic* PAL$_f$ *is consistent.* $\Box$

Let us note that the set of all theorems in PAL$_f$ is closed under the generalization rule

$$\frac{\alpha}{\Box K\alpha} \quad \text{for} \quad K \in V_p.$$

LEMMA 8.3. *The following formulas are theorems in* PAL$_f$:

$$\Box K(\alpha \vee \beta) \equiv \Box K\alpha \vee \Box K\beta,$$
$$\Diamond K(\alpha \wedge \beta) \equiv \Diamond K\alpha \wedge K\beta,$$
$$\sim \Box K\alpha \equiv \Diamond K \sim \alpha,$$

where K is a program variable and α, β are arbitrary formulas. $\Box$

COROLLARY 8.1. *There are formulas which are functionally valid and which are not valid in every structure.* $\Box$

Let T_f be a consistent theory based on PAL$_f$. We shall construct a model of such a theory. Let $\mathscr{F}$ be an arbitrary Q-filter in the Lindenbaum algebra of that theory (cf. § 7 of this chapter).

We shall consider a proper semantic structure $\mathfrak{M}_{\mathscr{F}}$

$$\mathfrak{M}_{\mathscr{F}} = \langle W_0, \mathscr{I} \rangle,$$

where

W_0 is a set of valuations v_{pref}, $\text{pref} \in (\{\Diamond K\}_{K \in V_p} \cup \{\Box K\}_{K \in V_p})^*$
$v_{\text{pref}}(q) = 1$ iff $\|\text{pref } q\| \in \mathscr{F}$, for every propositional variable q, and
$\mathscr{I}$ is a functional interpretation of the program variables in W_0 such that

$$\mathscr{I}(K) = \{(v_{\text{pref}}, v_{\text{pref} \Box K}): \text{for every prefix pref}\}, \quad K \in V_p.$$

The following lemma holds.

LEMMA 8.4. *For every formula α and every prefix c*

$$||c\alpha|| \in \mathscr{F} \quad \text{iff} \quad \mathfrak{M}_{\mathscr{F}}, v_c \models \alpha.$$

PROOF. The proof is by induction on the complexity of the formula. The basic step of induction follows immediately from the definition of $\mathfrak{M}_{\mathscr{F}}$. Assume that Lemma 8.4 holds for all formulas which are submitted to α_0.

1. Suppose α_0 is of the form $\square K\alpha$ where $K \in V_p$;

$$\mathfrak{M}_{\mathscr{F}}, v_c \models \square K\alpha \quad \text{iff} \quad \mathfrak{M}_{\mathscr{F}}, v_{c\,\square K} \models \alpha.$$

By the inductive assumption $||c\ \square K\alpha|| \in \mathscr{F}$.

2. Let α_0 be of the form $(\alpha \vee \beta)$.

$$\mathfrak{M}_{\mathscr{F}}, v_c \models (\alpha \vee \beta) \quad \text{iff} \quad \mathfrak{M}_{\mathscr{F}}, v_c \models \alpha \quad \text{or} \quad \mathfrak{M}_{\mathscr{F}}, v_c \models \beta.$$

By the inductive hypothesis, this is equivalent to $||c\alpha|| \in \mathscr{F}$ or $||c\beta|| \in \mathscr{F}$. But the formula $(c\alpha \vee c\beta) \equiv c(\alpha \vee \beta)$ is a theorem in T, thus $||c(\alpha \vee \beta)|| \in \mathscr{F}$.

3. Consider a formula α_0 of the form $\square$ **while** γ **do** M **od** β. By Lemma 2.3 we have $\mathfrak{M}_{\mathscr{F}}, v_c \models \square$ **while** γ **do** M **od** β, i.e. iff there exists an i_0 such that $\mathfrak{M}_{\mathscr{F}}, v_c \models \square(\textbf{if } \gamma \textbf{ then } M \textbf{ fi})^{i_0}(\sim\gamma \wedge \beta)$. By the inductive hypothesis this is equivalent to the following:

$$(\exists i_0)||\square(\textbf{if } \gamma \textbf{ then } M \textbf{ fi})^{i_0}(\sim\gamma \wedge \beta)|| \in \mathscr{F}.$$

By the definition of the Q-filter $\mathscr{F}$ we have

$$\mathfrak{M}_{\mathscr{F}}, v_c \models \alpha_0 \quad \text{iff} \quad \underset{i \in N}{\text{l.u.b.}}||c\square(\textbf{if } \gamma \textbf{ then } M \textbf{ fi})^i(\sim\gamma \wedge \beta)|| \in \mathscr{F}$$

and from Lemma 7.2.

$$\mathfrak{M}_{\mathscr{F}}, v_c \models \alpha_0 \quad \text{iff} \quad ||c\square \textbf{ while } \gamma \textbf{ do } M \textbf{ od } \beta|| \in \mathscr{F}.$$

4. Consider a formula α_0 of the form $\square$ **either** M_1 **or** M_2 **ro** β.

By the definition of interpretation and the value of the formula $\mathfrak{M}_{\mathscr{F}}, v_c \models \square$ **either** M_1 **or** M_2 **ro** β iff $\mathfrak{M}_{\mathscr{F}}, v_c \models \square M_1\beta$ and $\mathfrak{M}_{\mathscr{F}}, v_c \models \square M_2\beta$. By the inductive hypothesis and the properties of the Q-filter we have

$$\square \textbf{ (either } M_1 \textbf{ or } M_2 \textbf{ ro } \beta)_{\mathfrak{M}_{\mathscr{F}}}(v_c) = \mathbf{1} \quad \text{iff}$$
$$||c\ \square M_1\beta|| \in \mathscr{F} \quad \text{and} \quad ||c\ \square M_2\beta|| \in \mathscr{F}.$$

Since

$$\text{PAL}_f \vdash (\square M_1\beta \wedge \square M_2\beta) \equiv \square \textbf{ either } M_1 \textbf{ or } M_2 \textbf{ ro } \beta$$

then from Lemma 8.3 we have

$$\mathfrak{M}_{\mathscr{F}}, v_c \models \alpha_0 \quad \text{iff} \quad ||c \,\square\, \textbf{either } M_1 \textbf{ or } M_2 \textbf{ ro } \beta|| \in \mathscr{F}.$$

5. Let α_0 be of the form $\sim\alpha$. Then $\mathfrak{M}_{\mathscr{F}}, v_c \models \sim\alpha$ iff $||c\alpha|| \notin \mathscr{F}$. Since the filter is prime and the formula $(\square K \sim \alpha \equiv \sim \square K\alpha)$ is a theorem in PAL_f, we have $\mathfrak{M}_{\mathscr{F}}, v_c \models \sim\alpha$ iff $||c \sim \alpha|| \in \mathscr{F}$.

The remaining cases can be dealt with analogously. $\square$

THEOREM 8.5. *For every formula* α, α *is a theorem of* PAL_f *iff* α *is functionally valid.*

PROOF. By the Adequacy Theorem 4.1, if α is a PAL_f theorem, then α is functionally valid.

Suppose that α is a functionally valid formula and α is not a theorem in PAL_f. By Theorem 7.1, $||\sim\alpha|| \neq 0$. By Lemma 7.3 there exists a Q-filter $\mathscr{F}$ in the Lindenbaum algebra $F/\approx$ such that $||\sim\alpha|| \in \mathscr{F}$. Let us construct the set of valuations v_c and the interpretation $\mathscr{I}$ as defined above for this filter. By Lemma 8.4, $\mathfrak{M}_{\mathscr{F}}, v_\emptyset \models \sim\alpha$, and therefore α is not functionally valid. $\square$

THEOREM 8.6. *The theory* $T = \langle L_0, C, A \rangle$ *based on* PAL_f *is consistent iff* T *has a model.*

PROOF. The one-way implication is obvious. Assume that T is consistent, i.e. that there exists a formula α such that α is not a theorem in T. Thus $||\sim\alpha|| \neq 0$ and there exists a Q-filter $\mathscr{F}$ such that $||\sim\alpha|| \in \mathscr{F}$. The semantic structure $\mathfrak{M}_{\mathscr{F}}$ defined as above is a model of T. Indeed, if $\beta \in A$, then for every prefix c,

$$A \vdash c\beta.$$

Hence $||c\beta|| \in \mathscr{F}$. By Lemma 8.4 $\mathfrak{M}_{\mathscr{F}}, v_c \models \beta$ for every valuation v_c in $\mathfrak{M}_{\mathscr{F}}$, i.e. $\mathfrak{M}_{\mathscr{F}} \models \beta$. $\square$

As a consequence of the above theorems we have the following Completeness Theorem:

THEOREM 8.7. *For any consistent theory* T_f *based on* PAL_f, *the following conditions are equivalent:*
  (i) α *is a theorem of* T;
  (ii) α *is valid in every proper functional model of* T_f;

(iii) α *is valid in every normalized functional model of* T_f;

(iv) α *is valid in every functional model of* T_f.

PROOF.

(i) → (ii) by the Adequacy Theorem 5.3.

To prove that (ii) implies (i), assume that α is not a theorem in T_f. Thus $||\sim\alpha|| \neq 0$. A Q-filter $\mathscr{F}$ such that $||\sim\alpha|| \in \mathscr{F}$ therefore exists. Let us consider the proper model $\mathfrak{M}_{\mathscr{F}}$ connected with $\mathscr{F}$. From Lemma 8.4, for every prefix pref,

$$\mathfrak{M}_{\mathscr{F}}, v_{\text{pref}} \models \sim\alpha \quad \text{iff} \quad ||\text{pref} \sim\alpha|| \in \mathscr{F}.$$

In particular,

$$\text{non } \mathfrak{M}_{\mathscr{F}}, v_{\varnothing} \models \alpha,$$

and therefore it is not true that α is valid in every proper model of T_f.

(iii) → (i)   since the canonical model $\mathfrak{M}_{\mathscr{F}}$ is normalized.

(i)   → (iv) by Adequacy Theorem 8.5.

(iv) → (ii) and (iv) → (iii) are both obvious.   □

9. PARTIAL FUNCTIONAL INTERPRETATIONS

In § 8 of this chapter, where the simplest version of propositional algorithmic logic was described, the meaning of the program variable was a total function.

We now study another version of PAL, in which every interpretation of the program variable is a partial function. We shall prove the Completeness Theorem in a new way; the models constructed here are no longer proper models.

Let us denote by PAL$_{\text{pf}}$ a deductive system based on the axioms and rules described in § 5 of this chapter with one new axiom scheme

$$(\lozenge K\alpha \Rightarrow \square K\alpha)$$

for every program variable K and every formula α. Our aim in this section is to prove the following property:

For every set of formulas Z and every formula α

$$Z \vdash_{\text{pf}} \alpha \quad \text{iff} \quad Z \models_{\text{pf}} \alpha.$$

$\models_{\text{pf}}$ means that we shall consider only structures $\mathfrak{M}_{\text{pf}} = \langle S, \mathscr{I}, w \rangle$ in which for all $K \in V_p$ and $s \in S$, $K_{\mathfrak{M}}(s)$ is an at most one-element set. In other words, the meaning of a program variable is a partial function.

First of all, let us note that in every structure $\mathfrak{M}_{pf}$ the formula

$$(\lozenge K\alpha \Rightarrow \square K\alpha)$$

is valid for every $K \in V_p$ and $\alpha \in F$ (cf. Lemma 3.4).

As an immediate consequence of Lemma 5.2 and the above obser-vation we obtain the following lemma.

LEMMA 9.1. *For every theory T based on* PAL_{pf} *and every formula α,
if $T \vdash_{pf} \alpha$, then $T \models_{pf} \alpha$.*   □

Let T be a consistent theory based on PAL_{pf}. We shall construct a model of T in the Lindenbaum algebra of that theory (cf. § 7 of this chapter). By a *canonical structure* of a theory T we shall mean a semantic structure $\mathfrak{M}_0'$ such that

$$\mathfrak{M}_0 = \langle QF, \mathscr{I}_0, w_0 \rangle,$$

where:

QF is the family of all Q-filters in the Lindenbaum algebra of the theory T,

for every program variable K, $\mathscr{I}_0(K) = \{(\mathscr{F}_1, \mathscr{F}_2) \in QF^2 : ||\lozenge K \text{ true}|| \in \mathscr{F}_1$ and, for every α, if $||\square K\alpha|| \in \mathscr{F}_1$ then $||\alpha|| \in \mathscr{F}_2\}$;

w_0 is a function which to every Q-filter $\mathscr{F} \in QF$ assigns a valuation $v_{\mathscr{F}}$ such that for all $p \in V_0$

$$v_{\mathscr{F}}(p) = 1 \quad \text{iff} \quad ||p|| \in \mathscr{F}.$$

Let us consider the canonical model $\mathfrak{M}_0$ of a consistent theory T_{pf}, $\mathfrak{M}_0 = \langle QF, \mathscr{I}_0, w_0 \rangle$.

FACT 1. *QF is a non-empty set.*

FACT 2. *For every program variable K, every formula β and every Q-filter $\mathscr{F}$, if $||\lozenge K\beta|| \in \mathscr{F}$ then there exists a Q-filter $\mathscr{F}_1$ such that $\mathscr{F} K_{\mathfrak{M}_0} \mathscr{F}_1$ and $||\beta|| \in \mathscr{F}_1$.*

PROOF. Let us denote by $Z_{\mathscr{F}K}$ the following set

$$\{||\alpha|| : ||\square K\alpha|| \in \mathscr{F}\}.$$

$1°$ $Z_{\mathscr{F}K} \neq \varnothing$.

Indeed, since $||\lozenge K \text{ true}|| \in \mathscr{F}$ and $\vdash_{pf} (\lozenge K \text{ true} \Rightarrow \square K \text{ true})$ we have $||\square K \text{ true}|| \in \mathscr{F}$ and therefore $||\text{true}|| \in Z_{\mathscr{F}K}$.

$2°$ $Z_{\mathscr{F}K}$ is a filter.

Let us assume that $||(\alpha \wedge \beta)|| \in Z_{\mathscr{F}K}$. From the definition of $Z_{\mathscr{F}K}$, $||\square K(\alpha \wedge \beta)|| \in \mathscr{F}$. From the axiom $\square K(\alpha \wedge \beta) \equiv (\square K\alpha \wedge \square K\beta)$ we have $||(\square K\alpha \wedge \square K\beta)|| \in \mathscr{F}$. In fact $\mathscr{F}$ is a filter, thus $||\square K\alpha|| \in \mathscr{F}$ and $||\square K\beta|| \in \mathscr{F}$ and therefore $||\alpha|| \in Z_{\mathscr{F}K}$ and $||\beta|| \in Z_{\mathscr{F}K}$.

3° $Z_{\mathscr{F}K}$ is a maximal filter.

Suppose that $||\alpha|| \vee ||\beta|| \in Z_{\mathscr{F}K}$. Thus, $||(\alpha \vee \beta)|| \in Z_{\mathscr{F}K}$ and consequently, $||\square K(\alpha \vee \beta)|| \in \mathscr{F}$. From the axioms of PAL_{pf}, $\vdash_{pf} \lozenge K\delta \equiv \square K\delta$ for every formula δ and

$$\vdash \lozenge K(\alpha \vee \beta) \equiv (\lozenge K\alpha \vee \lozenge K\beta).$$

Hence, we have $||(\square K\alpha \vee \square K\beta)|| \in \mathscr{F}$. Since $\mathscr{F}$ is a maximal filter, $||\square K\alpha|| \in \mathscr{F}$ or $||\square K\beta|| \in \mathscr{F}$. As a consequence of this, $||\alpha|| \in Z_{\mathscr{F}K}$ or $||\beta|| \in Z_{\mathscr{F}K}$ for any two formulas α, β.

4° $Z_{\mathscr{F}K}$ is a Q-filter.

Suppose that

$$||\mathrm{pref} \; \square \; \mathbf{while} \; \gamma \; \mathbf{do} \; M \; \mathbf{od} \; \alpha|| \in Z_{\mathscr{F}K}$$

for some prefix pref, formulas $\gamma \in F_0$ and $\alpha \in F$ and a program $M \in \Pi$. Hence, $||\square K \, \mathrm{pref} \; \square \; \mathbf{while} \; \gamma \; \mathbf{do} \; M \; \mathbf{od} \; \alpha|| \in \mathscr{F}$. But $\mathscr{F}$ is a Q-filter, thus there exists an i such that $||\square K \, \mathrm{pref} \; \square \; (\mathbf{if} \; \gamma \; \mathbf{then} \; M \; \mathbf{fi})^i(\sim \gamma \wedge \alpha)|| \in \mathscr{F}$. By the definition of $Z_{\mathscr{F}K}$, $|| \, \mathrm{pref} \; \square \; (\mathbf{if} \; \gamma \; \mathbf{then} \; M \; \mathbf{fi})^i(\sim \gamma \wedge \alpha)|| \in Z_{\mathscr{F}K}$.

Analogously we can prove that if $||\mathrm{pref} \; \lozenge \; \mathbf{while} \; \gamma \; \mathbf{do} \; M \; \mathbf{od} \; \alpha|| \in Z_{\mathscr{F}K}$ then there exists an i such that $||\mathrm{pref} \; \lozenge \; (\mathbf{if} \; \gamma \; \mathbf{then} \; M \; \mathbf{fi})^i(\sim \gamma \wedge \alpha)|| \in Z_{\mathscr{F}K}$. Hence $Z_{\mathscr{F}K}$ is a Q-filter.

Observe that since $||\lozenge K\beta|| \in \mathscr{F}$ and $\vdash_{pf} (\lozenge K\beta \Rightarrow \square K\beta)$, then $||\beta|| \in Z_{\mathscr{F}K}$. By the definition of $\mathfrak{M}_0$, $(\mathscr{F}, Z_{\mathscr{F}K}) \in K_{\mathfrak{M}_0}$. Hence $Z_{\mathscr{F}K}$ is the required Q-filter. $\qquad \square$

FACT 3. *For every Q-filter $\mathscr{F} \in QF$ and every program variable K, $K_{\mathfrak{M}_0}(\mathscr{F})$ is an at most one-element set.*

PROOF. Suppose the contrary. Let $\mathscr{F}_1$, $\mathscr{F}_2 \in K_{\mathfrak{M}_0}(\mathscr{F})$. By the definition of the canonical structure and the proof of Fact 2,

$$\mathscr{F}_2 \supset Z_{\mathscr{F}K} \quad \text{and} \quad \mathscr{F}_1 \supset Z_{\mathscr{F}K}.$$

Since $Z_{\mathscr{F}K}$ is a Q-filter, it cannot be contained in any other Q-filter. Thus $\mathscr{F}_1 = \mathscr{F}_2 = Z_{\mathscr{F}K}$. $\qquad \square$

FACT 4. *If $||\Box K\alpha|| \in \mathcal{F}$ then, for every Q-filter $\mathcal{F}'$, $\mathcal{F}K_{\mathfrak{M}_0}\mathcal{F}'$ implies $||\alpha|| \in \mathcal{F}'$.*

This is an immediate consequence of Fact 2.   □

The following lemma is basic for our further considerations:

LEMMA 9.2. *For every formula α of a propositional algorithmic language and for every Q-filter $\mathcal{F} \in QF$,*

(∗) $\mathfrak{M}_0, \mathcal{F} \models \alpha$ iff $||\alpha|| \in \mathcal{F}$.

PROOF. The proof is by induction on the complexity of the formula α. For the base of induction (propositional variables) the proof of Lemma 9.2 follows immediately from the definition of the canonical structure $\mathfrak{M}_0$.

Assume that (∗) holds for all formulas that are submitted to a formula α (cf. Definition 3.3).

— Let us consider a formula $\Box K\beta$, where K is a program variable. Suppose that $\mathfrak{M}_0, \mathcal{F} \models \Box K\beta$.

By the definition of the structure $\mathfrak{M}_0$, $K_{\mathfrak{M}_0}(\mathcal{F}) \neq \emptyset$, and

(1)   for every $\mathcal{F}' \in K_{\mathfrak{M}_0}(\mathcal{F})$, $\mathfrak{M}_0, \mathcal{F}' \models \beta$.

By the inductive assumption $||\beta|| \in \mathcal{F}'$ and since $\mathcal{F}K_{\mathfrak{M}_0}\mathcal{F}'$, we have

(2) $||\Diamond K \text{ true}|| \in \mathcal{F}$.

Now suppose that $||\Box K\beta|| \notin \mathcal{F}$. The formula

$$(\sim \Box K\beta \equiv (\Diamond K \sim \beta \vee \sim \Diamond K \text{ true}))$$

is a theorem of PAL_{pf} and therefore

$$||(\Diamond K \sim \beta \vee \sim \Diamond K \text{ true})|| \in \mathcal{F}.$$

Since $\mathcal{F}$ is a maximal filter, we have

(3) $||\Diamond K \sim \beta|| \in \mathcal{F}$ or

(4) $||\sim \Diamond K \text{ true}|| \in \mathcal{F}$.

Case (4) is impossible because of (2).

Suppose (3). By Fact 2 there exists a Q-filter $\mathcal{F}''$ such that

$$\mathcal{F}K_{\mathfrak{M}_0}\mathcal{F}''   \text{and} ||\sim \beta|| \in \mathcal{F}''.$$

By (1) and the inductive assumption we have $||\beta|| \in \mathcal{F}''$, a contradiction. Hence, if $\mathfrak{M}_0, \mathcal{F} \models \Box K\beta$ then $||\Box K\beta|| \in \mathcal{F}$.

Conversely, if $||\Box K\beta|| \in \mathscr{F}$, then by Fact 3, for every Q-filter $\mathscr{F}'$ if $\mathscr{F}K_{\mathfrak{M}_0}\mathscr{F}'$ then $||\beta|| \in \mathscr{F}'$, and by Fact 2, $K_{\mathfrak{M}_0}(\mathscr{F}) \neq \varnothing$. By the inductive assumption $\mathfrak{M}_0, \mathscr{F}' \models \beta$ and therefore $\mathfrak{M}, \mathscr{F} \models \Box K\beta$.

— Consider the formula $\Diamond K\alpha$, for $K \in V_p$. Suppose that $\mathfrak{M}_0, \mathscr{F} \models \Diamond K\alpha$. By the definition of interpretation $\mathscr{I}_0$, there exists a Q-filter $\mathscr{F}'$ such that $\mathscr{F}K_{\mathfrak{M}_0}\mathscr{F}'$ and $\mathfrak{M}_0, \mathscr{F}' \models \alpha$. By the inductive assumption, we have

(5) $\mathscr{F}K_{\mathfrak{M}_0}\mathscr{F}'$ and $||\alpha|| \in \mathscr{F}'$.

Suppose that $||\Diamond K\alpha|| \notin \mathscr{F}$, then

(6) $|| \sim \Diamond K\alpha|| \in \mathscr{F}$.

Since $\mathscr{F}K_{\mathfrak{M}_0}\mathscr{F}'$, we have $||\Diamond K\,\mathbf{true}|| \in \mathscr{F}'$. Since $\vdash \Box K \sim \alpha$ $\equiv (\sim \Diamond K\alpha \wedge \Diamond K\,\mathbf{true})$, we have by (6) $||\Box K \sim \alpha|| \in \mathscr{F}$ and by (5) $|| \sim \alpha|| \in \mathscr{F}'$, a contradiction. Conversely, if $||\Diamond K\alpha|| \in \mathscr{F}$ then, by Fact 2 and the definition of the interpretation $\mathscr{I}_0$, $\mathfrak{M}_0, \mathscr{F} \models \Diamond K\alpha$.

The proof of other cases is similar to that of the analogous theorem for non-deterministic algorithmic logic (cf. Chapter VI).   □

THEOREM 9.3 (Model Existence Theorem). *For every consistent theory* $T_{\mathrm{pf}} = \langle L_0, C, A \rangle$ *based on* $\mathrm{PAL}_{\mathrm{pf}}$ *there exists a model of* T_{pf}.

PROOF. We shall prove that the canonical structure $\mathfrak{M}_0$ of the theory T_{pf} is a model of A.

If we let $\beta \in A$, then $||\beta|| \in \mathscr{F}$ for every Q-filter $\mathscr{F}$. By Lemma 9.2, $\mathfrak{M}_0, \mathscr{F} \models \beta$. Thus $\mathfrak{M}_0$ is a model of the set A.   □

The Completeness Theorem below is a simple consequence of the above considerations and the fact that $\mathfrak{M}_0$ is a normalized model.

THEOREM 9.4. *For every formula* α *of a consistent theory* T_{pf} $= \langle L, C, A \rangle$ *the following conditions are equivalent*:
  (i) α *is a theorem of* T_{pf}.
  (ii) α *is valid in every* pf-*model of* T_{pf}.
  (iii) α *is valid in every normalized structure* $\mathfrak{M}_{\mathrm{pf}}$ *which is a model of* T_{pf}.

PROOF. (i) implies (ii) by virtue of Theorem 5.3. To prove that (ii) implies (i), assume that α is valid in every model of T_{pf} and α is not a theorem. Then by Lemma 7.1, $|| \sim \alpha|| \neq \mathbf{0}$. Hence there exists a Q-filter $\mathscr{F}$

which contains $\| \sim \alpha \|$. From Theorem 9.3 the canonical structure $\mathfrak{M}_0$ is a model of A and from Lemma 9.2 formula α is not valid in $\mathscr{F}$. Thus $\mathfrak{M}_0$ is not a model of α. $\qquad\qquad\qquad\qquad\qquad\qquad\qquad$ $\square$

10. BOUNDED NON-DETERMINISM: THE COMPLETENESS THEOREM

In this section we shall consider another complete extension of PAL. Every program variable will now be interpreted as a relation which contains at most m pairs with the same first element.

We shall discuss in detail the case $m = 2$, i.e. we shall assume that in every semantic structure and for every state we can pass to at most two other states by means of an atomic program K, $K \in V_p$.

Let us denote by Ax_2 the following scheme

$$\left(\Diamond K(\alpha \wedge \beta) \wedge \Diamond K(\alpha \wedge \sim \beta) \Rightarrow \Box K\alpha \right)$$

where $K \in V_p$, $\alpha, \beta \in F$.

Lemma 10.1 below explains the meaning of this formula.

LEMMA 10.1. *Let $\mathfrak{M}$ be a normalized structure $\mathfrak{M} = \langle S, \mathscr{I}, w \rangle$. If $\mathfrak{M} \models \mathrm{Ax}_2$ then for all $s \in S$, $\mathrm{card}\left(\mathscr{I}(K)(s)\right) \leqslant 2$.*

PROOF. Let $\mathfrak{M}$ be a fixed normalized structure and

$$(\forall \alpha, \beta \in F) \ \mathfrak{M} \models \mathrm{Ax}_2.$$

Suppose that

$$\mathrm{card}\left(\mathscr{I}(K)(s)\right) > 2 \quad \text{for some state } s.$$

Let $s_1, s_2, s_3 \in K_{\mathfrak{M}}(s)$ and $s_1 \neq s_2, s_2 \neq s_3, s_1 \neq s_3$. There then exist formulas α, β such that

$$\alpha_{\mathfrak{M}}(s_1) \neq \alpha_{\mathfrak{M}}(s_2), \quad \alpha_{\mathfrak{M}}(s_1) = \alpha_{\mathfrak{M}}(s_3),$$
$$\beta_{\mathfrak{M}}(s_1) \neq \beta_{\mathfrak{M}}(s_3).$$

Let $\bar{\gamma}$ denote γ or $\sim \gamma$ depending on its value in the structure $\mathfrak{M}$ and state s_1,

$$\bar{\gamma} = \begin{cases} \gamma & \text{if } \mathfrak{M}, s_1 \models \gamma, \\ \sim \gamma & \text{if } \mathfrak{M}, s_1 \models \sim \gamma. \end{cases}$$

We now have

$$\mathfrak{M}, s_1 \models (\bar{\alpha} \wedge \bar{\beta}) \quad \text{and} \quad \mathfrak{M}, s_3 \models (\bar{\alpha} \wedge \sim \bar{\beta}).$$

Hence,

$$\mathfrak{M}, s \models \left(\Diamond K(\bar{\alpha} \wedge \bar{\beta}) \wedge \Diamond K(\bar{\alpha} \wedge \sim \bar{\beta}) \right),$$

and at the same time

$$\mathfrak{M}, s_2 \models \sim \bar{\alpha}, \quad \text{i.e.} \quad \mathfrak{M}, s \models \sim \Box K\bar{\alpha}.$$

Thus

$$\mathfrak{M}, s \models \sim \left(\left(\Diamond K(\bar{\alpha} \wedge \bar{\beta}) \wedge \Diamond K(\bar{\alpha} \wedge \sim \bar{\beta}) \right) \Rightarrow \Box K\bar{\alpha} \right),$$

a contradiction.   □

LEMMA 10.2. *If $\mathfrak{M}$ is a semantic structure $\mathfrak{M} = \langle S, \mathscr{I}, w \rangle$ such that for all $s \in S$, $\mathrm{card}\left(K_{\mathfrak{M}}(s)\right) \leqslant 2$, then $\mathfrak{M}$ is a model of Ax_2.*

PROOF. Since the value of the formula is defined in a unique way and any state cannot simultaneously satisfy both of the formulas $(\alpha \wedge \beta)$ and $(\alpha \wedge \sim \beta)$, Lemma 10.2 is obvious for $\mathrm{card}\left(K_{\mathfrak{M}}(s)\right) \leqslant 1$. Suppose that for some s, $\mathrm{card}\left(K_{\mathfrak{M}}(s)\right) = 2$. If one state of the set $K_{\mathfrak{M}}(s)$ satisfies $(\alpha \wedge \beta)$ and another one $(\alpha \wedge \sim \beta)$, then obviously $\Box K\alpha$ is also satisfied.   □

Let PAL_2 denote a propositional algorithmic logic which is an extension of PAL by the scheme Ax_2:

$$\left((\Diamond K(\alpha \wedge \beta) \wedge \Diamond K(\alpha \wedge \sim \beta)) \right) \Rightarrow \Box K\alpha), \qquad \alpha, \beta \in F, \quad K \in V_p.$$

Let T_2 be a consistent theory based on PAL_2 and let $\mathfrak{M}_0$ be the canonical structure for that theory, $\mathfrak{M}_0 = \langle QF, \mathscr{I}_0, w_0 \rangle$ (cf. § 9 of this chapter).

LEMMA 10.3. *The canonical structure for T_2 is a model of Ax_2.*

PROOF. By Lemma 10.2 it is sufficient to prove that $\mathrm{card}\left(K_{\mathfrak{M}_0}(\mathscr{F})\right) \leqslant 2$ for every Q-filter $\mathscr{F}$. Suppose that

$$K_{\mathfrak{M}_0}(\mathscr{F}) \supset \{\mathscr{F}_1, \mathscr{F}_2, \mathscr{F}_3\},$$

where $\mathscr{F}_1, \mathscr{F}_2, \mathscr{F}_3$ are different Q-filters. Hence, there exists a formula α such that

$$||\alpha|| \in \mathscr{F}_1 \quad \text{and} \quad ||\alpha|| \notin \mathscr{F}_2.$$

Since $\mathscr{F}_3$ is maximal, then either $||\alpha||$ or $||\sim \alpha||$ belongs to $\mathscr{F}_3$.

A. If $||\alpha|| \in \mathscr{F}_3$ then there exists a formula β such that

$$||\beta|| \notin \mathscr{F}_3 \quad \text{and} \quad ||\beta|| \in \mathscr{F}_1.$$

B. If $||\alpha|| \notin \mathscr{F}_3$ then there exists a formula β such that

$$||\beta|| \notin \mathscr{F}_3 \quad \text{and} \quad ||\beta|| \in \mathscr{F}_2.$$

These two possibilities are illustrated in the following diagram, Figure 10.1:

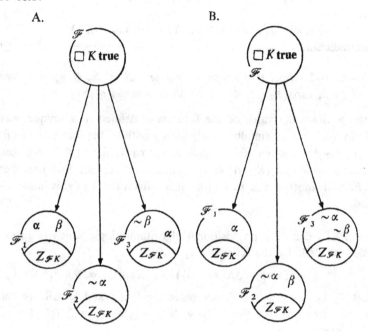

Fig. 10.1

Consider case A.

If $||\Box K\alpha|| \in \mathscr{F}$, then $||\alpha|| \in \mathscr{F}_1 \cap \mathscr{F}_2$ by the definition of the interpretation $\mathscr{I}_0$. This contradicts our assumptions. Thus, $||\Box K\alpha|| \notin \mathscr{F}$. Since $\mathscr{F}$ is a maximal filter, we have $|| \sim \Box K\alpha|| \in \mathscr{F}$.

By Ax_2 and the maximality of $\mathscr{F}$

(1) $|| \sim \Diamond K(\alpha \wedge \beta)|| \in \mathscr{F}$ or

(2) $|| \sim \Diamond K(\alpha \wedge \sim \beta)|| \in \mathscr{F}$.

Assuming (1), we have

$$||(\sim \Box K \text{ true} \vee \Box K \sim (\alpha \wedge \beta))|| \in \mathscr{F}.$$

Thus $||\Box K \sim (\alpha \wedge \beta)|| \in \mathscr{F}$ and in consequence

(3) $|| \sim (\alpha \wedge \beta)|| \in \mathscr{F}_1 \cap \mathscr{F}_2 \cap \mathscr{F}_3.$

But $||\sim\alpha|| \notin \mathscr{F}_1$, and $||\sim\beta|| \notin \mathscr{F}_1$, which contradicts (3).

Assuming (2), we have

$$||(\sim\Box K\ \textbf{true} \vee \Box K\sim(\alpha\wedge\sim\beta))|| \in \mathscr{F}.$$

Thus $||\Box K\sim(\alpha\wedge\sim\beta)|| \in \mathscr{F}$ and by the definition of $\mathscr{I}_0$,

$$||\sim(\alpha\wedge\sim\beta)|| \in \mathscr{F}_1\cap\mathscr{F}_2\cap\mathscr{F}_3.$$

But $||\sim\alpha|| \notin \mathscr{F}_3$ and $||\beta|| \notin \mathscr{F}_3$, a contradiction. Hence situation A is impossible.

Now let us consider case B.

After considerations similar to the above ones, we find that both of the assumptions

$$||\Box K\sim\alpha|| \in \mathscr{F} \quad \text{and} \quad ||\Box K\sim\alpha|| \notin \mathscr{F}$$

lead to a contradiction.

It is thus impossible to have three different Q-filters $\mathscr{F}_1, \mathscr{F}_2, \mathscr{F}_3$ such that

$$\mathscr{F} K_{\mathscr{I}_0}\mathscr{F}_i, \quad i = 1, 2, 3. \qquad \square$$

The following lemma is fundamental to our further discussion.

LEMMA 10.4. *If* $||\Diamond K\alpha|| \in \mathscr{F}$, *then there exists a Q-filter $\mathscr{F}'$ such that* $||\alpha|| \in \mathscr{F}'$ *and* $\mathscr{F} K_{\mathscr{I}_0}\mathscr{F}'$.

PROOF. Consider the set

$$Z_{\mathscr{F}K} = \{||\beta||: ||\Box K\beta|| \in \mathscr{F}\}.$$

$Z_{\mathscr{F}K}$ is a proper filter as was shown in § 9 of this chapter.

A. We claim that for every formula of the form

$$\text{pref} \bigcirc \textbf{while } \gamma \textbf{ do } M \textbf{ od } \beta$$

there is an index i such that

$$||\text{pref} \bigcirc \textbf{while } \gamma \textbf{ do } M \textbf{ od } \beta \Rightarrow \text{pref} \bigcirc(\textbf{if } \gamma \textbf{ then } M \textbf{ fi})^i(\sim\gamma\wedge\beta)||$$
$$\in Z_{\mathscr{F}K}.$$

We denote the antecedent by *while* for short and the succedent by *if*.

Suppose that for every i

(4) $\qquad ||(while \Rightarrow if^i)|| \notin Z_{\mathscr{F}K}.$

By the definition of $Z_{\mathscr{F}K}$

$$||\sim\Box K(while \Rightarrow if^i)|| \in \mathscr{F} \quad \text{for all } i.$$

By axiom Ax_2, for every formula δ

$$|| \sim \Diamond K((\textit{while} \Rightarrow \textit{if}^i) \wedge \delta) \vee \sim \Diamond K((\textit{while} \Rightarrow \textit{if}^i) \wedge \sim \delta)|| \in \mathscr{F}.$$

Let us take the formula *while* to be δ. We then have

$$||(\sim \Diamond K(\textit{if}^i \wedge \textit{while}) \vee \sim \Diamond K(\sim \textit{while}))|| \in \mathscr{F}.$$

Since $|| \Diamond K \textbf{ true}|| \in \mathscr{F}$ and since $\mathscr{F}$ is a maximal filter, we have for every natural number i, either

$$|| \Diamond K \textit{ if}^i|| \notin \mathscr{F} \quad \text{or} \quad || \Box K \textit{ while}|| \in \mathscr{F}.$$

Hence, either

(5) $|| \Box K \textit{ while}|| \in \mathscr{F}$ or

(6) $|| \Diamond K \textit{ if}^i|| \notin \mathscr{F}$ for every $i \in N$.

If (5), then by the properties of Q-filters there exists a natural number i such that $|| \Box K \textit{ if}^i|| \in \mathscr{F}$.

Hence $|| \textit{if}^i|| \in Z_{\mathscr{F}K}$ and since $\vdash (\textit{if}^i \Rightarrow (\textit{if}^i \vee \sim \textit{while}))$ we have $||(\textit{while} \Rightarrow \textit{if}^i)|| \in Z_{\mathscr{F}K}$, which contradicts (4).

If (6), and since $\mathscr{F}$ preserves all infinite operations, then

(7) $\underset{i \in N}{\text{l.u.b.}} || \Diamond K \textit{ if}^i|| \notin \mathscr{F}.$

By the properties of Lindenbaum algebra (cf. § 7 of this chapter) we have

$$\underset{i \in N}{\text{l.u.b.}} || \Diamond K \textit{ if}^i|| = || \Diamond K \textit{ while}||.$$

Thus by (7) $|| \sim \textit{while}|| \in Z_{\mathscr{F}K}$.

The formula

$$(\sim \textit{while} \Rightarrow (\textit{while} \Rightarrow \textit{if}^i)) \quad \text{for all } i \in N$$

is a theorem and $Z_{\mathscr{F}K}$ is a filter, hence $||\textit{while} \Rightarrow \textit{if}^i|| \in Z_{\mathscr{F}K}$, which contradicts (4). This proves supposition A.

  B. We shall now consider the set $Z_{\mathscr{F}K} \cup \{||\alpha||\}$.

This set has the finite intersection property (cf. Appendix A). So, if

$$||\beta_i|| \in Z_{\mathscr{F}K}, \ i \leqslant n \quad \text{and} \quad ||\beta_1|| \wedge \ ... \ \wedge ||\beta_n|| \wedge ||\alpha|| = \mathbf{0},$$

then

$$\vdash (\alpha \Rightarrow \sim (\beta_1 \wedge \ ... \ \wedge \beta_n)),$$

and consequently

$$\vdash \left(\Diamond K\alpha \Rightarrow \Diamond K \sim (\beta_1 \wedge \ldots \wedge \beta_n) \right).$$

Since $\|\Diamond K\alpha\| \in \mathscr{F}$, we have

$$\|\Diamond K \sim (\beta_1 \wedge \ldots \wedge \beta_n)\| \in \mathscr{F}.$$

Thus $\| \sim \Box K(\beta_1 \wedge \ldots \wedge \beta_n)\| \in \mathscr{F}$ in contradiction to $\|\beta_i\| \in Z_{\mathscr{F}K}$.

C. We can hence construct a proper filter which contains

$$Z_{\mathscr{F}K} \cup \{\|\alpha\|\}, \quad \text{cf. Appendix A.}$$

By the Kuratowski–Zorn Lemma (cf. Rasiowa and Sikorski, 1968) this filter can be extended to the maximal filter $\mathscr{F}'$.

This filter $\mathscr{F}'$ is a Q-filter, since from A, if

$$\text{l.u.b.} \| \operatorname{pref} \bigcirc (\textbf{if } \gamma \textbf{ then } M \textbf{ fi})^i (\sim \gamma \wedge \beta) \| \in \mathscr{F}'$$
$$\scriptstyle i \in N$$

then there exists an i such that

$$\| \operatorname{pref} \bigcirc (\textbf{if } \gamma \textbf{ then } M \textbf{ fi})^i (\sim \gamma \wedge \beta)\| \in \mathscr{F}'.$$

This proves Lemma 10.4. $\qquad\qquad\qquad\qquad\qquad\qquad\quad\Box$

We can now prove the following truth lemma:

LEMMA 10.5. *Let T_2 be a consistent theory based on the two-non-deterministic algorithmic logic PAL_2, and let $\mathfrak{M}_0$ be a canonical structure of T_2. For every Q-filter $\mathscr{F}$ in the Lindenbaum algebra of T_2 and every formula α,*

$$\|\alpha\| \in \mathscr{F} \quad \textit{iff} \quad \mathfrak{M}_0, \mathscr{F} \models \alpha.$$

The proof is by induction on the complexity of the formula α and is similar to the proof of Lemma 9.2. The fundamental step in this induction was proved in Lemma 10.4 in connection with the formula $\Diamond K\alpha$. $\qquad\qquad\qquad\qquad\qquad\qquad\qquad\qquad\qquad\Box$

COROLLARY. *The canonical structure $\mathfrak{M}_0$ of T_2 is a normalized two-non-deterministic model of T_2.* $\qquad\qquad\qquad\qquad\qquad\qquad\qquad\Box$

Using Lemma 10.5, we obtain the Model Existence Theorem:

THEOREM 10.6. *Theory T_2 is consistent iff there is a model of T_2.* $\square$

The following theorem asserts that the semantic consequence operation and the syntactic operation coincide.

THEOREM 10.7. (Completeness Theorem). *For every consistent theory T_2 based on PAL_2 the following conditions are equivalent*:
   (i) α *is a theorem of T_2*;
   (ii) α *is valid in every normalized two-non-deterministic model of T_2*;
   (iii) α *is valid in every two-non-deterministic model of T_2.*

PROOF.
(i) $\rightarrow$ (iii)   by the Adequacy Theorem 5.3 and Lemma 10.2.
(iii) $\rightarrow$ (ii) obvious.
To prove the theorem it is sufficient to verify that (ii) implies (i).
Suppose (ii) and non $T_2 \vdash \alpha$. Hence, by Lemma 7.1, $\| \sim \alpha \| \neq \mathbf{0}$, and from Lemma 7.3 we can construct a Q-filter $\mathscr{F}$ in the Lindenbaum algebra of that theory such that $\| \sim \alpha \| \in \mathscr{F}$. By Lemma 10.5, for the canonical structure $\mathfrak{M}_0$ of T_2 the following condition holds: $\mathfrak{M}_0, \mathscr{F} \models \sim \alpha$, in contradiction to (ii), since $\mathfrak{M}_0$ is a normalized two-non-deterministic model of T_2. Thus (ii) $\rightarrow$ (i), and Theorem 10.7 holds. $\square$

At the beginning of this section it was proved that two-non-deterministic structures can be characterized by formulas in algorithmic propositional language. It would be interesting to know whether the language assumed here allows us to characterize m-non-deterministic structures.

By an *m-non-deterministic structure* we shall mean a semantic structure $\mathfrak{M} = \langle S, \mathscr{I}, w \rangle$ such that $\mathrm{card}(K_{\mathfrak{M}}(s)) \leqslant m$ for all $s \in S$ and $K \in V_p$.

The following lemma provides an answer to our problem. For each natural number m there is a set of formulas Z in the propositional algorithmic language which satisfy the following condition: for every normalized semantic structure $\mathfrak{M}$, $\mathfrak{M} \models Z$ iff $\mathfrak{M}$ is m-non-deterministic.

LEMMA 10.8. *Let m be a fixed natural number and let Ax_m be the set of all formulas of the following form*:

$$\bigcap_{i=0}^{m-1} \Diamond K(\alpha_1^{n^i} \wedge \ldots \wedge \alpha_k^{n^i_k}) \Rightarrow \square K \left(\bigcup_{i=0}^{m-1} (\alpha_1^{n^i} \wedge \ldots \wedge \alpha_k^{n^i_k}) \right)$$

where $k = [\log m] + 1$, $(n^i_1 \ldots n^i_k)$ is a binary representation of i, $K \in V_p$, $\alpha_j \in F$, $j \leqslant k$, α^0 denotes α and α^1 denotes $\sim \alpha$. Then the following conditions hold:

    (i) *if $\mathfrak{M} \models Ax_m$ and $\mathfrak{M}$ is normalized, then $\mathfrak{M}$ is an m-non-deterministic structure;*

    (ii) *if $\mathfrak{M}$ is an m-non-deterministic structure, then*
$$\mathfrak{M} \models Ax_m.$$

PROOF. Let $\mathfrak{M} = \langle S, \mathscr{I}, w \rangle$ be a normalized structure and $\mathfrak{M} \models Ax_m$. Assume that the theorem does not hold, i.e. for some K and s, $\text{card}(K_{\mathfrak{M}}(s)) > m$. Let $s_0, \ldots, s_{m-1}, s_m$ be elements of $K_{\mathfrak{M}}(s)$. Since $\mathfrak{M}$ is normalized, there are formulas which distinguish these states. Let $k = [\log(m+1)] + 1$. Let $\alpha_1, \ldots, \alpha_k$ be formulas such that for any two states s_i, s_j there is a formula α_{ij} which is satisfied by s_i and is not satisfied by s_j. Let

$$\mathfrak{M}, s_0 \models (\alpha_1 \wedge \ldots \wedge \alpha_k),$$
$$\mathfrak{M}, s_1 \models (\alpha_1 \wedge \ldots \wedge \sim \alpha_k),$$
$$\mathfrak{M}, s_2 \models (\alpha_1 \wedge \ldots \wedge \sim \alpha_{k-1} \wedge \alpha_h) \ldots \text{etc.}$$

The state s_m does not satisfy any of the first m conjunctions. Thus $\bigcap_{i=0}^{m-1} \lozenge K(\alpha_1^{n^i_1} \wedge \ldots \wedge \alpha_k^{n^i_k})$ is satisfied in s and s_m does not satisfy $\bigcup_{i=0}^{m-1} (\alpha_1^{n^i_1} \wedge \ldots \wedge \alpha_k^{n^i_k})$, in contradiction to $\mathfrak{M} \models Ax_m$. This proves the first part of Lemma 10.8.

We now assume that $\mathfrak{M}$ is an m-non-deterministic structure. Thus for all $s \in S$

$$\text{card}(K_{\mathfrak{M}}(s)) \leqslant m.$$

If $\text{card}(K_{\mathfrak{M}}(s)) < m$ for some s, then the antecedents in the formulas from Ax_m are not satisfied by s. Hence Ax_m is valid in s.

Suppose that $\text{card}(K_{\mathfrak{M}}(s)) = m$ and for the formula

$$\delta_i = (\alpha_1^{n^i_1} \wedge \ldots \wedge \alpha_k^{n^i_k})$$

we have

$$\mathfrak{M}, s \models \bigcap_{i=0}^{m-1} \lozenge K \delta_i.$$

No two formulas δ_i, δ_j, $i \neq j$, can be satisfied by the same state. Hence for every $s' \in K_{\mathfrak{M}}(s)$ there exists i, such that $\mathfrak{M}, s' \models \delta_i$. Thus $\square K \bigcup_{i=0}^{m-1} \delta_i$ holds in s. Hence $\mathfrak{M} \models Ax_m$. $\qquad\square$

Note that we can also describe the strict degree of non-determinism by a set of formulas.

Assume that, for a program variable K,

$$X = \{\Box K \text{ true}, (\bigcap_{i=0}^{m-1} \Diamond K \delta_i \equiv \Box K \bigcup_{i=0}^{m-1} \delta_i)\},$$

where δ_i is the same as in the lemma above. Then for any normalized structure $\mathfrak{M} = \langle S, \mathscr{I}, w \rangle$

$$\mathfrak{M} \models X \quad \text{iff} \quad (\forall s \in S) \ \text{card}\big(\mathscr{I}(K)(s)\big) = m.$$

Consider the algorithmic logic PAL_m which arises from PAL by adding a scheme of axioms Ax_m. From the Adequacy Lemma 5.5 and Lemma 10.8 this logic is adequate.

LEMMA 10.9. *If α is a theorem of algorithmic theory based on PAL_m, then α is valid in every m-non-deterministic model of T_m.*   □

Adapting the procedure described in §§ 9 and 10, we can generalize the theorems obtained previously.

I. The canonical structure of T_m is a normalized m-non-deterministic model.

II. T_m is consistent iff T_m possesses a model.

III. For any consistent theory T_m the following conditions are equivalent:

$T_m \vdash \alpha$;

$\mathfrak{M} \models \alpha$ for all normalized m-non-deterministic models $\mathfrak{M}$ of T_m;

$\mathfrak{M} \models \alpha$ for all m-non-deterministic models $\mathfrak{M}$ of T_m.

We can also consider a mixed system such that from the point of view of one variable it is m-non-deterministic and from the point of view of another variable it is n-non-deterministic.

Let $\bar{m} = (m_1, m_2, \ldots)$ be an infinite sequence of natural numbers and let us assume that $K_1, K_2, \ldots$ is the sequence of all program variables in the algorithmic language L_0. We shall say that $\mathfrak{M} = \langle S, \mathscr{I}, w \rangle$ is an $\bar{m}$-*non-deterministic structure* if for every program variable K_i and for every $s \in S$, $\text{card}\big(\mathscr{I}(K_i)(s)\big) \leqslant m_i$.

We shall consider the propositional algorithmic logic $\text{PAL}_{\bar{m}}$ as an extension of PAL by the set of schemes

$$Ax_{m_i}(K_i) \quad \text{for } i = 1, 2, \ldots$$

Namely, for every program variable K_i which is m_i-non-deterministic we shall assume one scheme of axioms $Ax_{m_i}(K_i)$. It is obvious that properties I, II, III hold for any theory $T_{\vec{m}}$ based on $PAL_{\vec{m}}$.

I. $T_{\vec{m}}$ is consistent iff $T_{\vec{m}}$ possesses a model,

II. The canonical structure of $T_{\vec{m}}$ is a normalized $\vec{m}$-non-deterministic model,

III. For every consistent theory $T_{\vec{m}}$ the following holds

$$T_{\vec{m}} \vdash \alpha \quad \text{iff} \quad T_{\vec{m}} \models \alpha. \qquad \square$$

11. ELIMINATION OF BOUNDED NON-DETERMINISTIC PROGRAM VARIABLES

We shall prove that non-deterministic program variables can be eliminated by deterministic ones. For example, if K is a program variable satisfying axiom Ax_2

$$\big(\Diamond K(\alpha \wedge \beta) \wedge \Diamond K(\alpha \wedge \sim\beta) \Rightarrow \Box K\alpha\big),$$

then we can replace K by the non-deterministic program

either K_1 or K_2 ro

with two program variables K_1, K_2 which satisfy the axioms

$$(\Diamond K_1 \alpha \Rightarrow \Box K_1 \alpha), \quad (\Diamond K_2 \alpha \Rightarrow \Box K_2 \alpha).$$

In this way each m-non-deterministic theory T_m can be transformed to a partial function theory T_{pf}.

We shall construct a mapping which assigns formula α' in T_{pf} to every formula α in T_m with the following property

$$T_m \vdash_m \alpha \quad \text{iff} \quad T_{pf} \vdash_{pf} \alpha'.$$

Let T_m be a fixed consistent m-non-deterministic theory

$$T_m = \langle L_0, C, A \rangle.$$

To every program variable $K \in V_p$ of the language L_0 let us assign m program variables $K_1, K_2, \ldots, K_m$ which do not belong to L_0.

The propositional algorithmic language based on the set of program variables $V_p' = \{K_1, \ldots, K_m\}_{K \in V_p}$ and the same set of propositional variables will be denoted by L_0'.

Let α be a formula and let M be a program scheme in the language L_0. We shall write α' to denote a formula in L_0' and M' to denote a program scheme in L_0' which are the results of simultaneous replacement of all occurrences of K in α and in M by a program scheme of the form

> **either** K_1 **or**
>> **either** K_2 **or**
>>> ...
>>> **either** K_{m-1} **or** K_m **ro**
>> **ro**
>> ...
> **ro.**

For short, $(K_1$ **or** K_2 **or** ... **or** $K_m)$.

Let $\mathfrak{M}$ be a semantic m-non-deterministic structure

$$\mathfrak{M} = \langle S, \mathscr{I}, w \rangle$$

for the language L_0. We shall construct a new structure $\mathfrak{M}'$—a partial function structure —for the language L_0' in the following way:

$$\mathfrak{M}' = \langle S, \mathscr{I}', w \rangle:$$

If $\mathscr{I}(K) = \{(s, s_i): i = 1, ..., k\}$, then

$$\mathscr{I}'(K_j) = \{(s, s_j)\} \quad \text{for } j = 1, ..., k,$$

$$\mathscr{I}'(K_j) = \{(s, s_k)\} \quad \text{for } m \geqslant j > k.$$

If $\mathscr{I}(K)(s) = \varnothing$, then $\mathscr{I}'(K_j)(s) = \varnothing$ for $j = 1, ..., m$. This transformation is illustrated in Figure 11.1.

Let $m = 3$.

It is obvious from the definition that

$$K_{\mathfrak{M}}(s) = (K_1 \text{ or } ... \text{ or } K_m)_{\mathfrak{M}}(s)$$

for all $K \in V_p$ and every state s. This equality can be generalized to any program scheme. Let us first note that every computation $\mathcal{O}$ of the program scheme M in the structure $\mathfrak{M}$ at the initial state s can be transformed into a computation $\mathcal{O}'$ of the program M' in the structure $\mathfrak{M}'$ at the same initial state. This transformation is described as follows:

1. Let us put the program

$$(K_1 \text{ or } ... \text{ or } K_n)$$

in the place of K.

The structure $\mathfrak{M}$

The structure $\mathfrak{M}'$

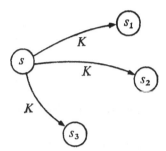

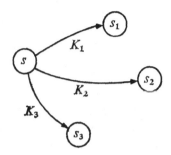

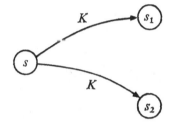

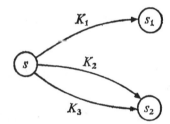

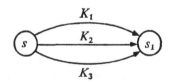

Fig. 11.1

2. Let us replace any two configurations of the form

$$\langle s; K, \text{Rest} \rangle,$$
$$\langle s_i; \text{Rest} \rangle,$$

by the following sequence of configurations:

$\langle s; (K_1 \text{ or } ... \text{ or } K_n), \text{ Rest} \rangle,$

$\langle s; (K_2 \text{ or } ... \text{ or } K_n), \text{ Rest} \rangle,$

. . . .

$\langle s; (K_i \text{ or } ... \text{ or } K_n), \text{ Rest} \rangle,$

$\langle s; K_i, \text{ Rest} \rangle,$

$\langle s_i; \text{ Rest} \rangle.$

3. The sequence obtained in this way is a computation $\mathcal{O}'$ of the program M' in the structure $\mathfrak{M}'$.

The converse transformation is obviously also possible. Moreover:

(i) The computation $\mathcal{O}$ is infinite iff the computation $\mathcal{O}'$ is infinite.

(ii) $\bar{s}$ is the result of $\mathcal{O}$ iff $\bar{s}$ is the result of $\mathcal{O}'$.

COROLLARY 11.1. *For every program scheme M and every formula* α

$$M^{\mathfrak{M}}(s) = M'_{\mathfrak{M}}(s),$$

$$\alpha_{\mathfrak{M}}(s) = \alpha'_{\mathfrak{M}}(s). \qquad \square$$

LEMMA 11.2. *A theory* $T_m = \langle L_0, C_m, A \rangle$ *has a model iff the theory* $T_{pf} = \langle L'_0, C_{pf}, A' \rangle$ *has a model.*

PROOF. Let $\mathfrak{M}$ be a model of T_m. By the corollary, $\mathfrak{M}'$ is a model of T_{pf}. Conversely, if $\mathfrak{M}'$ is a model of T_{pf} then the structure $\mathfrak{M} = \langle S, \mathcal{I}, w \rangle$, where $\mathcal{I}(K)(s) = \bigcup_{i=1}^{m} K_{i\mathfrak{M}'}(s)$, $K \in V_p$, $s \in S$, $K_i \in V'_p$, is a model of T_m. $\qquad \square$

From the Model Existence Theorem for T_{pf} and T_m we infer that an m-non-deterministic theory $T_m = \langle L_0, C_m, A \rangle$ is consistent iff the corresponding partial function theory $T_{pf} = \langle L_0, C_{pf}, A' \rangle$ is consistent.

Analogously, by the Completeness Theorems for T_{pf} and T_m we have the following theorem:

THEOREM 11.3. *For every formula* $\alpha \in L_0$, α *is a theorem of the m-non--deterministic theory* $T_m = \langle L_0, C_m, A \rangle$ *iff* α' *is a theorem of the partial function theory* $T_{p.} = \langle L'_0, C_{pf}, A' \rangle$. $\qquad \square$

12. YANOV SCHEMES

The original language of Yanov schemes (cf. Yanov, 1959) is different from that used here. We shall adapt the orthography of Yanov schemes to the syntactical patterns of this chapter.

Let us assume the following definition:

By a *Yanov scheme* we shall mean a program scheme in a propositional language for which every program variable has an associated carrier which is fixed and finite. (From this definition we can associate a finite carrier to every program scheme).

A natural interpretation of a Yanov scheme consists of a relation K_Y associated with every program variable K such that for every two valuations v, v' of propositional variables

$$(v, v') \in K_Y \quad \text{iff} \quad v = v' \text{ off } \text{Car}(K),$$

where $\text{Car}(K)$ is the carrier of K. Let us call this interpretation the *Yanov interpretation*.

By a *computation of a Yanov scheme* M with a given valuation v we shall understand a maximal sequence of configurations $c_0, c_1, c_2, \ldots$ such that $c_i \underset{Y}{\mapsto} c_{i+1}$ and $c_0 = \langle v; M \rangle$. The relation $\underset{Y}{\mapsto}$ is defined as in § 1 of this chapter. Let us mention here one step of this definition:

$$\langle v; K, \text{ Rest} \rangle \underset{Y}{\mapsto} \langle v'; \text{ Rest} \rangle,$$

where $K \in V_p$ and $(v, v') \in K_Y$.

Although all program variables are interpreted in a similar fashion (every program can change its variables in any possible way), their carriers may differ and this is why we cannot treat a Yanov scheme as an algorithm with a single program variable.

REMARK. There is a natural correspondence between Yanov scheme and non-deterministic programs (cf. Chapter VI).

For a fixed program variable K, let $\text{Car}(K) = \{q_1, \ldots, q_n\}$. The set of all possible valuations of these variables has 2^n elements. Let us consider the corresponding set of sequences of atomic formulas **true** and **false**, i.e. the set $\{(i_1^j, \ldots, i_n^j)\}_{j \leqslant 2^n}$ where $i_k^j \in \{\text{true}, \text{false}\}$ for $j \leqslant 2^n$ and $k \leqslant n$.

For a given sequence $(i_1^j, \ldots, i_n^j)$ let M_j denote the program

$$\textbf{begin } q_1 := i_1^j; \ldots; q_n := i_n^j \textbf{ end.}$$

Let M be a non-deterministic program of the following form

$$(M_1 \text{ or } M_2 \text{ or } \ldots \text{ or } M_{2^n})$$

It is easy to see that the sets of all results of M and K are equal for any given valuation. In conclusion we have the following result.

For every Yanov scheme M we can construct a non-deterministic program M' (with assignment instructions and without program variables) such that the behaviour of M and M' will be the same, i.e. the trees of the possible computations will be equal.   □

We can consider a logic of Yanov schemes as a propositional algorithmic theory with the set of specific axioms Yax, i.e. the set of all formulas of the forms:

$$\Box K \text{ true},$$

$$(\Diamond Kq \Rightarrow q), \quad (\Diamond K{\sim}q \Rightarrow {\sim}q) \quad \text{for all } q \notin \mathrm{Car}(K),$$

$$\bigcap_{j=0}^{2^n-1} \Diamond K(q_1^{m_1^j} \wedge \ldots \wedge q_n^{m_n^j}),$$

where $(m_1^j, \ldots, m_n^j)$ is a binary representation of the number j and q_i^0 is q and q_i^1 is q_i for $i \leqslant n$ and K is a program variable such that $\mathrm{Car}(K) = \{q_1, \ldots, q_n\}$.

We shall now prove the following lemma:

LEMMA 12.1.

(i) *Every semantic proper structure with a Yanov interpretation is a model of Yax.*

(ii) *If $\mathscr{I}$ is an interpretation of program variables such that $\mathfrak{M} = \langle W, \mathscr{I} \rangle$ is a proper model for Yax, then $\mathscr{I}$ is a Yanov interpretation.*

PROOF.

(i) Let us consider a proper semantic structure $\mathfrak{M} = \langle W, Y \rangle$, where Y is a Yanov interpretation of program variables.

The first two axioms are valid since by the definition of K_Y, $\mathfrak{M}, v \models q$ iff $\mathfrak{M}, v' \models q$ for all $q \notin \mathrm{Car}(K)$, all $v' \in K_Y(v)$ and for every valuation $v \in W$. The third axiom is also valid for every $v \in W$ since all possible changes of the values of variables from $\mathrm{Car}(K)$ are admissible as a result of K_Y.

(ii) Suppose $\mathfrak{M} = \langle W, \mathscr{I} \rangle$ is a model of Yax. If $(v, v') \in K_{\mathfrak{M}}$, then $v = v'$ off $\mathrm{Car}(K)$ since for $q \in \mathrm{Car}(K)$ $v'(q) = 1$ implies by the second axiom $v(q) = 1$ and $v'(q) = 0$ implies $v(q) = 0$.

Conversely, let us assume that $v = v'$ off $\mathrm{Car}(K)$ and let $q_1, ..., q_k$ be all variables from the set $\mathrm{Car}(K)$ such that $q_i(v') = 0$ for $i \leqslant k$. Consider the formula β of the form

$$(\sim q_1 \wedge \sim q_2 \wedge ... \wedge \sim q_k \wedge q_{k+1} \wedge ... \wedge q_n).$$

By the third axiom $\mathfrak{M}, v \models \Diamond K\beta$. Hence there exists v'' such that $v'' \in K_\mathfrak{M}(v)$ and $v'' \models \beta$, i.e. for all $i, v''(q_i) = v(q_i)$. By the previously proved implication

$$v'' = v \text{ off } \mathrm{Car}(K).$$

Thus $v'' = v'$ and therefore $v' \in K_\mathfrak{M}(v)$.   □

13. APPLICATION OF PAL IN MICROPROGRAMMING

Propositional algorithmic logic seems well-suited to the analysis of microprograms. In this section we present a small example of a microprogram and its transformation to another, more efficient microprogram which performs the same operation of multiplication of integers. We shall work within the frame of a theory of registers defined later. Before we present it let us recall the structure of a simple arithmetic unit. It will serve as a basis for future intuitions.

The unit consists of four registers and an adder, as shown in Figure 13.1.

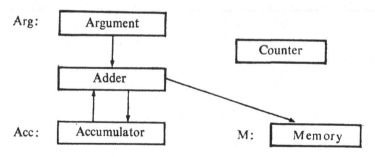

Fig. 13.1

The microoperations of the unit are

$$\mathrm{Acc} := \mathrm{Acc} + \mathrm{Arg},$$

shift Acc and M to left (to right), check whether the last bit of M is 0, check if Acc and M contains only zeros, subtract 1 from the counter, add 1 to the counter, test if counter contains 0, etc.

This physical model gives us an insight into the formal theory presented below. We shall imagine a collection of registers. Each register can contain an infinite sequence of bits (binary digits)

$$... d_3 d_2 d_1 d_0 d_{-1} d_{-2} d_{-3} ...$$

The set of program variables will represent microoperations on registers; for every i, j we have the following program variables:

a_{ij}—add the content of the register R_i to the register R_j,
$\qquad R_j := R_i + R_j$,
l_i —shift R_i to the left, $R_i := 2 \times R_i$,
r_i —shift R_i to the right, $R_i := R_i/2$,
o_i —put 0 into R_i, $R_i := 0$,
s_i —add 1 to R_i, $R_i := R_i + 1$,
p_i —subtract 1 from R_i, $R_i := R_i - 1$.

For every i we have the two propositional variables:

z_i—check if R_i contains only zeros,
e_i—check if R_i contains zeros on all non-positive positions.

The schemes of programs can in these circumstances be interpreted as microprograms. The algorithmic formulas need not contain the modality signs $\square$ and $\lozenge$ since we assume that the actions are deterministic, i.e. instead of $\square M\alpha$ or $\lozenge M\alpha$ we shall simply write $M\alpha$. We shall operate with axioms having the form of equalities of microprograms. The equality $M = M'$ should be conceived as the scheme $M\alpha \equiv M'\alpha$ for every formula α.

Below we present the schemes of axioms of our theory T of registers:

$$l_i a_{ij} = a_{ij}^2 l_i, \quad a_{ij}^2 r_j = r_j a_{ij}, \quad s_i p_i = p_i s_i = \mathrm{Id},$$
$$a_{ij} l_j = l_j a_{ij}^2, \quad a_{ij} r_i = r_i a_{ij}^2, \quad l_i r_i = r_i l_i,$$
$$l_i s_i^2 = s_i l_i, \quad l_i p_i^2 = p_i l_i,$$
$$r_i s_i = s_i^2 r_i, \quad p_i^2 r_i = r_i p_i,$$
$$r_i o_i = l_i o_i = s_i o_i = p_i o_i = o_i = a_{ji} o_i,$$
$$l_i z_i \equiv z_i \equiv r_i z_i,$$
$$\textbf{while } \sim z_i \textbf{ do } p_i \textbf{ od true} \Rightarrow \sim s_i z_i,$$
$$e_i \equiv \textbf{while } \sim z_i \textbf{ do } p_i^2 \textbf{ od true}.$$

Moreover, we assume that for different indices the operations commute, e.g.

$$a_{ij}p_k = p_k a_{ij} \quad \text{for } k \neq i \text{ and } k \neq j,$$
$$p_i l_j = l_j p_i \quad \text{for } i \neq j.$$

LEMMA 13.1. *The following formulas are provable in the theory T of registers*:

(1)   (**while** $\sim z_i$ **do** p_i **od true** $\Rightarrow s_i$ **while** $\sim z_i$ **do** p_i **od true**),

(2) $(e_i \Rightarrow \sim p_i e_i)$.

PROOF. By axioms

$$\text{(\textbf{while} } \sim z_i \text{ \textbf{do} } p_i \text{ \textbf{od true} } \Rightarrow s_i \sim z_i),$$
$$s_i p_i = \text{Id}$$

and, by classical propositional calculus, we have

$$\text{(\textbf{while} } \sim z_i \text{ \textbf{do} } p_i \text{ \textbf{od true} } \Rightarrow (s_i \sim z_i \wedge s_i p_i \text{ \textbf{while} } \sim z_i \text{ \textbf{do} } p_i$$
$$\text{\textbf{od true}})).$$

By axioms of PAL (cf. Chapter V, § 5)

$$((\sim s_i z_i \wedge s_i p_i \text{ \textbf{while} } \sim z_i \text{ \textbf{do} } p_i \text{ \textbf{od true}}) \Rightarrow s_i \text{ \textbf{while} } \sim z_i \text{ \textbf{do} } p_i$$
$$\text{\textbf{od true}}).$$

Hence by law of syllogism we have proved formula (1).

To prove the second implication (2) observe first that for arbitrary $j > 0$, $(z_i \Rightarrow s_i^j \sim z_i)$ is the consequence of (1) and of axiom

$$\text{(\textbf{while} } \sim z_i \text{ \textbf{do} } p_i \text{ \textbf{od true} } \Rightarrow s_i \sim z_i).$$

Hence using the following rule of inference

$$\frac{\alpha \Rightarrow \beta}{p_i^k \alpha \Rightarrow p_i^k \beta}$$

and axiom $s_i p_i = \text{Id}$ of theory T we obtain

$$(p_i^k z_i \Rightarrow \sim p_i^l z_i) \quad \text{for arbitrary natural numbers } k \neq l.$$

Hence for arbitrary k and l

$$\left(p_i \text{ (\textbf{if} } \sim z_i \text{ \textbf{then} } p_i^2 \text{ \textbf{fi}})^k z_i \Rightarrow \sim(\text{\textbf{if} } \sim z_i \text{ \textbf{then} } p_i^2 \text{ \textbf{fi}})^l z_i \right).$$

By ω-rule of PAL we obtain that for every $l \in N$,

$$((\text{\textbf{if} } \sim z_i \text{ \textbf{then} } p_i^2 \text{ \textbf{fi}})^l z_i \Rightarrow \sim p_i \text{ \textbf{while} } \sim z_i \text{ \textbf{do} } p_i^2 \text{ \textbf{od true}}).$$

Using once again ω-rule we obtain formula (2).   □

Let us consider the following microprogram M performing multiplication of registers R_1 and R_3, assuming that R_3 contains a natural number. The result of computation is placed in R_2:

$$M: \textbf{begin } o_2 \text{ ; } \textbf{while } \sim z_3 \textbf{ do } a_{12} \text{ ; } p_3 \textbf{ od}; \text{ } o_1 \text{ ; } o_3 \textbf{ end.}$$

The aim of this section is to improve the above program.

Assume the following denotations:

$$K_1 = \textbf{if } \sim z_3 \textbf{ then } a_{12} \text{ ; } p_3 \textbf{ fi,}$$
$$K_2 = \textbf{if } \sim z_3 \textbf{ then } a_{12}^2 \text{ ; } p_3^2 \textbf{ fi,}$$
$$K = \textbf{if } \sim e_3 \textbf{ then } a_{12} \text{ ; } p_3 \textbf{ fi.}$$

LEMMA 13.2. *For an arbitrary formula β and for an arbitrary natural number j the following formula is provable in T:*

(3) $\qquad \left(K_1^j(z_3 \wedge o_1 o_3 \beta) \equiv \textbf{if } \sim z_3 \textbf{ then } K; l_1 \text{ ; } r_3 \textbf{ fi } K_1^{\lfloor j/2 \rfloor}(z_3 \wedge o_1 o_3 \beta)\right),$

where $\lfloor 2j/2 \rfloor = \lfloor (2j+1)/2 \rfloor = j$.

PROOF. By axiom

$$\textbf{if } \sim z_3 \textbf{ then } K_1 \textbf{ else } \text{Id } \textbf{fi } \beta \equiv (z_3 \wedge K_1 \beta \vee \sim z_3 \wedge \beta),$$

we have

$$K_1^j(z_3 \wedge o_1 o_3 \beta)$$
$$\equiv \left(\sim z_3 \wedge a_{12} p_3 K_2^{\lfloor j/2 \rfloor}(z_3 \wedge o_1 o_3 \beta) \vee K_2^{\lfloor j/2 \rfloor}(z_3 \wedge o_1 o_3 \beta)\right).$$

Let us multiply the right hand-side of the above formula by $(e_3 \vee \sim e_3)$. Applying (2) and the following simple facts:

$$(\sim z_3 \wedge \sim e_3) \equiv \sim e_3,$$
$$((\textbf{if } \sim z_3 \textbf{ then } a_{12}^2 \text{ ; } p_3^2 \textbf{ fi})^{\lfloor j/2 \rfloor}(z_3 \wedge o_1 o_3 \beta) \Rightarrow e_3),$$
$$(a_{12} p_3(\textbf{if } \sim z_3 \textbf{ then } a_{12}^2 \text{ ; } p_3^2 \textbf{ fi})^{\lfloor j/2 \rfloor}(z_3 \wedge o_1 o_3 \beta) \Rightarrow \sim e_3)$$

we obtain

(4) $\qquad K_1^j(z_3 \wedge o_1 o_3 \beta) \equiv \textbf{if } \sim e_3 \textbf{ then } a_{12}; p_3 \textbf{ fi } K_2^{\lfloor j/2 \rfloor}(z_3 \wedge o_1 o_3 \beta).$

By axioms of the theory T of registers

$$(z_3 \wedge o_1 o_3 \beta) \equiv l_1 r_3(z_3 \wedge o_1 o_3 \beta)$$

and

$$a_{12}^2 p_3^2 l_1 r_3 \beta \equiv l_1 r_3 a_{12} p_3 \beta$$

for arbitrary formula β. Hence

$$K_2^{\lfloor j/2 \rfloor} l_1 r_3 \beta \equiv l_1 r_3 K_1^{\lfloor j/2 \rfloor} \beta.$$

Applying the obtained equivalence to the formula (4) we have

$$K_1^j (z_3 \wedge o_1 o_3 \beta)$$
$$\equiv \text{if} \sim e_3 \text{ then } a_{12}; p_3 \text{ fi } l_1 r_3 K_1^{\lfloor j/2 \rfloor} (z_3 \wedge o_1 o_3 \beta).$$

However

$$\left(z_3 \wedge K l_1 r_3 K_1^{\lfloor j/2 \rfloor} (z_3 \wedge o_1 o_3 \beta) \right) \equiv K_1^{\lfloor j/2 \rfloor} (z_3 \wedge o_1 o_3 \beta).$$

Thus

$$K_1^j (z_3 \wedge o_1 o_3 \beta) \equiv \text{if} \sim z_3 \text{ then } K; l_1; r_3 \text{ fi } K_1^{\lfloor j/2 \rfloor} (z_3 \wedge o_1 o_3 \beta).$$

$$\square$$

Let j be a natural number and let $2^{k-1} \leqslant j < 2^k$ for some $k \in N$, i.e. $\lfloor \log j \rfloor = k - 1$. By Lemma 13.2 and simple induction on l, $1 \leqslant l \leqslant k$, we have

$$K_1^j (z_3 \wedge o_1 o_3 \beta) \equiv (\text{if} \sim z_3 \text{ then } K; l_1; r_3 \text{ fi})^l K_1^{\lfloor j/2^l \rfloor} (z_3 \wedge o_1 o_3 \beta).$$

Hence for $l = k$, i.e. for $l = \lfloor \log j \rfloor + 1$

$$(5) \qquad K_1^j (z_3 \wedge o_1 o_3 \beta) \equiv (\text{if} \sim z_3 \text{ then } K; l_1; r_3 \text{ fi})^{\lfloor \log j \rfloor + 1} (z_3 \wedge o_1 o_3 \beta).$$

Applying twice the ω-rule of algorithmic logic we obtain

$$M \beta \equiv o_2 \, (\text{while} \sim z_3 \text{ do } K; l_1; r_3 \text{ od } (o_1 o_3 \beta)).$$

The final conclusion is that program M is equivalent to the following program;

```
begin
    o₂;
    while ~z₃ do
        if ~e₃ then a₁₂; p₃ fi;
    l₁; r₃
    od;
    o₁; o₃
end.
```

It is not difficult to observe by (5) that the complexity of the last microprogram is much better than the original one (the obtained microprogram is frequently implemented in computers). It requires $\lfloor \log R_3 \rfloor + 1$ steps in comparison with the R_3 steps of the original algorithm.

BIBLIOGRAPHIC REMARKS

The first result in propositional logic of programs belongs to Yanov (1959), who proved that the equivalence of program schemes is decidable. Many papers devoted to schematology have developed Yanov's ideas; it is impossible to quote all of them. The next important step was when Glushkov (1965) introduced algorithmic algebras. The same ideas and many new results were proposed by Fisher and Ladner (1979) in their paper introducing PDL—a propositional dynamic logic. Since 1977, when this paper appeared, many authors have studied the propositional logics of programs: Segerberg, Gabbay, Chlebus, Berman, Parikh, Kozen, Harel, Meyer, Valiev, Vakarelov, Passy, Mirkowska, Pratt. This list does not exhaust the names of all contributors to the field.

The results reported in this chapter are mainly from Mirkowska (1981) except for Section 13 which is based on an example from Glushkov *et al.* (1978).

CHAPTER VI

NON-DETERMINISM IN ALGORITHMIC LOGIC

In this chapter we shall deal with non-deterministic **while**-programs. Among many reasons for introducing non-determinism let us mention concurrency, whose semantics requires some non-deterministic actions. We shall study the semantic properties of non-deterministic programs, and also the non-deterministic logic NAL. The basis of our considerations is the algorithmic logic of deterministic **while**-programs and the propositional algorithmic logic PAL. In fact, every propositional tautology of PAL is a scheme of a tautology of non-deterministic algorithmic logic. On the other hand, NAL is a natural extension of algorithmic logic.

In contrast to the deterministic case, a non-deterministic program can have various computations. Thus we shall interpret a program as a tree in which every path represents one way of going through the program during the evaluation of its result. Hence a non-deterministic program can have many different results. We are therefore obliged to change our intuition connected with the algorithmic formula $K\alpha$.

There are two natural interpretations: to consider all results of all computations, or to consider a particular result.

Both interpretations are worthy of investigation. For this reason we shall introduce two modal constructions to the set of formulas $\Diamond M\alpha$ and $\Box M\alpha$, where M is a non-deterministic program and α is a formula. The informal meaning is as follows:

$\Diamond M\alpha$—it is possible that after performing M the formula α holds,

$\Box M\alpha$—it is necessary that after performing M the formula α holds. (We have already met these constructions in PAL.)

Formulas of this kind can easily express properties of programs like termination, correctness, etc., and properties of data structures.

In thic chapter we shall present a Hilbert-style axiomatization; it is also possible to construct a Gentzen-type axiomatization. The logic presented, NAL, is complete in the sense that the semantic and syntactic consequence operations determine the same sets of consequences.

However, the axiomatization has an infinitary character, since, following the arguments presented for AL (see Chapter II, § 4), we can prove that the semantic consequence is not compact.

1. NON-DETERMINISTIC **while**-PROGRAMS AND THEIR SEMANTICS

Let us assume that we are given a fixed alphabet in which V is a set of individual and propositional variables, P is a set of predicates, and Φ is a set of functors. On the basis of this alphabet we are going to construct a non-deterministic algorithmic language L and in particular the most important element of L—the notion of a non-deterministic program.

DEFINITION 1.1. *By a non-deterministic program we shall mean any expression M such that*:

(i) *M is an assignment instruction, $(x := \tau)$ or $(q := \gamma)$, where x is an individual variable, q is a propositional variable, τ is a term and γ is an open formula (for the notion of term or open formula see Chapter* II, § 1), *or*

(ii) *M is of the form* if γ then M_1 else M_2 fi, begin M_1 ; M_2 end, while γ do M_1 od, *where M_1, M_2 are arbitrary non-deterministic programs and γ is an open formula, or*

(iii) *M is of the form* either M_1 or M_2 ro, *where M_1, M_2 are arbitrary non-deterministic programs.*   □

Hence the set of all non-deterministic **while**-programs is an extension of the set of deterministic programs defined in Chapter II, § 1. We shall denote this set by Π.

EXAMPLE. Let empty be a one-argument predicate and let left and right be one-argument functors. The following expression is then an example of a non-deterministic program:

> while ~empty(x) do
>> either $(x := \text{left}(x))$ or $(x := \text{right}(x))$ ro
>
> od.   □

Let $\mathfrak{A}$ be a data structure

$$\mathfrak{A} = \langle A, \{\psi_{\mathfrak{A}}\}_{\psi \in \Phi}, \{\varrho_{\mathfrak{A}}\}_{\varrho \in P} \rangle$$

in which, for every n-argument predicate ϱ, $\varrho_{\mathfrak{A}}$ is an n-argument relation in A and, for every n-argument functor ψ, $\psi_{\mathfrak{A}}$ is an n-argument operation in A.

The given data structure $\mathfrak{A}$ determines the interpretation of open formulas and terms as defined in Chapter II, § 2. The interpretation of non-deterministic programs will be defined in a way similar to that presented in PAL (cf. Chapter V, § 1).

DEFINITION 1.2. *By a tree of possible computations of a program M in the structure $\mathfrak{A}$ from the initial valuation v we mean a tree* $\mathrm{Comp}(M, v, \mathfrak{A})$ *such that the configuration* $\langle v; M \rangle$ *is the root of the tree and*:

(i) *If a configuration* $\langle v'; \mathbf{if} \ \gamma \ \mathbf{then} \ M_1 \ \mathbf{else} \ M_2 \ \mathbf{fi}, \mathrm{Rest} \rangle$ *is a vertex of* Comp, *then the unique son of this vertex is* $\langle v'; M_1, \mathrm{Rest} \rangle$ *in the case* $\mathfrak{A}, v' \models \gamma$ *and* $\langle v'; M_2, \mathrm{Rest} \rangle$ *in the case* $\mathfrak{A}, v' \models \sim\gamma$ (Rest *denotes a sequence of programs*).

(ii) *If the configuration* $\langle v'; \mathbf{begin} \ K; M \ \mathbf{end}, \mathrm{Rest} \rangle$ *is a vertex of the tree* Comp, *then the unique son of this vertex is* $\langle v'; K, M, \mathrm{Rest} \rangle$.

(iii) *If the configuration* $\langle v'; \mathbf{while} \ \gamma \ \mathbf{do} \ M \ \mathbf{od}, \mathrm{Rest} \rangle$ *is a vertex of* Comp, *then the unique son of this vertex is* $\langle v'; \mathrm{Rest} \rangle$ *in the case* $\mathfrak{A}, v' \models \sim\gamma$ *and is* $\langle v'; M, \mathbf{while} \ \gamma \ \mathbf{do} \ M \ \mathbf{od}, \mathrm{Rest} \rangle$ *in the case* $\mathfrak{A}, v' \models \gamma$.

(iv) *If the configuration* $\langle v'; \mathbf{either} \ M_1 \ \mathbf{or} \ M_2 \ \mathbf{ro}, \mathrm{Rest} \rangle$ *is a vertex of* Comp, *then the left son of this vertex is* $\langle v; M_1, \mathrm{Rest} \rangle$ *and the right son is* $\langle v'; M_2, \mathrm{Rest} \rangle$.

(v) *If the configuration* $\langle v'; (x := w), \mathrm{Rest} \rangle$ *is in* Comp, *then the unique son of this vertex is* $\langle v''; \mathrm{Rest} \rangle$ *where* $v''(z) = v'(z)$ *for* $z \neq x$ *and* $v''(x) = w_{\mathfrak{A}}(v')$.

(vi) *If the configuration* $\langle v'; \ \rangle$ *is a vertex of* Comp, *then it is a leaf of* D, *i.e., has no sons*.

Every path of the tree $\mathrm{Comp}(M, v, \mathfrak{A})$ *is called a computation of a program M in the structure $\mathfrak{A}$ at the initial valuation v.*

If $\langle v'; \ \rangle$ *is a leaf of the tree* Comp, *then the valuation v' is called the result of the corresponding computation.* □

LEMMA 1.1. *Let K be a program of the form* $\mathbf{while} \ \gamma \ \mathbf{do} \ M \ \mathbf{od}$. *If all computations of K at the initial valuation v in a data structure $\mathfrak{A}$ are finite, then there exists a common upper bound of the length of the computations.*

PROOF. Let Comp be a tree of all possible computations of the program K starting from the valuation v in $\mathfrak{A}$. Suppose on the contrary that for every natural number b, there exists a path in Comp of length n. Since the degree of any vertex in Comp is equal to 1 or 2, by König's Lemma (Kuratowski and Mostowski, 1967) there exists an infinite path in the tree Comp, contrary to the assumption.   □

Let us remark that the set of all finite computations of the program M determines a binary relation $M_{\mathfrak{A}}$ in the set of all valuations of a data structure $\mathfrak{A}$ such that

$$(v, v') \in M_{\mathfrak{A}} \quad \text{iff} \quad v' \text{ is a result of a computation of } M \text{ from the valuation } v \text{ in the structure } \mathfrak{A}.$$

The relation $M_{\mathfrak{A}}$ is called the *interpretation* of a program M in the structure $\mathfrak{A}$.

Hence, the interpretation of a program **begin** K; M **end** is a composition of the interpretations of K and of M; the interpretation of a program **either** K **or** M **ro** is the set-theoretical sum of the interpretations of K and M and the interpretation of **while** γ **do** M **od** in $\mathfrak{A}$ is

$$\bigcup_{i \in N} (\textbf{if } \gamma \textbf{ then } M \textbf{ fi})^{i}_{\mathfrak{A}} \circ \{(v, v): \mathfrak{A}, v \models \sim\gamma\}.$$

Let $K_{\mathfrak{A}}(v)$ denote the set of all results of the program K at the valuation v in the structure $\mathfrak{A}$, $K_{\mathfrak{A}}(v) = \{v': (v, v') \in K_{\mathfrak{A}}\}$. The following lemma gives a characterization of this set according to the structure of the program.

LEMMA 1.2. *For arbitrary programs K, M and an arbitrary valuation in a data structure $\mathfrak{A}$ the following equalities hold*:

$$(\textbf{begin } K;\ M \textbf{ end})_{\mathfrak{A}}(v) = \bigcup_{v' \in K_{\mathfrak{A}}(v)} M_{\mathfrak{A}}(v'),$$

$$(\textbf{if } \gamma \textbf{ then } K \textbf{ else } M \textbf{ fi})_{\mathfrak{A}}(v) = \begin{cases} K_{\mathfrak{A}}(v) & \text{if } \mathfrak{A},\ v \models \gamma, \\ M_{\mathfrak{A}}(v) & \text{if } \mathfrak{A},\ v \models \sim\gamma, \end{cases}$$

$$(\textbf{either } K \textbf{ or } M \textbf{ ro})_{\mathfrak{A}}(v) = K_{\mathfrak{A}}(v) \cup M_{\mathfrak{A}}(v),$$

$$(\textbf{while } \gamma \textbf{ do } M \textbf{ od})_{\mathfrak{A}}(v)$$
$$= \bigcup_{i \in N} (\textbf{if } \gamma \textbf{ then } M \textbf{ fi})^{i}_{\mathfrak{A}}(v) \cap \{v': \mathfrak{A}, v' \models \sim\gamma\}.$$

For the proof see the similar considerations which have been presented in PAL (cf. Chapter V, § 2).   □

2. PROPERTIES OF NON-DETERMINISTIC PROGRAMS

We shall begin our considerations from a description of a formalized non-deterministic algorithmic language and its semantics, since the formulas of this language will represent the properties of the programs.

DEFINITION 2.1. *By a formula of non-deterministic algorithmic language we shall understand every expression α such that:*

(i) α *is a propositional variable, or α is an elementary formula* (*cf. Chapter* II, § 1),

(ii) α *is of the form* $(\exists x)\beta(x)$, $(\forall x)\beta(x)$, *where x is an individual variable,*

(iii) α *is of the form* $(\beta \vee \delta)$, $(\beta \wedge \delta)$, $(\beta \Rightarrow \delta)$, $\sim\beta$,

(iv) α *is of the form* $\square M\beta$, $\lozenge M\beta$,

(v) α *is of the form* $\bigsqcup M\beta$, $\bigsqcap M\beta$, $\bigvee M\beta$, $\bigwedge M\beta$ (*the signs* $\bigsqcap$, $\bigsqcup$, $\bigvee$, $\bigwedge$ *will be called iteration quantifiers*), *where δ, β are arbitrary formulas and M is an arbitrary non-deterministic program.* $\square$

The set of all formulas will be denoted by F. The sets of terms, formulas, and non-deterministic programs determine the non-deterministic algorithmic language L.

We shall define below the semantics of the language under consideration.

Let $\mathfrak{A}$ be a fixed data structure for L. The semantics of non-deterministic programs has been defined in § 1 of this chapter. Hence it remains to define the semantics of formulas. However, the formulas constructed by means of the classical connectives $\wedge$, $\vee$, $\sim$, $\Rightarrow$, and quantifiers $\exists$, $\forall$ are interpreted in the usual way (see Chapter II, § 1) and therefore need not be mentioned here.

Thus for an arbitrary valuation v in the data structure $\mathfrak{A}$ we assume

$$\mathfrak{A}, v \models \lozenge M\alpha \quad \text{iff} \quad (\exists v' \in M_{\mathfrak{A}}(v))\; \mathfrak{A}, v' \models \alpha,$$

$$\mathfrak{A}, v \models \square M\alpha \quad \text{iff} \quad (\forall v' \in M_{\mathfrak{A}}(v))\; \mathfrak{A}, v' \models \alpha \text{ and all computations of } M \text{ at the valuation } v \text{ in } \mathfrak{A} \text{ are finite,}$$

$$\mathfrak{A}, v \models \bigsqcup M\alpha \quad \text{iff} \quad (\exists i \in N)\; \mathfrak{A}, v \models \square M^i\alpha,$$

$$\mathfrak{A}, v \models \bigsqcap M\alpha \quad \text{iff} \quad (\forall i \in N)\; \mathfrak{A}, v \models \square M^i\alpha,$$

$$\mathfrak{A}, v \models \bigvee M\alpha \quad \text{iff} \quad (\exists i \in N)\; \mathfrak{A}, v \models \lozenge M^i\alpha,$$

$$\mathfrak{A}, v \models \bigwedge M\alpha \quad \text{iff} \quad (\forall i \in N)\; \mathfrak{A}, v \models \lozenge M^i\alpha.$$

REMARK. If M is a deterministic program then the formulas $\Diamond M\alpha$ and $\Box M\alpha$ are equivalent. Moreover, every formula α of a non-deterministic algorithmic language in which the instruction **either—or—ro** and classical quantifiers do not occur is equivalent to an algorithmic formula α' which is obtained by replacing all subformulas of the form $\Box M\beta, \Diamond M\beta, \bigsqcup M\beta, \sqcap M\beta, \bigvee M\beta, \bigwedge M\beta$ by the corresponding expressions $M\beta, \bigcup M\beta, \bigcap M\beta$ of AL, i.e., $\mathfrak{A}, v \models \alpha$ iff $\mathfrak{A}, v \models \alpha'$ for an arbitrary data structure $\mathfrak{A}$ and valuation v. $\Box$

It follows directly from the definition of semantics that algorithmic formulas can describe the properties of computations. For example the formula $\Box M$ **true** descibes the stop property of the program M, since for an arbitrary data structure $\mathfrak{A}$ and every valuation v

$$\mathfrak{A}, v \models \Box M \text{ true} \quad \text{iff} \quad \text{all computations of the program } M$$
$$\text{at the valuation } v \text{ in } \mathfrak{A} \text{ are finite.}$$

There are some variants of this formula which also express interesting properties:

    $\sim \Diamond M$ **true**—all computations of the program M are infinite,
    $\Diamond M$ **true**—there exists a finite computation,
    $\sim \Box M$ **true**—there exists an infinite computation.

EXAMPLE 2.1. Let us consider the following program M

    M: **while** b **do**
          either $x := x+1$ **or** $x := x-1$ **ro**;
          either $b := $ **true or** $b := $ **false ro**;
     od

The formulas $\Diamond M$ **true** and $\Box M$ **true** are both valid in the data structure of integers since both infinite and finite computations are possible. $\Box$

One of the most important properties of programs is *correctness*. In the case of non-deterministic programs the partial correctness property (cf. Chapter II, § 3) and the correctness property have different variants:

(1) $(\alpha \Rightarrow \Diamond M\beta)$—if an input data satisfies the condition α, then there exists a finite computation of M starting from this data whose result satisfies condition β,

(2) $((\alpha \wedge \Diamond M \text{ true}) \Rightarrow \Diamond M\beta)$—if an input data satisfies condition α and there exists a finite computation then one of the results of M satisfies property β,

(3) $((\alpha \wedge \Box M \text{ true}) \Rightarrow \Diamond M\beta)$—if an input data satisfies condition α and all computations of M from this data are finite, then there exists a result of M which satisfies β,

(4) $(\alpha \Rightarrow \Box M\beta)$—if an input data satisfies condition α then, all computations of M are finite and all results satisfy β,

(5) $((\alpha \wedge \Box M \text{ true}) \Rightarrow \Box M\beta)$—if an input data satisfies condition α and all computations of M are finite, then all results satisfy property β.

EXAMPLE 2.2. Let M be a non-deterministic program and let $\mathfrak{A}$ be a data structure of real numbers.

M: **begin**

            either $c := a$ **or** $c := b$ **ro**

            while $|b-a| > \varepsilon \wedge |f(c)| > \varepsilon$ **do**

            $x := (a+b)/2;$

            either $a := c$ **or** $b := c$ **ro**

        od

    end.

Program M is correct (in the sense of (1)) with respect to the input formula $f(b) \cdot f(a) \leqslant 0$ and the output formula **true** since for every valuation v in $\mathfrak{A}$

$$\mathfrak{A}, v \models (f(b) \cdot f(a) < 0 \Rightarrow \Diamond M \text{ true}),$$

and is not correct in the sense of (4) since

$$\mathfrak{A}, v \models (f(b) \cdot f(a) < 0 \wedge \sim \Box M \text{ true}). \qquad \Box$$

In the case where a program is of the form **while** γ **do** K **od** we can construct formulas which determine the length of the computation:

(6) $\Box(\text{if } \gamma \text{ then } M \text{ fi})^i \sim \gamma$—the number of iterations of the program M in every computation of the program **while** γ **do** M **od** is at most i,

(7) $\Diamond(\text{if } \gamma \text{ then } M \text{ fi})^i \gamma$—there exists a computation of the program **while** γ **do** M **od** such that the number of iterations of M is at least i.

The last property we shall mention has a different character: it expresses that a program satisfies some condition throughout the computation. We shall say that such a condition is an *invariant* of the program. To show that a formula α is an invariant of a program M we shall introduce a recursive definition of the expression ⊔ $M\alpha$:

$$\sqcup\, s\alpha \equiv (\alpha \wedge \Diamond s\alpha),$$

$$\sqcup \text{ if } \gamma \text{ then } K \text{ else } K' \text{ fi}\alpha \equiv ((\gamma \wedge \sqcup\, K\alpha) \vee (\sim\gamma \wedge \sqcup\, K'\alpha)),$$

$$\sqcup \text{ begin } K;\ K' \text{ end } \alpha \equiv (\sqcup\, K\alpha \wedge \sim\Diamond K(\sim \sqcup\, K'\alpha)),$$

$$\sqcup \text{ either } K \text{ or } K' \text{ ro } \alpha \equiv (\sqcup\, K\alpha \wedge \sqcup\, K'\alpha),$$

$$\sqcup \text{ while } \gamma \text{ do } K \text{ od } \alpha \equiv \left(\alpha \wedge \sim \bigvee \text{ if } \gamma \text{ then } K \text{ fi}(\gamma \wedge \sim \sqcup\, K\alpha)\right)$$

where s is an assignment instruction, K, K' are programs and γ is an open formula.

LEMMA 2.1. *For every data structure $\mathfrak{A}$ and every valuation v, $\mathfrak{A}, v \models \sqcup M\alpha$ iff the formula α is satisfied by every valuation of every computation of the program M starting from the valuation v in $\mathfrak{A}$.*

PROOF. The proof is by induction on the length of the program M.

It is obvious that Lemma 2.1 holds for assignment instructions, since there is a unique computation of such a program.

Suppose the lemma holds for the programs K and K' (the induction hypothesis).

Let us consider the program M of the form **either** K **or** K' **ro**. By definition

$$\mathfrak{A}, v \models \sqcup\, M\alpha \quad \text{iff} \quad \mathfrak{A}, v \models \sqcup\, K\alpha \quad \text{and} \quad \mathfrak{A}, v \models \sqcup\, K'\alpha.$$

Hence by the inductive assumption every valuation which occurs in a computation of K or in a computation of K' from the valuation v in $\mathfrak{A}$ satisfies the formula α. Since every computation of **either** K **or** K' **ro** is either a computation of K or a computation of K', every valuation of every computation of M satisfies α.

Similar considerations for the programs **begin** K; K' **end**, **if** γ **then** K **else** K' **fi** are omitted.

Let us consider the program M of the form **while** γ **do** K **od** and let Comp be the tree of all possible computations of M at the initial valuation v in $\mathfrak{A}$. Suppose that for some vertex $\langle \bar{v}; \ldots \rangle$ of the tree Comp, $\mathfrak{A}, \bar{v} \models \sim\alpha$. Let us consider a path going through this vertex. Assume that we have made exactly i iterations of K on this path such that all the valuations obtained satisfy the property α. Hence there exists a valuation $\bar{\bar{v}} \in (\text{if } \gamma \text{ then } K \text{ fi})^i_{\mathfrak{A}}(v)$ such that $\mathfrak{A}, \bar{\bar{v}} \models \gamma$ and $\bar{\bar{v}}$ occurs in a computation of the program K from the valuation $\bar{v}$ or $\mathfrak{A}, v \models \sim \alpha$. Thus by the induction hypothesis

$$\mathfrak{A}, v \models (\gamma \vee \sim \sqcup K\alpha),$$

and consequently

$$\mathfrak{A}, v \models (\sim \alpha \vee \Diamond(\text{if } \gamma \text{ then } K \text{ fi})^i (\gamma \wedge \sim \sqcup K\alpha)).$$

From the definition of semantics we obtain

$$\mathfrak{A}, v \models (\sim \alpha \vee \bigvee \text{ if } \gamma \text{ then } K \text{ fi} (\gamma \wedge \sim \sqcup K\alpha)).$$

The above considerations can easily be converted so as to show that $\mathfrak{A}, v \models (\sim \alpha \vee \bigvee \text{ if } \gamma \text{ then } K \text{ fi} (\gamma \wedge \sim \sqcup K\alpha))$ implies the existence of a computation of **while** γ **do** K **od** in which not every valuation satisfies the formula α.

This will complete the proof of Lemma 2.1.   □

REMARK. The set of all invariants of a given program M creates a (distributive) lattice, since if α, β are two invariants of M, $(\alpha \vee \beta)$ and $(\alpha \wedge \beta)$ are also invariants of M.   □

3. THE SUBSTITUTION THEOREM

In this section we aim to show that the tautologies of propositional algorithmic logic are schemes of tautologies of non-deterministic algorithmic logic. The replacement of atomic formulas and atomic program schemes by formulas and programs of non-deterministic algorithmic logic NAL applied to a tautology of propositional algorithmic logic PAL$_{pf}$ gives a tautology of NAL, or the resulting expression does not belong to NAL.

Let α be a formula of PAL and let s ba a substitution of the form

$$(1) \qquad (q_1/\alpha_1, \ldots, q_n/\alpha_n, K_1/M_1, \ldots, K_m/M_m),$$

where $q_i \in V_0$, for $i = 1, \ldots, n$, $K_i \in V_p$ for $j = 1, \ldots, m$, α_j are formulas of NAL and M_j are deterministic programs of NAL. By $\overline{s\alpha}$ we shall mean the expression obtained from the formula α by the simultaneous replacement of any variable q_i by the formula α_i and any program variable K_j by the program M_j. Analogously, we shall denote by $\overline{sM}$ the expression obtained from the program scheme M by the simultaneous replacement of any variable q, by the formula α_j and of any program variable K_j by the deterministic program M_j.

For every data structure $\mathfrak{A}$ of NAL, every valuation of individual variables in $\mathfrak{A}$, and every substitution s of form (1), let us define the set W_0 of valuations of propositional variables $v_{s\mathfrak{A}v}$ as follows:

$$v_{s\mathfrak{A}v}(q_i) = (\overline{sq}_i)_{\mathfrak{A}}(v), \quad i = 1, \ldots, n,$$

$$v_{s\mathfrak{A}v}(q) = 1 \quad \text{for all } q \notin \{q_1, \ldots, q_n\}.$$

Let $\mathscr{I}$ denote an interpretation of program variables such that

$$\mathscr{I}(K) = \{(v_{s\mathfrak{A}v}, v_{s\mathfrak{A}v'}): v' \in \overline{sK}_{\mathfrak{A}}(v)\}$$

for $K \in \{K_1, \ldots, K_m\}$ and $K_{\mathscr{I}} = \varnothing$ for all other program variables. Denote by $\mathfrak{M}$ the semantic structure $\langle W_0, \mathscr{I} \rangle$.

We can now formulate the following fundamental lemma.

LEMMA 3.1. *For every substitution s of the form* (1), *every data structure $\mathfrak{A}$ of non-deterministic algorithmic language, every valuation of individual variables v, every formula α and program scheme M of* PAL$_{\text{pf}}$, *if $\overline{s\alpha}$ is a well-formed formula and $\overline{sM}$ is a well-formed program of* NAL, *then the following holds*:
  (i) $\overline{s\alpha}_{\mathfrak{A}}(v) = \alpha_{\mathfrak{M}}(v_{s\mathfrak{A}v})$,
  (ii) $\pmb{v'} \in \overline{sM}_{\mathfrak{A}}(v)$ iff $v_{s\mathfrak{A}v'} \in M_{\mathfrak{M}}(v_{s\mathfrak{A}v})$.

The proof of Lemma 3.1 is by induction on the complexity of the formula α and of the program M.

We shall use the following definition.

A program scheme M *is of less complexity* than a program scheme N iff the pair (M, N) belongs to the transitive closure of the relation given below:

$$((\text{if } \gamma \text{ then } M_1 \text{ fi})^i, \text{ while } \gamma \text{ do } M_1 \text{ od}) \quad \text{for all } i \in N,$$
$$(M_i, \text{ if } \gamma \text{ then } M_1 \text{ else } M_2 \text{ fi}),$$
$$(M_i, \text{ either } M_1 \text{ or } M_2 \text{ ro}),$$
$$(M_i, \text{ begin } M_1; M_2 \text{ end}),$$

where γ is a propositional classical formula and M_1 and M_2 are program schemes of PAL.

PROOF OF LEMMA 3.1. By definition of the valuation $v_{s\mathfrak{A}v}$ and interpretation $\mathscr{I}$ the lemma holds for all open classical propositional formulas and for all program variables of PAL.

Inductive assumption: Lemma 3.1 holds for all formulas that are submitted to the formula α' and all program schemes that are of less complexity than M'.

Let $\mathfrak{A}$ and s be a fixed data structure of NAL and a fixed substitution of the form (1) respectively. We shall discuss different forms of the formula α' and the program M' such that $\overline{s\alpha'}$ and $\overline{sM'}$ are a well-formed formula and a well-formed program, respectively.

1p. Let M' be a program scheme of the form **begin** $M_1 ; M_2$ **end**. Hence

$$v' \in \overline{sM'_{\mathfrak{A}}}(v) \quad \text{iff} \quad v' \in \overline{s \text{ begin } M_1 ; \ M_2 \text{ end}}_{\mathfrak{A}}(v).$$

By the definition of the semantics we have $v' \in \overline{sM'_{\mathfrak{A}}}(v)$ iff there exists a valuation of individual variables v'' such that $v'' \in \overline{sM_{1\mathfrak{A}}}(v)$ and $v' \in \overline{sM_{2\mathfrak{A}}}(v'')$. By the inductive hypothesis the last sentence is equivalent to: There exists a valuation of propositional variables $v_{s\mathfrak{A}v''}$ such that

$$v_{s\mathfrak{A}v''} \in M_{1\mathfrak{M}}(v_{s\mathfrak{A}v}) \quad \text{and} \quad v_{s\mathfrak{A}v'} \in M'_{2\mathfrak{M}}(v_{s\mathfrak{A}v''}).$$

Hence, from the definition of interpretation

$$v_{s\mathfrak{A}v'} \in \text{begin } M_1 ; \ M_2 \text{ end}_{\mathfrak{M}}(v_{s\mathfrak{A}v}).$$

2p. Let M' be a program scheme of the form **if** γ **then** M_1 **else** M_2 **fi**. From the definition of semantics of NAL we have

$$\begin{aligned} v' \in \overline{sM'_{\mathfrak{A}}}(v) \quad \text{iff} \quad & v' \in \overline{sM_{1\mathfrak{A}}}(v) \text{ and } \overline{s\gamma}_{\mathfrak{A}}(v) = \mathbf{1} \text{ or} \\ & v' \in \overline{sM_{2\mathfrak{A}}}(v) \text{ and } \overline{s\gamma}_{\mathfrak{A}}(v) = \mathbf{0}. \end{aligned}$$

By the inductive assumption we have

$$v_{s\mathfrak{A}v} \in M_{1\mathfrak{M}}(v_{s\mathfrak{A}v}) \quad \text{and} \quad \gamma_{\mathfrak{M}}(v_{s\mathfrak{A}v}) = \mathbf{1}$$

or

$$v_{s\mathfrak{A}v'} \in M_{2\mathfrak{M}}(v_{s\mathfrak{A}v}) \quad \text{and} \quad \sim\gamma_{\mathfrak{M}}(v_{s\mathfrak{A}v}) = \mathbf{1}.$$

Thus by the definition of an interpretation $v_{s\mathfrak{A}v'} \in M_{\mathfrak{M}}(v_{s\mathfrak{A}v})$.

3p. Consider the program scheme $M' = $ **either** M_1 **or** M_2 **ro**. By the definition

$$\begin{aligned} v' \in \overline{s \text{ (either } M_1 \text{ or } M_2 \text{ ro)}}_{\mathfrak{A}}(v) \quad \text{iff} \quad & v' \in \overline{sM_{1\mathfrak{A}}}(v) \text{ or} \\ & v' \in \overline{sM_{2\mathfrak{A}}}(v). \end{aligned}$$

This is equivalent (by the inductive hypothesis) to:

$$v_{s\mathfrak{A}v'} \in M_{1\mathfrak{M}}(v_{s\mathfrak{A}v}) \quad \text{or} \quad v_{s\mathfrak{A}v'} \in M_{2\mathfrak{M}}(v_{s\mathfrak{A}v}).$$

Hence

$$v' \in \overline{sM_\mathfrak{A}^i}(v) \quad \text{iff} \quad v_{s\mathfrak{A}v} \in (\text{either } M_1 \text{ or } M_2 \text{ ro})_{\mathfrak{M}}(v_{s\mathfrak{A}v}).$$

4p. Consider the program scheme $M = \textbf{while } \gamma \textbf{ do } M \textbf{ od}$. By the semantic properties of non-deterministic algorithmic logic NAL we have

$$v' \in \overline{sM_\mathfrak{A}^i}(v) \quad \text{iff} \quad \text{there exists an } i_0 \in N \text{ such that}$$
$$v' \in \overline{s(\textbf{if } \gamma \textbf{ then } M \textbf{ fi})_\mathfrak{A}^{i_0}}(v).$$

By the inductive assumption this is equivalent to the statement that there exists an i_0 such that $v_{s\mathfrak{A}v'} \in (\textbf{if } \gamma \textbf{ then } M \textbf{ fi})_{\mathfrak{M}}^{i_0}(v_{s\mathfrak{A}v})$ and therefore

$$v_{s\mathfrak{A}v'} \in (\textbf{while } \gamma \textbf{ do } M \textbf{ od})_{\mathfrak{M}}(v_{s\mathfrak{A}v}).$$

Now let us consider the formulas.

1f. Let us assume that α' is of the form $\Diamond K\alpha$, where $K \in V_p$. By the definition of semantics

$$\overline{s \Diamond K\alpha_\mathfrak{A}}(v) = 1 \quad \text{iff} \quad \text{there exists a finite computation of the}$$
$$\text{program } s\overline{K}_\mathfrak{A} \text{ at the initial valuation } v$$
$$\text{such that its result } v' \in \overline{sK_\mathfrak{A}}(v) \text{ satisfies}$$
$$s\alpha.$$

By the inductive hypothesis, there exists a successful computation of the program K such that

$$v_{s\mathfrak{A}v'} \in K_{\mathfrak{M}}(v_{s\mathfrak{A}v}) \quad \text{and} \quad \mathfrak{M}, v_{s\mathfrak{A}v'} \models \alpha.$$

Hence

$$\mathfrak{A}, v \models \overline{s \Diamond K\alpha} \quad \text{iff} \quad \mathfrak{M}, v_{s\mathfrak{A}v} \models \Diamond K\alpha.$$

2f. Consider the formula α' of the form $\Box K\alpha$, where $K \in V_p$. By the definition of a semantic we have

$$\mathfrak{A}, v \models \overline{s \Box K\alpha} \quad \text{iff} \quad \text{all computations of the program } s\overline{K}_\mathfrak{A}$$
$$\text{are finite and for all } v' \in \overline{sK_\mathfrak{A}}(v), \mathfrak{A},$$
$$v' \models \overline{s\alpha}.$$

By the inductive assumption for the program variable K and for the formula α we have

$$\mathfrak{A}, v \models \overline{s \Box K\alpha} \quad \text{iff} \quad \text{all computations of the program}$$
$$\text{scheme } K \text{ are successful and for all}$$
$$v_{s\mathfrak{A}v'} \in K_{\mathfrak{M}}(v_{s\mathfrak{A}v}) \text{ we have } \mathfrak{M}, v_{s\mathfrak{A}v} \models \alpha.$$

By the definition of the value of the formula in PAL

$$A, v \models s \; \square \; K\alpha \quad \text{iff} \quad \mathfrak{M}, v_{s\mathfrak{A}v} \models \square K\alpha.$$

3f. Let us consider a formula α' of the form $\square$ **either** M_1 **or** M_2 **ro** α. By the properties of semantics we have

$$\mathfrak{A}, v \models \overline{s\square \; \textbf{either} \; M_1 \; \textbf{or} \; M_2 \; \textbf{ro} \; \alpha} \quad \text{iff}$$
$$\mathfrak{A}, v \models \overline{s\square M_1 \alpha} \quad \text{and} \quad \mathfrak{A}, v = \overline{s\square M_2 \alpha}.$$

Hence, by the inductive hypothesis,

$$\mathfrak{M}, v_{s\mathfrak{A}v} \models \square M_1 \alpha \quad \text{and} \quad \mathfrak{M}, v_{s\mathfrak{A}v} \models \square M_2 \alpha,$$

and therefore $\mathfrak{M}, v_{s\mathfrak{A}v} \models \square$ **either** M_1 **or** M_2 **ro** α.

4f. Suppose now that α' is of the form $\Diamond$ **while** γ **do** M **od** β. By the definition of semantics we have

$$\mathfrak{A}, v \models \overline{s \; \Diamond \; \textbf{while} \; \gamma \; \textbf{do} \; M \; \textbf{od} \; \beta} \quad \text{iff}$$
$$\text{l.u.b.} \overline{\big(s\Diamond \, (\textbf{if} \; \gamma \; \textbf{then} \; M \, \textbf{fi})^i (s\beta \wedge \sim s\gamma)\big)}_{\mathfrak{A}}(v) = \mathbf{1}.$$
$$\scriptstyle i \in N$$

Hence, by the inductive hypothesis,

$$\mathfrak{A}, v \models \overline{s\alpha'} \quad \text{iff} \quad \text{l.u.b.} \big(\Diamond \, (\textbf{if} \, \gamma \, \textbf{then} \, M \, \textbf{fi})^i (\sim \gamma \wedge \beta)_{\mathfrak{M}}(v_{s\mathfrak{A}v}) \big) = \mathbf{1}.$$
$$\scriptstyle i \in N$$

By Lemma 2.3 from Chapter V we have

$$\mathfrak{A}, v \models \overline{s\alpha'} \quad \text{iff} \quad \mathfrak{M}, v_{s\mathfrak{A}v} \models \Diamond \; \textbf{while} \; \gamma \; \textbf{do} \; M \; \textbf{od} \; \beta.$$

The proof of the remaining cases runs analogously. $\square$

The following theorem is our goal in this section.

THEOREM 3.2. *For every formula α of PAL and for every substitution s of the form (1), if $\overline{s\alpha}$ is a well-formed formula of NAL and α is a tautology of PAL_{pt} then the formula $\overline{s\alpha}$ is a tautology of NAL.*

PROOF. Let α be a tautology of PAL_{pt} and let $\overline{s\alpha}$ be a well-formed formula of NAL for some substitution s.

Suppose that $\mathfrak{A}, v \models \sim \overline{s\alpha}$ for some fixed data structure $\mathfrak{A}$ of a non-deterministic algorithmic logic and valuation v of individual variables and $\mathfrak{M}$ the corresponding to $\mathfrak{A}$ semantic structure. From Lemma 3.1 of this chapter,

$$\mathfrak{A}, v \models \overline{s\alpha} \quad \text{iff} \quad \mathfrak{M}, v_{s\mathfrak{A}v} \models \alpha.$$

Hence $\alpha_{\mathfrak{M}}(v_{s\mathfrak{A}v}) = \mathbf{0}$, and therefore α is not a propositional tautology, a contradiction. $\square$

4. NON-DETERMINISTIC ALGORITHMIC LOGIC

In this section we shall introduce the deductive system called *Non-deterministic Algorithmic Logic* (NAL), which enables us to character-ize syntactically the notion of tautology. As a result of the PAL—com-pleteness theorem (cf. Chapter V, § 9) and of the Substitution Theorem (cf. § 3 of this chapter) all instances of axioms of PAL which are non-deterministic formulas are tautologies of NAL. This justifies the adop-tion of the following set of axioms and reference rules.

Ax1–Ax11—axioms of the classical propositional calculus (cf. Chap-ter II, § 5).

$$\Diamond s\gamma \equiv \overline{s\gamma}, \qquad\qquad \Box s\gamma, \equiv \overline{s\gamma}$$
$$\Diamond M(\alpha \vee \beta) \equiv (\Diamond M\alpha \vee \Diamond M\beta), \quad \Box M(\alpha \wedge \beta) \equiv (\Box M\alpha \wedge \Box M\beta),$$
$$\bigvee M\alpha \equiv (\alpha \vee \bigvee M(\Diamond M\alpha)), \quad \bigsqcup M\alpha \equiv (\alpha \vee \bigsqcup M(\Box M\alpha)),$$
$$\bigwedge M\alpha \equiv (\alpha \wedge \bigwedge M(\Diamond M\alpha)), \quad \bigsqcap M\alpha \equiv (\alpha \wedge \bigsqcap M(\Box M\alpha)),$$
$$s\big((\exists x)\alpha(x)\big) \equiv (\exists y)s\big((x := y)\alpha(x)\big),$$

where y is an individual variable not occurring in s,
$$(\sim \Diamond M\alpha \Rightarrow \Box M \sim \alpha), \quad \Box M \text{ true} \Rightarrow (\Diamond M \sim \alpha \equiv \; \sim \Box M\alpha'),$$
$$\Diamond(x := \tau)\alpha(x) \equiv (\exists x)\alpha(x) \quad \text{for every term } \tau,$$
$$(\forall x)\alpha(x) \equiv \; \sim(\exists x)\alpha(x),$$
$$\Diamond \text{ begin } M; \; M' \text{ end } \alpha \equiv \Diamond M(\Diamond M'\alpha),$$
$$\Box \text{ begin } M; \; M' \text{ end } \alpha \equiv \Box M(\Box M'\alpha),$$
$$\Diamond \text{ if } \gamma \text{ then } M \text{ else } M' \text{ fi } \alpha \equiv \big((\gamma \wedge \Diamond M\alpha) \vee (\sim\gamma \wedge \Diamond M'\alpha)\big),$$
$$\Box \text{ if } \gamma \text{ then } M \text{ else } M' \text{ fi } \alpha \equiv \big((\gamma \wedge \Box M\alpha) \vee (\sim\gamma \wedge \Box M'\alpha)\big),$$
$$\Diamond \text{ while } \gamma \text{ do } M \text{ od } \alpha$$
$$\equiv \big((\sim\gamma \wedge \alpha) \vee (\gamma \wedge \Diamond M(\Diamond \text{ while } \gamma \text{ do } M \text{ od } \alpha))\big),$$
$$\Box \text{ while } \gamma \text{ do } M \text{ od } \alpha$$
$$\equiv \big((\sim \gamma \wedge \alpha) \vee (\gamma \wedge \Box M(\Box \text{ while } \gamma \text{ do } M \text{ od } \alpha))\big),$$
$$\Diamond \text{ either } M \text{ or } M' \text{ ro } \alpha \equiv (\Diamond M\alpha \vee \Diamond M'\alpha),$$
$$\Box \text{either } M \text{ or } M' \text{ ro } \alpha \equiv (\Box M\alpha \wedge \Box M'\alpha).$$

In the above schemes of formulas α, β are arbitrary formulas, γ is an open formula, M and M' are arbitrary programs and s is an assignment instruction.

The set of inference rules contains all rules of PAL and some rules which characterize the classical and iteration quantifiers.

Rules

$$\frac{\alpha,\ (\alpha \Rightarrow \beta)}{\beta} \quad \textit{modus ponens,}$$

$$\frac{((x := y)\alpha(x) \Rightarrow \beta)}{((\exists x)\alpha(x) \Rightarrow \beta)}, \quad \text{where } y \text{ is an individual variable occurring neither in } \alpha \text{ nor in } \beta,$$

$$\frac{(\alpha \Rightarrow \beta)}{(\lozenge M\alpha \Rightarrow \lozenge M\beta)}, \qquad\qquad \frac{(\alpha \Rightarrow \beta)}{(\square M\alpha \Rightarrow \square M\beta)},$$

$$\frac{\{(\lozenge M'(\lozenge M^i\alpha) \Rightarrow \beta)\}_{i \in N}}{(\lozenge M'(\bigvee M\alpha) \Rightarrow \beta)}, \qquad \frac{\{(\lozenge M'(\square M^i\alpha) \Rightarrow \beta)\}_{i \in N}}{(\lozenge M'(\sqcup M\alpha) \Rightarrow \beta)},$$

$$\frac{\{(\beta \Rightarrow \square M'(\lozenge M^i\alpha))\}_{i \in N}}{(\beta \Rightarrow \square M'(\bigwedge M\alpha))}, \qquad \frac{\{(\beta \Rightarrow \square M'(\square M^i\alpha))\}_{i \in N}}{(\beta \Rightarrow \square M'(\sqcap M\alpha))},$$

$$\frac{\{(\lozenge M'(\lozenge (\text{if } \gamma \text{ then } M \text{ fi})^i(\alpha \wedge \sim\gamma)) \Rightarrow \beta)\}_{i \in N}}{(\lozenge M'(\lozenge \text{ while } \gamma \text{ do } M \text{ od } \alpha) \Rightarrow \beta)},$$

$$\frac{\{(\lozenge M'(\square(\text{if } \gamma \text{ then } M \text{ fi})^i(\sim\gamma \wedge \alpha)) \Rightarrow \beta)\}_{i \in N}}{(\lozenge M'(\square \text{ while } \gamma \text{ do } M \text{ od } \alpha) \Rightarrow \beta)}.$$

Note that some of the inference rules have infinitely many premises. This is an effect of the non-compactness of the semantic consequence operation.

DEFINITION 4.1. *By the non-deterministic algorithmic logic* NAL *we shall understand a system* $\langle L, C \rangle$, *where L is a non-deterministic algorithmic language and C is a syntactic consequence determined by the axioms and rules mentioned above.*

By the non-deterministic theory we shall understand a formal system $\langle L, C, A \rangle$ *based on* NAL *such that A is a set of formulas of non-deterministic algorithmic language L.*   □

The notions of a theorem, of a model, and of consistency are very like those of algorithmic logic (see Chapter II, Definitions 5.2, 6.1, 6.2) and therefore are not presented here.

LEMMA 4.1. *If* α *is a theorem of a non-deterministic theory T, then* α *is valid in every model of that theory.*

PROOF. Let $T = \langle L, C, A \rangle$ and let $\mathfrak{M}$ be a data structure for L.

For an arbitrary valuation v, $\mathfrak{M}, v \models \bigwedge M\alpha$ is equivalent by the definition of semantics (see § 2) to the following:

$\mathfrak{M}, v \models \Diamond M^i \alpha$ for every natural number i.

Hence $\mathfrak{M}, v \models \alpha$ and $\mathfrak{M}, v \models \Diamond M^i(\Diamond M\alpha)$ for every $i \geqslant 0$. The latter formula is equivalent to $\mathfrak{M}, v \models (\alpha \wedge \bigwedge M(\Diamond M\alpha))$. As a consequence of the above considerations the formula

$$\bigwedge M\alpha \equiv (\alpha \wedge \bigwedge M(M\alpha))$$

is valid in every data structure for L, i.e. it is a tautology.

In a similar way we can prove that all axioms of NAL are valid in every data structure (see also Chapter II, § 5 and Chapter V, § 5).

Moreover, we claim that the inference rules go from valid premises to a valid conclusion.

Let us check the last sentence in the case of the rule

$$\frac{\{(\Diamond M'(\square M^i\alpha) \Rightarrow \beta)\}_{i \in N}}{(\Diamond M'(\bigsqcup M\alpha) \Rightarrow \beta)} \ .$$

Assume that $\mathfrak{M}$ is a model of T and that all formulas $(\Diamond M'(\square M^i\alpha) \Rightarrow \beta)$ are valid in $\mathfrak{M}$. Suppose that for some valuation $\hat{v}$

$$\mathfrak{M}, \hat{v} \models \Diamond M'(\bigsqcup M\alpha) \quad \text{and} \quad \mathfrak{M}, \hat{v} \models \sim\beta.$$

Hence there exists a finite computation of M' such that its result satisfies the formula $\bigsqcup M\alpha$. By the definition of semantics it follows that $\mathfrak{M}, v \models \square M^i\alpha$ for a certain $i \in N$ and a certain valuation $v \in M_{\mathfrak{M}}^v(\hat{v})$.

Thus non $\mathfrak{M}, \hat{v} \models (\Diamond M'(\square M^i\alpha) \Rightarrow \beta)$ contrary to the assumption. $\square$

The most important theorem of this section is the Model Existence Theorem. The proof of the theorem makes use of the Rasiowa–Sikorski algebraic method (see Chapter III, § 2).

THEOREM 4.2 (Model Existence Theorem). *A non-deterministic algorithmic theory is consistent if and only if it has a model.*

PROOF. One implication is obvious. We shall present below a sketch of the proof that if a theory $T = \langle L, C, A \rangle$ is consistent then there exists a model of the set A.

(1) The first step is to construct the Lindenbaum algebra $F/\approx$ of the theory (cf. Chapter III, § 1).

(2) Since the theory T is consistent, the Lindenbaum algebra is a non-degenerate Boolean algebra and moreover

$$||\bigvee M\alpha|| = \text{l.u.b.}_{i \in N} ||\Diamond M^i\alpha||, \quad ||\bigwedge M\alpha|| = \text{g.l.b.}_{i \in N} ||\Diamond M^i\alpha||,$$

$$||\bigsqcup M\alpha|| = \text{l.u.b.}_{i \in N} ||\Box M^i\alpha||, \quad ||\bigsqcap M\alpha|| = \text{g.l.b.}_{i \in N} ||\Box M^i\alpha||,$$

$$||(\exists x)\alpha(x)|| = \text{l.u.b.}_{\tau \in T} ||(x := \tau)\alpha(x)||,$$

$$||(\forall x)\alpha(x)|| = \text{g.l.b.}_{\tau \in T} ||(x := \tau)\alpha(x)||.$$

(3) Let Q denote the set of all infinite operations mentioned in (2). By the Rasiowa–Sikorski Lemma (Rasiowa and Sikorski, 1968) for every non-zero element a of the Lindenbaum algebra there exists a Q-filter V such that $a \in V$ (see Appendix A).

(4) Let $\mathfrak{M}$ be a data structure in the set of all terms of the language L such that

$$(\tau_1, \ldots, \tau_n) \in \varrho_{\mathfrak{M}} \quad \text{iff} \quad ||\varrho(\tau_1, \ldots, \tau_n)|| \in V,$$

$$\psi_{\mathfrak{M}}(\tau_1, \ldots, \tau_n) - \psi(\tau_1, \ldots, \tau_n),$$

for an arbitrary n-argument predicate ϱ, an arbitrary n-argument functor ψ and arbitrary terms $\tau_1, \ldots, \tau_n$ of the non-deterministic language L.

(5) By induction on the length of the formula α we can prove that

$$\mathfrak{M}, v_0 \models \alpha \quad \text{iff} \quad ||\alpha|| \in V,$$

where v_0 is a valuation such that $v_0(x) = x$ for all individual variables x and $v(q) = \mathbf{1}$ iff $||q|| \in V$ for all propositional variables.

(6) It follows by (5) that $\mathfrak{M}$ is a model of the set of specific axioms A. □

The last theorem of this section characterizes the connections between the syntactic and the semantic consequence operations.

THEOREM 4.3 (The Completeness Theorem). *For every consistent non-deterministic algorithmic theory the following conditions are equivalent:*

(i) *α is a theorem of T;*

(ii) *α is valid in every model of T.*

In other words, $A \vdash \alpha$ iff $A \models \alpha$ for an arbitrary set A. We shall omit the proof since it is very similar to the proof of the Completeness Theorem of algorithmic logic. □

5. CERTAIN METAMATHEMATICAL RESULTS

The aim of this section is to generalize some results obtained in algorithmic logic.

THEOREM 5.1 (Downward Skolem-Löwenheim Theorem). *If a non-deterministic algorithmic theory has a model, then it has an enumerable model.*   □

This is an immediate effect of the construction presented in the proof of the Model Existence Theorem (cf. Theorem 4.2).

As a consequence of the Completeness Theorem we have the following fact:

THEOREM 5.2 (on deduction). *If α is a closed formula of non-deterministic algorithmic language, then for an arbitrary set of formulas A and a formula β*

$$A \vdash (\alpha \Rightarrow \beta) \quad iff \quad A \cup \{\alpha\} \vdash \beta.$$ □

LEMMA 5.3 *For an arbitrary formula α which does not contain any* **while**-*instruction or quantifiers there exists an open formula γ such that*

$$(1) \qquad \mathfrak{A}, v \models \gamma \quad iff \quad \mathfrak{A}, v \models \alpha$$

for an arbitrary valuation v and an arbitrary data structure $\mathfrak{A}$.

PROOF. The lemma holds trivially for open formulas. Let us assume that (1) holds for all formulas which are submitted to the formula α (see Appendix B) and let us consider the formula $\alpha = \Box M\beta$.

If M is an assignment instruction $(x := w)$, then by the induction hypothesis there exists an open formula β such that for an arbitrary data structure $\mathfrak{A}$

$$\mathfrak{A} \models (\beta' \equiv \beta).$$

Hence by the rule $\dfrac{(\alpha \Rightarrow \beta)}{(\Box M\alpha \Rightarrow \Box M\beta)}$ we have $\mathfrak{A} \models (\Box M\beta' \equiv \Box M\beta)$.

Thus by the axioms of NAL,

$$\mathfrak{A} \models \Box(x := w)\beta \equiv \overline{(x := w)\beta'}$$

which completes the proof since $\overline{(x := w)\beta'}$ is an open formula obtained from β' by the simultaneous replacement of all occurrences of x by the expression w.

If M is of the form **either** M_1 **or** M_2 **ro**, then by the Completeness

Theorem we have

$$\mathfrak{A} \models \Box \text{ either } M_1 \text{ or } M_2 \text{ ro } \beta \equiv (\Box M_1 \beta \wedge \Box M_2 \beta).$$

By the induction hypothesis there exist open formulas γ_1 and γ_2 such that for an arbitrary data structure $\mathfrak{A}$,

$$\mathfrak{A} \models \Box M_1 \beta \equiv \gamma_1 \quad \text{and} \quad \mathfrak{A} \models \Box M_2 \beta \equiv \gamma_2.$$

Hence

$$\mathfrak{A} \models \Box \text{ either } M_1 \text{ or } M_2 \text{ ro } \beta \equiv (\gamma_1 \wedge \gamma_2).$$

We shall omit the easy next steps of induction. $\qquad \Box$

THEOREM 5.4. *Let K be a program without the* **while**-*operation and let γ be an open formula.*

(i) *The formula $\bigvee K \gamma$ is a tautology iff there exists a natural number n such that the formula $\bigvee_{i \leqslant n} \Diamond K^i \gamma$ is a tautology.*

(ii) *The formula $\bigsqcup K \gamma$ is a tautology iff there exists a natural number n such that the formula $\bigvee_{i \leqslant n} \Box K^i \gamma$ is a tautology.*

PROOF. Since the proofs in cases (i) and (ii) are essentially the same we shall discuss case (ii) only. Moreover one implication is obvious by the definition of semantics.

Let H_m denotes the formula $\bigvee_{i \leqslant m} \Box K^i \gamma$ and suppose H_m is not a tautology for arbitrary $m \in N$. For arbitrary natural number i, the formula $\Box K^i \gamma$ is equivalent to an open formula, say β_i, cf. Lemma 5.3. Let us put $H'_m = \bigvee_{i \leqslant m} \beta_i$. Hence for an arbitrary data structure $\mathfrak{A}$

$$\mathfrak{A} \models H_m \equiv H'_m \quad \text{for every } m \in N.$$

For every $m \in N$, let H''_m be the formula obtained from H'_m by the simultaneous replacement of all elementary formulas of the form $\varrho(\tau_1, \ldots, \tau_n)$ that occur in H'_m by the corresponding propositional variables $q_{\varrho(\tau_1, \ldots, \tau_n)}$ which do not occur in any formula H'_m (different propositional variables correspond to different elementary formulas).

The formulas H''_m satisfy the following condition:

$$\models H''_m \quad \text{implies} \quad \models H'_m \quad \text{for } m \in N.$$

By the assumption, H''_m is not a tautology for arbitrary m, hence the set W^m of all valuations which do not satisfy the formula H''_m restricted to the set $V(H''_m)$ is a finite non-empty set. Moreover, it follows easily from the construction that, if $n > m$ then for every $v \in W^n$ there exists a valuation $v' \in W^m$ such that $v = v'$ off$(V - V(H''_m))$, i.e. $v(z) = v'(z)$ for $z \in V(H''_m)$. The set $\bigcup_{m \in N} W^m$ creates a tree such that the elements of W^m are on the $(m+1)$ level of the tree and a valuation v on the $(m+1)$ level is a son of the valuation $\hat{v}$ on the m level if and only if $v = \hat{v}$ off$(V - V(H''_m))$.

Since the degree of any vertex in the tree is finite (the set W^m is finite for every $m \in N$), then by König's Lemma (cf. Kuratowski and Mostowski, 1967) there exists an infinite path $\varnothing, v_0, v_1, \ldots$ such that $v_j \in W^j$, $j \in N$. Let us denote by v_∞ a valuation such that

$$v_\infty = v_m \text{ off } (V - V(H'')) \quad \text{for every } m \in N.$$

Thus for every natural number m, $H''_m(v_\infty) = 0$.

Let $\mathfrak{A}$ be a data structure in the set of all terms such that

$$(\tau_1, \ldots, \tau_n) \in \varrho_\mathfrak{A} \quad \text{iff} \quad v_\infty(q_{\varrho(\tau_1, \ldots, \tau)_n}) = 1,$$

$$\psi_\mathfrak{A}(\tau_1, \ldots, \tau_n) = \psi(\tau_1, \ldots, \tau_n),$$

for an arbitrary n-argument predicate ϱ and an arbitrary n-argument functor ψ.

Let $\bar{v}$ be a valuation in $\mathfrak{A}$ such that

$$\bar{v}(x) = x \quad \text{for all individual variables } x,$$

$$\bar{v}(q) = v_\infty(q) \quad \text{for all propositional variables } q.$$

From the above construction we have

$$\text{non } \mathfrak{A}, \bar{v} \models H'_m \quad \text{for every } m \in N$$

and therefore

$$\text{non } \mathfrak{A}, \bar{v} \models H_m \quad \text{for all } m \in N.$$

By the definition of semantics.

$$\text{l.u.b.}_{i \in N} (\square K^i \gamma)_\mathfrak{A}(\bar{v}) = 0.$$

Hence $\mathfrak{A}, \bar{v} \models \sim \bigsqcup K\gamma$, and therefore $\bigsqcup K\gamma$ is not a tautology. $\square$

As a result of Theorem 5.4 we have the following.

THEOREM 5.5. *If a program M of the form*

begin M_1 ; while γ do M_2 od end,

where M_1, M_2 do not contain any **while**-instruction, does not diverge in any data structure, then there exists a common upper bound on the length of all computations of that program.

For the proof see Lemma 3.6 from Chapter III.   □

6. ON ISOMORPHISM OF DATA STRUCTURES

Let $\mathfrak{A}$ and $\mathfrak{B}$ be arbitrary data structures for the language L,

$$\mathfrak{A} = \langle A, \{\varphi_\mathfrak{A}\}_{\varphi\in\Phi}, \{\varrho_\mathfrak{A}\}_{\varrho\in P}\rangle, \quad \mathfrak{B} = \langle B, \{\varphi_\mathfrak{B}\}_{\varphi\in\Phi}, \{\varrho_\mathfrak{B}\}_{\varrho\in P}\rangle$$

Let h be an isomorphism of data structures $\mathfrak{A}$ and $\mathfrak{B}$, i.e., let h be a one-to-one mapping such that

$$h: A \xrightarrow[\text{onto}]{} B$$

and for every n-argument functor φ,

(1) $h\big(\varphi_\mathfrak{A}(a_1, \ldots, a_n)\big) = \varphi_\mathfrak{B}\big(h(a_1), \ldots, h(a_n)\big)$

and for every n-argument predicate ,

(2) $(a_1, \ldots, a_n) \in \varrho_\mathfrak{A}$ iff $\big(h(a_1), \ldots, h(a_n)\big) \in \varrho_\mathfrak{B}$,

where $a_1, \ldots, a_n$ are arbitrary elements of A.

For an arbitrary valuation v in the data structure $\mathfrak{A}$ we shall denote by hv a valuation in $\mathfrak{B}$ such that

$$hv(x) = h\big(v(x)\big) \quad \text{for every individual variable } x,$$
$$hv(q) = v(q) \quad \text{for every propositional variable } q.$$

LEMMA 6.1. *If h is an isomorphism of $\mathfrak{A}$ and $\mathfrak{B}$, then for every term τ, every open formula γ and an arbitrary valuation v in the structure $\mathfrak{A}$*

(3) $h\big(\tau_\mathfrak{A}(v)\big) = \tau_\mathfrak{B}(hv),$

(4) $\mathfrak{A}, v \models \gamma$ iff $\mathfrak{B}, hv \models \gamma.$

The proof is by induction on the length of term τ and formula γ and is an easy consequence of definitions (1) and (2).   □

Let $\text{Comp}(M, v, \mathfrak{A})$ be a tree of all possible computations of the program M starting from the valuation v in the data structure $\mathfrak{A}$ and let $\text{Comp}(M, hv, \mathfrak{B})$ be a tree of all possible computations of the program M starting from the valuation hv in the data structure $\mathfrak{B}$ (cf. § 1).

We shall denote by h' a mapping which to every configuration $\langle v; \text{Rest} \rangle$ of the tree $\text{Comp}(M, v, \mathfrak{A})$ assigns a configuration $\langle hv'; \text{Rest} \rangle$.

LEMMA 6.2. *If h is an isomorphism from $\mathfrak{A}$ onto $\mathfrak{B}$, then h' is an isomorphism from the tree $\text{Comp}(M, v, \mathfrak{A})$ onto the tree $\text{Comp}(M, hv, \mathfrak{B})$.*

PROOF. Let us denote by D a tree which is an image of $\text{Comp}(M, v, \mathfrak{A})$ under the mapping h'. We shall prove that $D = \text{Comp}(M, hv, \mathfrak{B})$.

Let us note first that the configuration $\langle hv; M \rangle$ is a root of both trees.

Suppose that the trees D and $\text{Comp}(M, hv, \mathfrak{B})$ are both identical to the level n.

Let σ_n be a configuration on the level n in the tree D and $\sigma_n = \langle h(v_n); K, \text{Rest} \rangle$. We shall now consider the different forms of the program K.

$1°$ If K is an assignment instruction $(x := \tau)$ then the unique son of σ_n in D is a configuration $\sigma = \langle h(v_{n+1}); \text{Rest} \rangle$, where v_{n+1} is the result of performing K at the valuation v_n, by the construction of the tree $\text{Comp}(M, v, \mathfrak{A})$ and by the definition of h'.

By the induction hypothesis $\sigma_n \in \text{Comp}(M, hv, \mathfrak{B})$. Hence the unique son of σ_n is configuration σ' such that $\sigma' = \langle v'; \text{Rest} \rangle$ and $v' = (x := \tau)_{\mathfrak{B}}(hv_n)$. However, by (3) $hv_{n+1} = v'$ and therefore $\sigma' = \sigma$, i.e., $\sigma \in \text{Comp}(M, hv, \mathfrak{B})$,

$2°$ If K is of the form **if then** K_1 **else** K_2 **fi**, then the next configuration depends on the value of the formula at the valuation v_n. Suppose $\mathfrak{A}, v_n \models \gamma$. The configuration $\sigma = \langle hv_n; K_1, \text{Rest} \rangle$ is then an element of the $(n+1)$ level of the tree D. By the induction hypothesis σ_n is in the tree $\text{Comp}(M, hv, \mathfrak{B})$ and the configuration $\langle hv_n; K_1, \text{Rest} \rangle$ is its unique son whenever $\mathfrak{B}, hv_n \models \gamma$. However, $\mathfrak{B}, hv_n \models \gamma$ iff $\mathfrak{A}, v_n \models \gamma$, by (4). Thus $\sigma \in \text{Comp}(M, hv, \mathfrak{B})$,

$3°$ The remaining forms of the program K can be discussed analogously.

As a consequence it follows that all configurations that occur on the $(n+1)$ level of the tree D are on the $(n+1)$ level of the tree $\text{Comp}(M, hv, \mathfrak{B})$.

Conversely, we can prove that all $(n+1)$ level vertices of $\text{Comp}(M, hv, \mathfrak{B})$ occur on the $(n+1)$ level of D.

Thus

$$D = \text{Comp}(M, hv, \mathfrak{B}),$$

by the induction principle. $\square$

As a corollary of Lemma 6.2 we obtain the following fact:

LEMMA 6.3. *If h is an isomorphism from $\mathfrak{A}$ onto $\mathfrak{B}$, then for an arbitrary program K and an arbitrary valuation v in $\mathfrak{A}$*

(i) $v' \in K_{\mathfrak{A}}(v)$ *iff* $hv' \in K_{\mathfrak{B}}(hv)$,

(ii) *there exists an infinite computation of K starting from the valuation v in the structure $\mathfrak{A}$ iff there exists an infinite computation of K starting from hv in the structure $\mathfrak{B}$.* □

THEOREM 6.4. *If h is an isomorphism from $\mathfrak{A}$ onto $\mathfrak{B}$, then for every formula α of the language L*

$$\mathfrak{A}, v \models \alpha \quad \text{iff} \quad \mathfrak{B}, hv \models \alpha,$$

where v is an arbitrary valuation in $\mathfrak{A}$.

The proof is by induction on the length of formula α and follows immediately from Lemma 6.1 and Lemma 6.3. □

DEFINITION 6.1. *We shall say that the two data structures $\mathfrak{A}$ and $\mathfrak{B}$ are algorithmically equivalent iff for every formula α*

$$\mathfrak{A} \models \alpha \quad \text{iff} \quad \mathfrak{B} \models \alpha.$$ □

The following corollary is a consequence of Theorem 6.4.

COROLLARY. *Every two isomorphic data structures are algorithmically equivalent.* □

7. ON THE EQUIVALENCE OF NON-DETERMINISTIC PROGRAMS

DEFINITION 7.1. *We shall say that two non-deterministic programs are equivalent, $K \sim M$ for short, whenever they determine the same relations in every data structure.* □

EXAMPLE 7.1. The following programs M, K are equivalent:
(i)

```
M: either                         K: if γ then
       if γ then M' else M'' fi          either M' or K' ro
   or                                else
       if γ then K' else K'' fi          either M'' or K'' ro
   ro,                             fi;
```

(ii)

        M: **either** K: **begin**

                begin K'; M' **end**   **either** K' **or** K'' **ro**;

            or M'

                begin K''; M' **end**   **end**.

            ro,   □

From the practical point of view the above definition is not very useful, since two programs which in fact compute the same function are not equivalent if they make use of different auxiliary variables.

EXAMPLE 7.2. The following programs are not equivalent in the sense of Definition 7.1:

        M: **either**

                while γ **do** K **od**

            or

                while γ' **do** K' **od**

            ro,

        M': **begin**

                either $q := $ **true or** $q := $ **false ro**;

                while $(\gamma \wedge q) \vee (\gamma' \vee \sim q)$ **do**

                      if q **then** K **else** K' **fi**

                od

            end,

where q is a propositional variable not occurring in K, K' and γ.

Moreover, let us note that Definition 7.1 does not capture the difference if one program has infinite computation and the other has not.

The programs

        K: $x := 1$;

        M: **either**

               $x := 1$

            or

                while $x \geqslant 1$ **do** $x := x+1$ **od**

            ro

are equivalent in the sense of relation $\sim$ although M has an infinite computation, while the unique computation of K is finite.   □

Hence we shall modify Definition 7.1 to avoid the disadvantages mentioned above.

DEFINITION 7.2. *The two programs K and M are equivalent up to the set of variables X, $K \sim M$ off X for short, iff for an arbitrary data structure $\mathfrak{A}$ and an arbitrary valuation v:*

(i) *there exists an infinite computation of K from the valuation v in $\mathfrak{A}$ iff there exists an infinite computation of M from the valuation v in $\mathfrak{A}$.*

(ii) $K_{\mathfrak{A}} = M_{\mathfrak{A}}$ *off X, i.e.,*

$$(v, v') \in K_{\mathfrak{A}} \quad implies$$
$$(\exists v'')(v, v'') \in M_{\mathfrak{A}} \quad and \quad v' = v'' \text{ off } X$$

and

$$(v, v') \in M_{\mathfrak{A}} \quad implies$$
$$(\exists \bar{v})(v, \bar{v}) \in K_{\mathfrak{A}} \quad and \quad v' = \bar{v} \text{ off } X. \qquad \square$$

LEMMA 7.1. *Let X, Y be arbitrary sets of variables and $K_1 \sim K_2$ off X and $M_1 \sim M_2$ off Y. The following properties are then valid:*

(i) *If $V(\gamma) \cap X = \emptyset$, then*

if γ **then** K_1 **else** M_1 **fi** $\sim$ **if** γ **then** K_2 **else** M_2 **fi** off $(X \cup Y)$.

(ii) *If X is a set of variables inessential (cf. Chapter III, § 6) for M_1 and M_2, then*

begin K_1 ; M_1 **end** $\sim$ **begin** K_2 ; M_2 **end** off $(X \cup Y)$.

(iii) *If $V(\gamma) \cap X = \emptyset$ and X is a set of variables inessential for K_1 and K_2, then*

while γ **do** K_1 **od** $\sim$ **while** γ **do** K_2 **od** off X.

(iv) **either** K_1 **or** M_1 **ro** $\sim$ **either** K_2 **or** M_2 **ro** off $(X \cup Y)$.

PROOF. We shall consider case (iv) since the proofs of the remaining cases are similar to those presented in Chapter III, § 6.

Suppose there exists an infinite computation of the program **either** K_1 **or** M_1 **ro** from the valuation v in the structure $\mathfrak{A}$. Hence there exists an infinite computation of K_1 or of M_1 starting from v in $\mathfrak{A}$. By the assumption there exists an infinite computation of K_2 or of M_2 starting from v in $\mathfrak{A}$ and therefore there exists an infinite computation of **either** K_2 **or** M_2 **ro** from the valuation v.

Suppose $(v, v') \in ($**either** K_1 **or** M_1 **ro**$)_{\mathfrak{A}}$. By the definition of semantics $(v, v') \in K_{1\mathfrak{A}}$ or $(v, v') \in M_{1\mathfrak{A}}$. Thus, by the assumption, either there exists a valuation v_1 such that $v' = v_1$ off X and $(v, v_1) \in K_{2\mathfrak{A}}$ or there exists a valuation v_2 such that $v_2 = v'$ off Y and $(v, v_2) \in M_{2\mathfrak{A}}$. Hence there exists a $v'' \in$(either K_2 or M_2 ro)$_\mathfrak{A}$ (v) and $v'' = v'$ off $(X \cup Y)$.

The converse implications are abviously true also.   □

LEMMA 7.2. *For arbitrary sets of variables* X, Y *and for arbitrary programs* K, M, M', *if* $K \sim M$ *off* X *and* $M \sim M'$ *off* Y *then* $K \sim M'$ *off* $(X \cup Y)$

For the proof see Lemma 6.2 from Chapter III.   □

Let us adopt the same definition of the normal form of programs as in the deterministic case. Hence, we shall say that a program M *is in the normal form* iff.

$$M = \textbf{begin } M_1 \text{ ; } \textbf{while } \gamma \textbf{ do } M_2 \textbf{ od end}$$

where γ is an open formula and M_1 and M_2 are programs in which the **while**-operation does not occur.

The following theorem is a generalization of the theorem on the normal form of deterministic programs (cf. Chapter III, § 6):

THEOREM 7.3. *For every program* M *we can find in an effective way a program* M' *in the normal form such that* $M \sim M'$ *off* X, *where* X *is a set of inessential variables for* M *and for* M'.

PROOF. The proof is by induction on the length of the program. It proceeds analogously to those presented in Chapter III, § 6.

In the case of an **either**-program the theorem follows immediately from the fact that the programs M and M' are equivalent up to $\{q\}$, where $q \notin V(M_1) \cup V(M_2) \cup V(K_1) \cup V(K_2)$.

M: **either**
     begin
         M_1 ;
         while γ_1 **do** M_2 **od**
     end
    or
     begin
         K_1 ;
         while γ_2 **do** K_2 **od**
     end
   ro,

M': **begin**

  either $q :=$ **true or** $q :=$ **false ro**;

  if q **then** M_1 **else** K_1 **fi**;

  while $(q \wedge \gamma_1) \vee (\sim q \wedge \gamma_2)$ **do**

    if q **then** M_2 **else** K_2 **fi**

  od

 end.   ☐

Let X be a set of variables inessential for K and for M and let F_X denote the set of all formulas α such that $V(\alpha) \cap X = \emptyset$.

LEMMA 7.4. *If $K \sim M$ off X, then for an arbitrary formula $\alpha \in F_X$*
(1) $\vdash \Diamond M\alpha \equiv \Diamond K\alpha$ *and* $\vdash \Box M\alpha \equiv \Box K\alpha$.

PROOF. Let us observe first that condition (1) is equivalent to

$\vdash \Diamond M\alpha \equiv \Diamond K\alpha$ and $\vdash \Box M$ **true** $\equiv \Box K$ **true**.

By the completeness result the latter condition is equivalent to

$\models \Diamond M\alpha \equiv \Diamond K\alpha$ and $\models \Box M$ **true** $\equiv \Box K$ **true**.

Since the results of K and M may differ in at most X and since by the assumption we consider formulas which do not contain variables X, then for an arbitrary data structure $\mathfrak{A}$ and for an arbitrary valuation v

$\mathfrak{A}, v \models \Diamond K\alpha$ iff $\mathfrak{A}, v \models \Diamond M\alpha$

and

$\mathfrak{A}, v \models \Box K$ **true**   iff $\mathfrak{A}, v \models \Box M$ **true**.

This completes the proof of Lemma 7.4.   ☐

LEMMA 7.5. *If $K \sim M$ off X then for an arbitrary formula $\alpha \in F_X$*

$\vdash \Diamond M\alpha$ *iff* $\vdash \Diamond K\alpha$,

$\vdash \Box M\alpha$ *iff* $\vdash \Box K\alpha$.

This follows immediately from Lemma 7.4.   ☐

For any program K, let $\mathrm{PC}_X(K)$ denote the partial correctness theory of K such that

$$\mathrm{PC}_X(K) = \{(\alpha, \beta) \in F^2 : \ \vdash \big((\alpha \wedge \Diamond K \text{ **true**}) \Rightarrow \Diamond K\beta\big)\}.$$

As the next consequence of Lemma 7.4 we find that if two programs are equivalent then their partial correctness theories are equivalent.

LEMMA 7.6. *If $K \sim M$ off X, then for arbitrary formulas $\alpha, \beta \in F_X$*

$$\vdash ((\alpha \wedge \Diamond K \text{ true}) \Rightarrow \Diamond K\beta) \quad iff$$
$$\vdash ((\alpha \wedge \Diamond M \text{ true}) \Rightarrow \Diamond M\beta),$$

i.e., $\mathrm{PC}_X(K) = \mathrm{PC}_X(M)$.

PROOF. Suppose that $(\alpha, \beta) \notin \mathrm{PC}_X(M)$ and $K \sim M$ off X. By the definition we have

$$\text{non} \vdash ((\alpha \wedge \Diamond M \text{ true}) \Rightarrow \Diamond M\beta).$$

This implies by the completeness theorem that there exists a data structure $\mathfrak{A}$ and a valuation v such that

$$\text{non } \mathfrak{A}, v \models ((\alpha \wedge \Diamond M \text{ true}) \Rightarrow \Diamond M\beta).$$

Hence

(2) $\mathfrak{A}, v \models \alpha$ and $\mathfrak{A}, v \models (\Diamond M \text{ true} \wedge \sim \Diamond M\beta)$.

Thus, by the Completeness Theorem and Lemma 7.4, $\mathfrak{A}, v \models \sim \Diamond K\beta$, and therefore, by (2) non $\mathfrak{A}, v \models ((\alpha \wedge \Diamond K \text{ true}) \Rightarrow \Diamond K\beta)$. Hence $((\alpha \wedge \Diamond K \text{ true}) \Rightarrow \Diamond M\beta)$ is not a theorem of NAL and moreover $(\alpha, \beta) \notin \mathrm{PC}_X(K)$. As a consequence $\mathrm{PC}_X(K) \subset \mathrm{PC}_X(M)$.

Analogously it can be proved that $\mathrm{PC}_X(M) \subset \mathrm{PC}_X(K)$. $\square$

A similar reasoning can be followed for some other versions of partial and total correctness theories. Let us assume the following notation:

$$\mathrm{Cor}X_{\Diamond}(K) = \{(\alpha, \beta) \in F_X^2 : \vdash (\alpha \Rightarrow \Diamond K\beta)\},$$
$$\mathrm{Cor}X_{\Box}(K) = \{(\alpha, \beta) \in F_X^2 : \vdash (\alpha \Rightarrow \Box K\beta)\},$$
$$\mathrm{PCX}_{\Diamond\Diamond}(K) = \{(\alpha, \beta) \in F_X^2 : \vdash ((\alpha \wedge \Diamond K \text{ true}) \Rightarrow \Diamond K\beta)\},$$
$$\mathrm{PCX}_{\Box\Diamond}(K) = \{(\alpha, \beta) \in F_X^2 : \vdash ((\alpha \wedge \Box K \text{ true}) \Rightarrow \Diamond K\beta)\},$$
$$\mathrm{PCX}_{\Diamond\Box}(K) = \{(\alpha, \beta) \in F_X^2 : \vdash ((\alpha \wedge \Diamond K \text{ true}) \Rightarrow \Box K\beta)\},$$
$$\mathrm{PCX}_{\Box\Box}(K) = \{(\alpha, \beta) \in F_X^2 : \vdash ((\alpha \wedge \Box K \text{ true}) \Rightarrow \Box K\beta)\}.$$

LEMMA 7.7. *For arbitrary programs K, M and for an arbitrary partial or total correctness theory* Th *as defined above, if $K \sim M$ off* Th *then* $\mathrm{Th}(K) = \mathrm{Th}(M)$. $\square$

The last problem we shall consider in this section is the following: Is it possible to express by a formula that two programs are equivalent with respect to a set of variables X?

The answer is positive in the case where the non-deterministic language L contains the predicate $=$ (cf. Chapter III, § 7).

Assume that $V(K) = \{z_1, ..., z_n, q_1, ..., q_m\}$, $V(M) = V(K) \cup X$ and $X \cap V(K) = \emptyset$, where z_i is an individual variable for $i \leqslant n$ and q_j is a propositional variable for $j \leqslant m$. Let $K(\overline{yp})$ be a copy of the program K which is obtained by the simultaneous replacement of all occurrences of z_i by y_i for $i \leqslant n$ and all occurrences of q_j by p_j for $j \leqslant m$. Moreover, let

$$\{y_1 \ ... \ y_n, \ p_1, \ ..., \ p_m\} \cap X = \emptyset.$$

LEMMA 7.8. *Let L be a non-deterministic language with equality. Then*

$$K \sim M \text{ off } X \quad iff \quad \vdash \ \square K \text{ true} \ \equiv \ \square M \text{ true} \quad and$$

$$\vdash \bigwedge_{i \leqslant n} \Diamond K(\overline{yp}) \, \Diamond M(y_i = z_i) \wedge \bigwedge_{j \leqslant m} \Diamond K(\overline{yp}) \, \Diamond M(p_j \equiv q_j). \qquad \square$$

BIBLIOGRAPHIC REMARKS

For the motivations of non-determinism in logics of programs see Harel and Pratt (1978c). The results of this chapter concerning the NAL were proved in Mirkowska (1980, 1980b). Dynamic logic, cf. Harel (1979), is another approach to the non-deterministic programs.

CHAPTER VII

PROBLEMS AND THEORIES INSPIRED BY THE LOGLAN PROJECT

This chapter differs in character from those preceding it. It presents problems of semantics which grow up during work on the design and implementation of modern, very high level programming languages like SIMULA, ADA, LOGLAN and others. A sample of current research into the semantics of LOGLAN is presented below. The sections also vary in the degree of descriptive details. Some contain theories which are almost complete, others present problems which are still open. The reader will find a new mathematical model of concurrent computations, a theory of the notion of reference and a few remarks on other semantic problems. As we said earlier the content of this chapter reflects the status of present (1982) work on the formal specification of LOGLAN semantics.

The chapter presents a method for the formal specification of a very high level programming language. The method may be called axiomatic since it brings in axioms; it may be also called algorithmic because of the form of the axioms. It may be called modular since we factorize the semantics of the programming language into modules (or subsystems), then give every module a theory describing its properties and finally put all the constructed theories together in order to give a theory of the system of modules under consideration. This method is exemplified in the sections devoted to the notion of reference.

In 1976 a group working on the design of the LOGLAN programming language had to define the semantics of parallel processes. The models known from the literature did not seem to be adequate for the description of computational processes generated by statements of a very high-level programming language. Consequently, a new mathematical model of concurrent computations called the MAX model was invented. The model facilitates the analysis of programs, since for a given program K and initial state s of a computing system it defines the set of all possible computations which have their origin in the initial configuration $\langle s; K \rangle$ The model is not a scheduler's design, and it is not meant as a concept of implementation. The description of the model might give this impression, but the reader should not be mislead—this is an analytical model. It is easy to observe that the MAX model presented here differs from the ARB model consisting of arbitrary interleavings of atomic actions.

1. CONCURRENT PROGRAMS

A language of concurrent programs is determined by its sets of atomic instructions and open formulas. Informally, by a *concurrent program* we shall mean an expression constructed from atomic instructions and open formulas by means of the program connectives: composition, branching, iteration, non-deterministic choice, and parallel execution.

DEFINITION 1.1. *By a set of concurrent programs we shall understand the least set which contains all assignments and such that*:

1. *If $K_1, ..., K_n$ are programs then* **begin** $K_1 ; ... ; K_n$ **end,** **either** K_1 **or** ... **or** K_n **ro,** **cobegin** $K_1 \| ... \| K_n$ **coend** *are programs.*

2. *If γ is an open formula and K, M are programs, then* **if** γ **then** M **else** K **fi,** **while** γ **do** M **od,** *are programs.*   □

EXAMPLE. Assume that $+$, $-$, $/$ are two-argument functors and $>$, $=$ are two-argument predicates. The expression

> **cobegin**
> > **while** $x >$ eps **do** $x := b - a$; $a := a + x/2$ **od** $\|$
> > **while** $x >$ eps **do** $x := b - a$; $b := b - x/2$ **od**
> **coend**

is then a program.   □

REMARK. In this and in the subsequent examples we shall omit the superfluous parentheses **begin ... end.**   □

DEFINITION 1.2. *By a process we shall mean a maximal instruction contained in a concurrent program M of the form*

$$\textbf{cobegin } K_1 \parallel ... \parallel K_n \textbf{ coend,}$$

i.e., every program $K_1, K_2, ..., K_n$ is a process of the program M. □

2. MAX SEMANTICS

In this section we shall present the definition of MAX-semantics of the programming language introduced above.

The meaning of a concurrent program can be determined by means of the notion of computation. As in the sequential case, a computation is a sequence of configurations such that the consecutive configurations are in the relation of direct successorship. The definition of this relation is the most important point in the presentation of the semantics.

We assume that each process has a processor assigned to it. It is possible to imagine that the nature of these processors is not important and especially that we should ignore the real and the relative speed of the processors. Our main assertion is quite the opposite. In our view it is the nature of the processors which should be taken into consideration. The definition of a computation given below makes use of the assumption of eagerness of processors, that is, processors cannot refuse to make the next step in the computation. However, we cannot predict how long it will take to execute a step.

One of the most important notions in this section is the notion of a *conflict set*. We shall use the following definition.

DEFINITION 2.1. *Let I be a finite set of instructions which consists of a set A of assignment instructions, a set C of conditional instructions and a set W of iteration instructions.*

We shall say that the set I is a conflict set iff there exists a variable x occurring on the left-hand side of an assignment instruction of A which also occurs in another instruction of the set A or in a test formula in an instruction belonging to the set $C \cup W$. □

EXAMPLE 2.1. Let I_1, I_2 be sets of instructions such that

$$I_1 = \{x := y+z; \textbf{ while } x > 0 \textbf{ do } y := \tau \textbf{ od}\},$$
$$I_2 = \{x := y+z, \textbf{ while } y > 0 \textbf{ do } x := \tau \textbf{ od}\}.$$

The set I_1 is a conflict set and the set I_2 is a non-conflict set.   □

The notion of a state of a computation, i.e., a configuration, is different from the analogous notion for sequential computations.

By a configuration of a concurrent computation we shall mean an ordered pair consisting of a valuation of variables and a list of programs in which certain instructions are indicated by an asterisk or circle ∘. The intuitive meaning of these symbols is:

∘—the instruction is under execution,

∗—the instruction is ready, not yet started.

The initial configuration of a computation of a concurrent program K has the form $\langle v; *K \rangle$. To describe the notion of computation we shall give another definition of the notion of direct successorship (see Chapter II).

DEFINITION 2.2. *Let* $\mathfrak{A}$ *be a fixed data structure. The configuration* $\langle v'; M' \rangle$ *is a direct successor of the configuration* $\langle v; M \rangle$ *in the data structure* $\mathfrak{A}$ *iff the configuration* $\langle v'; M' \rangle$ *is obtained from* $\langle v; M \rangle$ *by means of the following non-deterministic algorithm*:

1. *Each mark* ∗ *which precedes the symbol* **begin**, **cobegin** *or* **either** *moves inside the program according to the following rules*:

$\quad$ ∗ **begin** $K_1 ; ...; K_n$ **end** → **begin** ∗ $K_1 ; ...; $ ∗ K_n **end**,

$\quad$ ∗ **cobegin** K_1 ||...|| K_n **coend** → **cobegin** ∗K_1 ||...|| ∗K_n **coend**,

$\quad$ ∗ **either** K_1 **or** ... **or** K_n **ro** → ∗K_i $\quad$ *for arbitrary* $i \leqslant n$.

Repeat the first step until each mark ∗ *precedes an assignment, conditional instruction or iterative instruction. Let* I *be the set of instructions marked* ∘ *or* ∗ *and let* $I_0 \subset I$ *be the set of instructions marked* ∘.

2. *Choose the maximal non-conflict set* J *(a non-deterministic choice) such that*

$$I_0 \subset J \subset I$$

and denote by ∘ *all instructions from the set* $J - I_0$.

3. *Choose an arbitrary non-empty set* $J' \subset J$ *(a non-deterministic choice). Let* $J' = A \cup C \cup W$, *where* A *is a set of assignments, C is the set of conditional instructions and W the set of iterative instructions.*

4. *Replace by v' the result of the simultaneous execution of all assignments from the set A, and replace instructions from set J' by their successors as in Table 2.1.*

Table 2.1

the instruction	replace by	iff
○ if γ then M_1 else M_2 fi	$*M_1$	$\mathfrak{A}, v \models \gamma$
○ if γ then M_1 else M_2 fi	$*M_2$	$\mathfrak{A}, v \models \sim\gamma$
○ while γ do M od	$*$ begin M; while γ do M od end	$\mathfrak{A}, v \models \gamma$
○ while γ do M od	$*$	$\mathfrak{A}, v \models \sim\gamma$
○ ($x := \tau$)	$*$	

5. *Mark out all empty instructions, i.e., all occurrences of* **begin** $*$ **end,**
cobegin $*||$ ... $||*$ **coend,** *replace by* $*$.   □

Let us recall that a direct successor of a configuration $\langle v; M\rangle$ is
any configuration $\langle v'; M'\rangle$ which can be obtained by the following
operations:

(1) moving marks $*$ inside,

(2) choosing a maximal non-conflict subset J of instructions (which
retains a remainder of previous choices),

(3) choosing a subset $J' \subset J$ of instructions that are to be completed in this step,

(4) execution of instructions from J',

(5) deleting empty instructions.

EXAMPLE 2.2. Consider the following configuration:

$$\left\langle v: \frac{x}{v_x}\bigg|\frac{y}{v_y}\,;\; *\text{ cobegin } x := \tau\,||\,y := \eta\,||\right.$$

$$\left. \text{if } \gamma(x) \text{ then } M_1 \text{ else } M_2 \text{ fi coend}\right\rangle.$$

Let us assume that the variable x does not occur in the expression
η and that the variable y occurs neither in the term τ nor in the formula γ.
Suppose that $\mathfrak{A}, v \models \gamma$.

This configuration has six different successors:

$$\left\langle \frac{x}{\tau(v)}\bigg|\frac{y}{v_y}\,;\; \text{cobegin } *||* \text{ if } \gamma(x) \text{ then } M_1 \text{ else } M_2 \text{ fi }||\right.$$

$$\left. \circ\, y := \eta \;\text{ coend}\right\rangle,$$

$$\left\langle \frac{x}{\tau(v)} \middle| \frac{y}{\eta(v)} ;\ \textbf{cobegin } * \ || \ * \textbf{ if } \gamma(x) \textbf{ then } M_1 \textbf{ else } M_2 \textbf{ fi} || \right.$$

$$\left. * \textbf{ coend} \right\rangle,$$

$$\left\langle \frac{x}{v_x} \middle| \frac{y}{\eta(v)} ;\ \textbf{cobegin } \circ x := \tau \ || \ * \textbf{ if } \gamma(x) \textbf{ then } M_1 \textbf{ else } M_2 \right.$$

$$\left. \textbf{fi} || \ * \textbf{ coend} \right\rangle,$$

$$\left\langle v ;\ \textbf{cobegin } * \ x := \tau \ || \ * M_1 \ || \circ y := \eta \textbf{ coend} \right\rangle,$$

$$\left\langle \frac{x}{v_x} \middle| \frac{y}{\eta(v)} ;\ \textbf{cobegin } * x := \tau || \circ \textbf{if } \gamma(x) \textbf{ then } M_1 \textbf{ else } M_2 \textbf{ fi} || \right.$$

$$\left. * \textbf{ coend} \right\rangle,$$

$$\left\langle \frac{x}{v_x} \middle| \frac{y}{\eta(v)} ;\ \textbf{cobegin } * x := \tau || * M_1 || * \textbf{ coend} \right\rangle. \qquad \square$$

DEFINITION 2.3. *The tree of possible computations of a program K at the valuation v_0 in a data structure $\mathfrak{A}$ is defined in the following way:*

The root of the tree is labelled by the configuration $\langle v_0 ; *K \rangle$, i.e., the initial configuration.

If the tree contains the node labelled by configuration $\langle v ; M \rangle$, then this node has as many direct successors as the configuration $\langle v ; M \rangle$ has and they are labelled by the successors of this configuration.

Any node labelled by $\langle v' ; * \rangle$ is a leaf.

Every maximal path of the above tree is called a possible computation of program K at valuation v_0. $\qquad \square$

In this way the semantics MAX is defined. The conflicts are ascertained at the level of instructions and a maximal non-conflict set of instructions is initialized at every step.

3. COMPARISON WITH SOME OTHER CONCEPTS OF CONCURRENCY

The main ideas of MAX semantics are

(i) Double non-deterministic choice (we do not assume that all instructions which start execution will finish in the same step).

(ii) The maximal set of non-conflict actions are taken into consideration in each step (we do not admit lazy processors).

Let us now call the reader's attention to the second non-deterministic choice in the definition of MAX semantics. One may think that this choice is not essntial; however, this is not the case. We shall discuss this problem briefly below.

Let us consider a modification of MAX semantics which is obtained by omitting step (3) of the definition (cf. Definition 2.2), i.e., all instructions chosen for execution will finish in this step (observe that marks are not necessary in this case). Thus, the relation of direct successorship is determined by the following steps:

(1) putting marks ∗ inside,

(2) choosing a maximal non-conflict set J of instructions,

(3) execution of all instructions from J,

(4) deleting empty instructions.

Let us call a semantics with the above direct successorship relation simple MAX semantics, or SMAX semantics for short. Below we shall indicate the difference between MAX and SMAX semantics.

EXAMPLE 3.1. Let M denote the program

cobegin $x := 1$; $x := 2$; $x := y \| y := 3$; $y := 4$ **coend**.

Let v be an arbitrary valuation in the data structure of natural numbers $\mathfrak{N}$. The only possible computation of the program M in SMAX semantics is as follows:

$$\langle v; \text{ \textbf{cobegin} } \ast x := 1; \quad x := 2; \quad x := y \| \ast y := 3; \quad y := 4$$
$$\text{\textbf{coend}} \rangle,$$

$$\left\langle \frac{x}{1} \middle| \frac{y}{3} ; \text{ \textbf{cobegin} } \ast x := 2; \quad x := y \| \ast y := 4 \text{ \textbf{coend}} \right\rangle,$$

$$\left\langle \frac{x}{2} \middle| \frac{y}{4} ; \text{ \textbf{cobegin} } \ast x := y \| \ast \text{ \textbf{coend}} \right\rangle,$$

$$\left\langle \frac{x}{4} \middle| \frac{y}{4} ; \ast \right\rangle.$$

One of the possible computations in MAX semantics is the following:

$$\langle v; \text{ \textbf{cobegin} } \ast x := 1; x := 2; x := y \| \ast y := 3; y := 4$$
$$\text{\textbf{coend}} \rangle,$$

$$\left\langle \frac{x}{1}\middle|\frac{y}{v(y)} \right.;\ \mathbf{cobegin}\ *x := 2;\quad x := y\ \|\ \circ\ y := 3;\quad y := 4$$

$$\left. \mathbf{coend} \right\rangle,$$

$$\left\langle \frac{x}{2}\middle|\frac{y}{3};\ \mathbf{cobegin}\ *x := y\ \|\ *y := 4\ \mathbf{coend} \right\rangle$$

$$\left\langle \frac{x}{3}\middle|\frac{y}{3};\ \mathbf{cobegin}\ *\ \|\ *y := 4\ \mathbf{coend} \right\rangle,$$

$$\left\langle \frac{x}{3}\middle|\frac{y}{4};\ * \right\rangle.$$

As a conclusion from the above we can observe the different algorithmic properties of program M in MAX and in SMAX semantics,

$$\mathfrak{R} \models_{\text{SMAX}} \square M(x = y) \quad \text{and} \quad \mathfrak{R} \models_{\text{MAX}} \Diamond M(x \neq y). \qquad \square$$

To complete these considerations, let us observe that any computation of a program in SMAX semantics is one of the possible computations in MAX semantics. Thus any result obtained for a program in SMAX semantics is also a result of that program in MAX semantics starting from the same valuation. We shall express our observation shortly as

(5)   SMAX $\subset$ MAX,

One of the simplest models of concurrency is known as ARB semantics. The relation of direct successorship differs from the analogous relation defined for SMAX semantics in the second step. We take an arbitrary set of non-conflict instructions instead of the maximal one.

To obtain the next configuration of a computation in an ARB semantics we ought to proceed by the following steps:

(6) putting marks $*$ inside the program,

(7) choosing an arbitrary set of non-conflict instructions J,

(8) execution of all instructions from J,

(9) deleting empty instructions.

It is a straightforward consequence of the definition that

(10)   SMAX $\subset$ ARB   and   MAX $\subset$ ARB.

To see the difference between MAX and ARB semantics, let us consider the following example.

EXAMPLE 3.2. Let p and q be propositional variables and let K be the program

$\mathbf{cobegin}\ p := \mathbf{false}\ \|\ q := \mathbf{true};\ p := q\ \mathbf{coend}.$

In Figures 3.1 and 3.2 we present the set of all possible computations of K in MAX semantics and in ARB semantics.

ARB   semantics

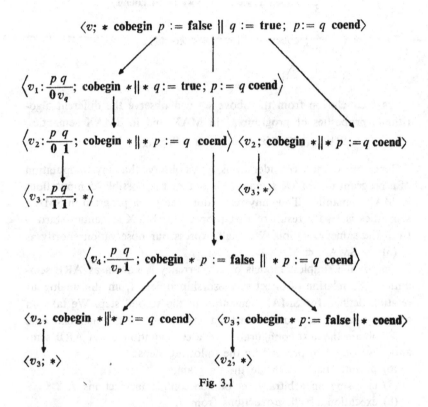

Fig. 3.1

Thus in ARB semantics it is possible that after the execution of K the formula p holds and it is possible that this formula does not hold after another execution of K, i.e.,

$$\models_{ARB} (\Diamond Kp \wedge \Diamond K{\sim}p).$$

In MAX semantics, however, it is necessary that after every execution of K the formula p holds, i.e.,

$$\models_{MAX} \Box Kp. \qquad\qquad\qquad\qquad\qquad\qquad\qquad \Box$$

MAX   semantics

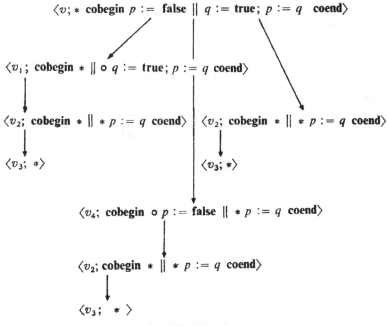

Fig. 3.2

REMARK. One may think of an ARB′ semantics with two non-deterministic choices:

(i) The first choice of arbitrary non-conflict instructions which start execution in one step.

(ii) The second choice of an arbitrary subset (of the previously chosen set) of instructions which will finish execution in this step.

Thus the only difference between MAX and ARB′ semantics would be in the word maximal (cf. Definition 2.2). However, as can easily be seen, in this case we can replace the two non-deterministic choices by only one. Hence   ARB′ = ARB.   □

In all the semantics discussed above we have not assumed any restrictions on the set of processors, i.e., the number of processors was potentially infinite. Obviously in practice any computer has a bounded number of processors. We shall discuss the consequences of this assumption below.

Suppose that the computer we are talking about has only n processors. Thus in all kinds of semantics we can consider at most n instructions to be executed simultaneously. Let us call such semantics ARB(n), SMAX(n), MAX(n), respectively.

LEMMA 3.1. *For every natural number i,* ARB(i) = ARB($i+1$).

PROOF. It is obvious by definition that every computation of an arbitrary program K in ARB(i) semantics is also a computation of the program K in ARB($i+1$) semantics. Hence ARB(i) $\subset$ ARB($i+1$).

Conversely, every computation of a program K in ARB($i+1$) semantics can be simulated in ARB(i) semantics.

Suppose $I_1, ..., I_{i+1}$ are instructions which have been chosen for execution in a configuration $\langle v; M \rangle$ of some computation in ARB($i+1$) semantics. Let $\langle v'; M' \rangle$ be the direct successor of $\langle v; M \rangle$,

$$\langle v; M \rangle \xrightarrow{\quad I_1 \& ... \& I_{i+1} \quad} \langle v'; M' \rangle.$$

By the definition of semantics, the instructions $I_1, ..., I_{i+1}$ create a non-conflict set, hence the result v' can be obtained in two steps:

1. By the execution of $I_1, ..., I_i$ in the first step, and
2. by the execution of I_{i+1} in the second step.

Thus, for an appropriate valuation v'' and an appropriate program M'' we can replace the above-mentioned fragment of computation by the following:

$$\langle v; M \rangle \xrightarrow{\quad I_1 \& ... \& I_i \quad} \langle v''; M'' \rangle \xrightarrow{\quad I_{i+1} \quad} \langle v'; M' \rangle.$$

Such a transformation applied to an ARB($i+1$) computation results in an ARB(i) computation with the same result. Hence ARB($i+1$) $\subset$ ARB(i).   □

As a conclusion from the above simple observation we have, for every natural number i,

$$\text{ARB}(1) = \text{ARB}(i),$$

Thus the semantics ARB(1) also called *multiplexing* and ARB semantics are not essentially different.

Observe that

$$\mathrm{ARB}(1) = \mathrm{MAX}(1) = \mathrm{SMAX}(1).$$

The same problem has a different solution for $\mathrm{SMAX}(n)$ semantics from the one it has for $\mathrm{MAX}(n)$ semantics.

EXAMPLE 3.3. Let p be a propositional variable and $\mathfrak{N}$ a data structure of natural numbers. The behaviour of the following program M is different in $\mathrm{SMAX}(n+1)$ semantics from that in $\mathrm{SMAX}(n)$ semantics.

> M: **cobegin**
> $p :=$ **false** $\|$
> $x_1 :- x_1 + 1;$ **while** p **do** $x_1 :- x_1 + 1$ **od** $\|...\|$
> $x_n := x_n + 1;$ **while** p **do** $x_n := x_n + 1$ **od**
> **coend.**

$\mathrm{SMAX}(n+1)$ *semantics*

$$\left\langle \frac{p\ \ x_1 \dots x_n}{1\ \ 0 \dots 0};\ * M \middle\rangle \right.,$$

$$\left\langle \frac{p\ \ x_1 \dots x_n}{0\ \ 1 \dots 1};\ \mathbf{cobegin} * \mathbf{while}\dots\|\dots\| * \mathbf{while}\dots\mathbf{coend} \right\rangle,$$

$$\left\langle \frac{p\ \ x_1 \dots x_n}{0\ \ 1 \dots 1};\ * \right\rangle.$$

Let v be a valuation in $\mathfrak{N}$ such that $v(x_i) = 1$ for $i \leqslant n$. Hence we have the following properties: for every $i \leqslant n$,

$$\mathfrak{N}, v \models_{\mathrm{SMAX}(n+1)} \square M(x_i = 1)$$

and

$$\mathfrak{N} \models_{\mathrm{SMAX}(n+1)} \square M\ \mathbf{true}.$$

$\mathrm{SMAX}(n)$ *semantics*

One of the possible computation of the program M in the data structure $\mathfrak{N}$ is as follows:

$$\left\langle \frac{p\ \ x_1 \dots x_n}{1\ \ 1 \dots 1};\ \mathbf{cobegin} * p := \mathbf{false} \| * \mathbf{while}\ p\ \mathbf{do}\ x_1 := x_1 + 1 \right.$$

$$\mathbf{od} \|\dots\| * \mathbf{while}\ p\ \mathbf{do}\ x_n := x_n + 1\ \mathbf{od}\ \mathbf{coend} \right\rangle,$$

$$\left\langle\frac{p\ \ x_1\ ...\ x_n}{1\ \ 1\ \ ...\ \ 1}; \textbf{cobegin}\ *p := \textbf{false}\ ||\ *x_1 := x_1 + 1;\ \textbf{while}\ p\ \textbf{do}\right.$$

$$\left. x_1 := x_1 + 1\ ||...\ ||\ *\ x_n := x_n + 1;\ \textbf{while}\ ...\ \textbf{coend}\right\rangle,$$

$$\left\langle\frac{p\ \ x_1\ ...\ x_n}{1\ \ 2\ \ ...\ \ 2}; \textbf{cobegin}\ *\ p := \textbf{false}\ ||*\ \textbf{while}\ p\ \textbf{do}\ x_1 := x_1 + 1\right.$$

$$\left.\textbf{od}\ ||\ ...\ ||\ *\ \textbf{while}\ p\ \textbf{do}\ x_n := x_n + 1\ \textbf{od}\ \textbf{coend}\right\rangle,$$

Hence, for every natural number i

$$\mathfrak{N}, v \models_{\mathrm{SMAX}(n)} \Diamond M(x_1 = ... = x_n = i)$$

$$\mathfrak{N} \models_{\mathrm{SMAX}(n)} (p \Rightarrow \sim \Box M\ \text{true}).$$

MAX(n) *semantics*

One of the possible computations of the program M in the structure $\mathfrak{N}$ is the following

$$\langle v; * M \rangle$$

$$\left\langle\frac{p\ \ x_1\ x_2\ ...\ x_n}{1\ \ 0\ \ 1\ \ ...\ \ 1}; \textbf{cobegin}\ *\ p := \textbf{false}\right.$$

$$\circ\ x_1 := x_1 + 1;\ \textbf{while}\ p\ \textbf{do}$$

$$x_1 := x_1 + 1\ \textbf{od}\ ||$$

$$*\ \textbf{while}\ p\ \textbf{do}\ x_2 := x_2 + 1\ \textbf{od}\ ||\ ...\ ||$$

$$\left. *\ \textbf{while}\ p\ \textbf{do}\ x_n := x_n + 1\ \textbf{od}\ \textbf{coend}\right\rangle$$

$$\left\langle\frac{p\ \ x_1\ x_2\ x_3\ ...\ x_n}{1\ \ 0\ \ 1\ \ 1\ \ ...\ \ 1}; \textbf{cobegin}\ *\ p := \textbf{false}||\right.$$

$$\circ\ x_1 := x_1 + 1;\ \textbf{while}\ ...\ ||$$

$$*\ x_2 := x_2 + 1;\ \textbf{while}\ ...\ ||\ ...\ ||$$

$$\left. *\ x_n := x_n + 1;\ \textbf{while}\ ...\ \textbf{coend}\right\rangle$$

$$\left\langle\frac{p\ \ x_1\ x_2\ x_3\ ...\ x_n}{1\ \ 1\ \ 2\ \ 2\ \ ...\ \ 2}; \textbf{cobegin}\ *\ p := \textbf{false}||\right.$$

$$*\ \textbf{while}\ p\ \textbf{do}\ x_1 := x_1 + 1\ \textbf{od}||\ ...\ ||$$

$$\left. *\ \textbf{while}\ p\ \textbf{do}\ x := x + 1\ \textbf{od}\ \textbf{coend}\right\rangle,$$

...

It is easy to see that every n-element sequence of natural numbers is a possible result of the program M in MAX(n) semantics.   □

The conclusion is

$$\mathrm{MAX}(n) \neq \mathrm{SMAX}(n) \quad \text{and} \quad \mathrm{MAX}(n) \neq \mathrm{MAX}(n+1),$$

$$\mathrm{SMAX}(n) \neq \mathrm{SMAX}(n+1).$$

4. A COMPARISON OF MAX AND ARB SEMANTICS IN THE CASE
OF PETRI NETS

The previous section has indicated the differences between MAX and
ARB semantics. Now we shall discuss the same problem on the basis
of Petri nets, to show that the distinction is deeper than could be seen
from the examples given. We shall start with the necessary definitions.

DEFINITION 4.1. *By a Petri net we understand a quintuple*

$$PN = \langle PL, TR, BACK, FOR, m_0 \rangle,$$

where PL *and* TR *are finite disjoint sets, and* BACK, FOR, *and* m_0
are functions such that

$$BACK: PL \times TR \to N,$$
$$FOR: TR \times PL \to N,$$
$$m_0: PL \to N.$$ □

Each Petri net PN can be regarded as a bipartite graph

$$G = (PL \cup TR, E),$$

where PL∪TR is the set of all vertices and the set of edges is equal
to $\{(p, t): BACK(p, t) > 0\} \cup \{(t, p): FOR(t, p) > 0\}$. The elements of
the set PL we shall call *places* and the elements of the set TR—*transitions*. An edge of the form (p, t) will be called an *in-arrow* of the transition t. The function BACK determines the capacity of in-arrows.
An edge of the form (t, p) will be called an *out-arrow* of the transition t. The function FOR determines the capacity of out-arrows. The
function m_0 is called the *initial marking* of the net PN.

EXAMPLE 4.1. The graph shown in Figure 4.1 (p. 312) is a Petri net
in which $p_1, ..., p_5$ are places and $t_1, ..., t_4$ are transitions. The initial
marking is described by dots at the corresponding places and the values
of functions BACK and FOR are indicated on the arcs.

DEFINITION 4.2. *We shall say that transitions* $t_1, ..., t_n$ *can fire
simultaneously at a marking* m *in a Petri net* PN = (PL, TR, BACK,
FOR, m_0) *iff for every place* $p \in$ PL *the following condition holds*:

$$m(p) \geqslant \sum_{i \leqslant n} BACK(p, t_i).$$

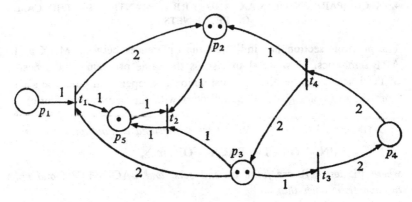

Fig. 4.1

If the condition does not hold, then we say that transitions $\{t_1, \dots, t_n\}$
are in conflict at the marking m.   □

DEFINITION 4.3. *Let m, m' be arbitrary markings. We shall say that
the marking m' is the result of the simultaneous firing of transitions*
$t_1, \dots, t_n$ *at the marking m in a Petri net PN iff*
(a) *the set* $t_1, \dots, t_n$ *of transitions can be fired simultaneously at the
marking m,*
(b) *for an arbitrary* $p \in PL$,

$$m'(p) = m(p) - \sum_{i \leqslant n} BACK(p, t_i) + \sum_{i \leqslant n} FOR(t_i, p).$$ □

DEFINITION 4.4. *A sequence of pairs* $\{(m_J, c_J)\}_{j \in J}$, *where* $J \subset N$,
will be called a firing sequence in SMAX semantics (ARB semantics) iff
(i) *for every* $j \in J$ *the set* c_J *is a maximal (an arbitrary) set of transitions that can fire at marking* m_J,
(ii) *the marking* m_{J+1}, *is the result (if it is defined) of firing* c_J *at* m_J. □

EXAMPLE 4.2. Let us return to Figure 4.1. Let m_0 be the initial marking described on the graph. There are three possible non-conflict sets
of transitions in marking m_0:

$$\{t_2\}, \quad \{t_3\} \quad \text{and} \quad \{t_2, t_3\}.$$

Observe that the transitions t_1 and t_4 cannot be fired at m_0.

Below we present three firing sequences in ARB semantics. The first example is a finite sequence:

$$\left\langle \frac{p_1\ p_2\ p_3\ p_4\ p_5}{0\ \ 2\ \ 2\ \ 0\ \ 1}\ ;\ \{t_2\} \right\rangle,$$

$$\left\langle \frac{p_1\ p_2\ p_3\ p_4\ p_5}{0\ \ 1\ \ 1\ \ 0\ \ 1}\ ;\ \{t_2\} \right\rangle,$$

$$\left\langle \frac{p_1\ p_2\ p_3\ p_4\ p_5}{0\ \ 0\ \ 0\ \ 0\ \ 1}\ ;\ \right\rangle.$$

Clearly, after two repetitive firings of the transition t_2 the net is dead—no transition can be fired.

The second example shows an infinite firing sequence such that the values of the marking functions are not bounded

$$\langle m_0, \{t_3\} \rangle, \langle m_1, \{t_4\} \rangle, \langle m_2, \{t_3\} \rangle, \langle m_3, \{t_4\} \rangle, \ldots$$

The sequence of consecutive markings is defined as follows:

$$m_{2i+1} : \frac{p_1\ \ p_2\ \ \ \ p_3\ \ \ \ p_4\ \ p_5}{0\ \ i+2\ \ i+1\ \ 2\ \ 1}\ ,$$

$$m_{2i} : \frac{p_1\ \ p_2\ \ \ \ p_3\ \ \ p_4\ p_5}{0\ \ i+2\ \ i+2\ \ 0\ \ 1}\ ,$$

where i is an arbitrary non-negative integer.

The third example presents an infinite cyclic firing sequence

$$\langle m_0, \{t_2, t_3\} \rangle, \langle m_1, \{t_4\} \rangle, \langle m_0, \{t_2, t_3\} \rangle, \langle m_1, \{t_4\} \rangle, \ldots,$$

where

$$m_1 : \frac{p_1\ p_2\ p_3\ p_4\ p_5}{0\ \ 1\ \ 0\ \ 2\ \ 1}\ . \qquad\qquad \Box$$

We observed earlier (cf. § 3 of this chapter) that the semantics MAX and ARB are different. Now we shall mention observations showing that this difference is essential.

Consider a net in Figure 4.2 (p. 314).

In SMAX semantics the net behaves as follows:

(1)   **if** $p > 0$ **then** t_1 **else** t_2 **fi.**

The meaning of the same net in ARB semantics would read

        if $p = 0$ **then** t_2 **else either** t_1 **or** t_2 **ro fi.**

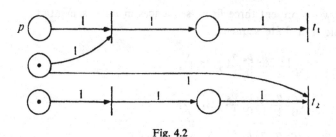

Fig. 4.2

A small modification of the net mentioned in Figure 4.2 gives a net (Figure 4.3) which analysed in SMAX semantics behaves as follows:

(2)   **begin while** $p > 0$ **do** t_1 **od;** t_2 **end.**

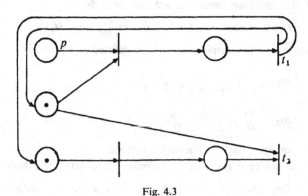

Fig. 4.3

As a consequence of the above observations, every partial recursive function can be computed by a Petri net with SMAX semantics. Hence, the stop problem for Petri nets with SMAX semantics is undecidable. On the other hand, it was proved by E. Meyr that the reachability problem, and therefore also the stop problem for Petri nets interpreted in ARB semantics are decidable. This suggests that the constructions (1) and (2) cannot be interpretations of any Petri net with ARB semantics. More generally, we can conjecture that there exists a Petri net with SMAX semantics which cannot be simulated by any Petri net with ARB semantics. An example of such a net was given by H. D. Burkhard (1983). The result is, that using the constructions (1) and (2), we can construct a net which interpreted in SMAX seman-

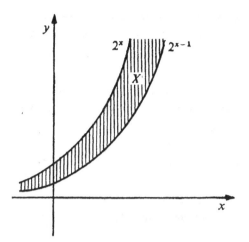

Fig. 4.4

tics, describes the set X presented in Figure 4.4. This example is very important since, by the Pumping Lemma (cf. Burkhard, 1981b), any set computable by a Petri net in ARB semantics contains an infinite linear subset. Obviously, the set described in Figure 4.4 does not contain an infinite line.

5. CRITICAL REMARKS CONCERNING MAX SEMANTICS

In this section we would like to consider the question of whether the MAX model adequately captures the phenomena of parallel distributed computations. We shall start our analysis with examples which shows an unexpected power of MAX semantics.

EXAMPLE 5.1. Consider the following program in the data structure of natural numbers $\mathfrak{N}$:

K: **begin**

    $p := $ **true**;

    cobegin while p **do** $x := x+1$ **od** $\|$ $p := $ **false coend**

  end.

Considered within ARB semantics, the program has the following properties:

$$\mathfrak{N} \models_{\mathrm{ARB}} (\forall k \in \omega)\, (\Diamond K(x > k) \land \sim \Box K \,\mathbf{true}).$$

Analysed in MAX semantics the program behaves fairly, i.e., both processes are active and therefore the program terminates.

$$\mathfrak{N} \models_{MAX} (k = x \Rightarrow \Box K(x = k \vee x = k+1)).$$ ☐

One can compare this example with the remarks of Dijkstra and Lamport (1980) that the termination of the program K means exactly the fairness property of the semantics.

EXAMPLE 5.2. Let us modify our previous example slightly:

> K_1 : **begin**
> > $p :=$ **true**;
> > **cobegin while** p **do** $x := x+1$ **od** $|| \ y := f(y)$;
> > > > $p :=$ **false coend**
>
> **end**.

We have as before

$$\mathfrak{N} \models_{ARB} (\forall k \in \omega) \ (\Diamond K_1(x > k) \wedge \sim \Box K_1 \ \text{true}),$$

and a similar behaviour can be observed in MAX semantics. The program K_1 admits an infinite computation in both semantics.

In ARB semantics there is a possibility of unfair choice: the second process can be delayed for ever.

In MAX semantics the second process will begin execution of the instruction $y := f(y)$. However, the completion of the instruction can be postponed *ad infinitum* by the second non-deterministic choice (cf. Definition 2.2). ☐

This minor disadvantage of the model can easily be overcomed. Suppose that with the initiation of an instruction one can associate a non-negative integer l, with the intention that, each time we make the second non-deterministic choice, the number l should be decreased or the instruction should be terminated. The instruction must be terminated (chosen) when l equals zero. The range of choice for the number l can reflect certain information about the executing system. If there is no limit for l, i.e., if l can be any non-negative integer then the formula

$$(\forall k) \ \Diamond K \ (x > k)$$

will be satisfied in MAX semantics.

Another possibility, that l is taken from a finite set of non-negative integers, seems closer to physical hardware. The number l then corresponds to the number of time units which are necessary in order to complete the instruction.

EXAMPLE 5.3. Suppose we wish to limit the class of computing systems only to those where the relative speed of processors is comparable. Assuming the formula

$$(p \wedge x = k) \Rightarrow \Box \text{ cobegin while } p \text{ do } x := x+1 \text{ od } \|$$
$$y := y \cdot y; p := \text{false coend } (x < k+5)$$

to be valid, we arrive at the conclusion that the formula expresses the following fact:

One multiplication takes no more than five additions.   □

We can end up with the following remark. Formulas like those in the above example can be regarded as axioms about computing systems. Such axioms can accept certain systems and reject others.

Another unexpected aspect of **MAX** semantics can be seen from the following discussion of the well-known case of the philosophers (cf. Figure 5.1).

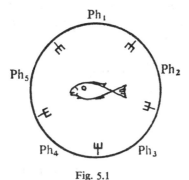

Fig. 5.1

Five philosophers are sitting around the table. There are five forks and a fish. Each philosopher alternately 'thinks' or 'eats'. We assume that eating is possible only when the philosopher has involved two forks. When passing from thinking to eating each philosopher must synchronize his actions with his two neighbours since the forks are shared.

Let us regard the system as a concurrent program written informally as follows:

cobegin

$$\overset{5}{\underset{i=1}{\|}} \ \text{Ph}_i : \textbf{do} \text{ 'think'; 'take one fork'; 'take another}$$

fork'; 'eat' **od**

coend.

We have used instructions which are not totally defined (e.g., 'take one fork') in order to leave more freedom in the system. Anyway, there is a possibility that each philosopher will take the fork on his left, and consequently no one will be able to continue. In the next section we shall discuss such cases.

The situation changes completely if we consider a slight modification of the program and analyse it in MAX semantics:

cobegin

$$\overset{5}{\underset{i=1}{\|}} \ \text{Ph}_i : \textbf{do} \text{ 'think'; 'take forks } f_{i-1}, \ f_{i \, \text{mod} \, 5}';' \text{ eat' } \textbf{od}$$

coend.

The change consists in treating the operation 'take forks f_{i-1}, $f_{i \, \text{mod} \, 5}$' as an atomic one. Due to this fact the analysis of the possible computations will show that no deadlock situation will occur. For every configuration there is a next one. This example might satisfy supporters of the MAX model, but in fact it can be regarded as a source of deep criticism of MAX semantics: it seems that our model is stronger than the system it is modelling.

6. LIBERAL SEMANTICS

In accordance with the criticism of § 5 of this chapter we are now going to change the structure of concurrent programs and their semantics to make them more natural. Moreover, the new semantics will be closer to that of the LOGLAN programming language.

The most important modification is that the semantics itself does not take care of conflicts. We allow all processes to work simultaneously if they are ready. However, it may appear that two processes will try to change the value of the same shared variable. We assume that the result is then undefined, i.e. the processes put a value in the shared variable but we do not know which value.

EXAMPLE 6.1. If two processes try to execute simultaneously $x := 1$ and $x := 2$, then as a result the value of x may become 1, 2, or any possible integer.   □

Obviously such a situation is not a desirable one. Hence, the language will be equipped with control variables, called *semaphores*, and special atomic actions on them, called lock and unlock, which allow us to solve conflicts properly. We assume that the programmer himself will take care of conflicts to avoid undefined results.

Let us study the picture presented in Figure 6.1. The semaphore keeps watch over one or more variables to exclude possible conflict actions.

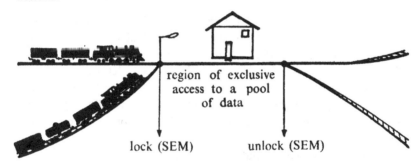

region of exclusive access to a pool of data

lock (SEM)   unlock (SEM)

Fig. 6.1

The intuitive meaning of the action lock(SEM) is to close the semaphore SEM, so that the other process cannot change the variables guarded by this semaphore. The meaning of the other action, unclock(SEM), is just the opposite—to open semaphore SEM when the previous process has finished its action.

The strict definition of the new concurrent language is as follows: Let L be an algorithmic language with the sets V_i and V_0 of individual and propositional variables. We assume that the language L contains the set of control variables *Sem*.

DEFINITION 6.1. *By the atomic program of the language L we shall understand*:

(i) *every assignment instruction and*
(ii) *every expression of the form*

   lock(SEM),   unlock(SEM),

where SEM *is an arbitrary contr 1 variable.*   □

DEFINITION 6.2. *By a set of concurrent programs we shall understand the least set of expressions which contains all atomic instructions and is closed with respect to the same formation rules which appear in Definition* 1.1.   □

EXAMPLE 6.2. The following expression is a concurrent program, where x is an individual variable, SEM is a control variable and 1, 2 are constants:

   cobegin lock(SEM); $x := 1$; unlock(SEM) || lock(SEM);
   $x := 2$; unlock(SEM) **coend.** □

Let $\mathfrak{A}$ be a data structure for the language L. The valuation in $\mathfrak{A}$ consists of three parts: the valuation of individual variables, the valuation of propositional variables and the valuation of control variables. The values of control variables will be 'open' and 'closed'.

By a *configuration of a concurrent computation* we shall mean an ordered pair consisting of a valuation of all variables and a list of programs in which certain instructions are marked by *, ∘ or ⊗. The intuitive meaning of these marks is:

   ∘ —the instruction is under execution,
   * —the instruction is ready for execution,
   ⊗ —the process is passivated (this sign will appear only before control instructions).

DEFINITION 6.3. *We shall say that the configuration $\langle v'; M' \rangle$ is a direct successor of a configuration $\langle v; M \rangle$ in* LIBERAL *semantics iff $\langle v'; M' \rangle$ is obtained by means of the following non-deterministic algorithm:*

1. *Each mark * moves inside the program as long as it precedes the basic instructions according to the rules mentioned in Definition* 2.2.

*Let I_* be the set of all instructions marked with *, but not control actions; Let IC_* be the set of all control actions marked with * and let $I_\circ$ be the set of all instructions marked with ∘. If the set $I_* \cup IC_* \cup I_\circ$ is empty then the configuration $\langle v; M \rangle$ does not have any direct successor.*

2. *Change the marks of all instructions from the set I_* into* ∘; *(all instructions, except control actions, start execution). For every* SEM ∈ *Sem take from the set IC_* only one instruction* lock(SEM) *or* unlock(SEM) *and change its mark into* ∘. *Let I' be the set of all control instructions marked with* ∘.

3. *Choose an arbitrary subset I'' of the set $I_* \cup I_\circ$ (the set of instructions which will finish execution).*

4. *Execute all instructions from the set I' and I''.*

The resulting configuration $\langle v'; M' \rangle$ is obtained by the simultaneous execution of all modifications displayed in Table 6.1 (p. 322). □

EXAMPLE 6.3. Let us consider the following program

> M: **cobegin**
> > lock(SEM); $x := 1$; unlock(SEM)|| lock(SEM);
> > **if** $\gamma(x)$ **then** M_1 **else** M_2 **fi**;
> > unlock(SEM)|| $y := 2$
>
> **coend.**

The reader is asked to compare this with Example 2.1.
The configuration $\langle v; *M \rangle$ has four direct successors:

$$\left\langle v: \frac{\text{SEM } x\, y}{\text{closed}} \; ; \textbf{cobegin} * x := 1; \; \text{unlock(SEM)}|| \right.$$
$$\otimes \text{lock(SEM)}; \textbf{if } \gamma(x) \textbf{ then } \dots ||$$
$$\left. \circ y := 2 \textbf{ coend} \right\rangle.$$

$$\left\langle v': \frac{\text{SEM } x\, y}{\text{closed } 2}; \textbf{cobegin} * x := 1; \; \text{unlock(SEM)}|| \right.$$
$$\otimes \text{lock(SEM)}; \textbf{if } \gamma(x) \textbf{ then}\dots||$$
$$\left. * \textbf{ coend} \right\rangle,$$

$$\langle v; \textbf{cobegin} \otimes \text{lock(SEM)}; x := 1; \; \text{unlock(SEM)}||$$
$$* \textbf{if } \gamma(x) \textbf{ then } M_1 \textbf{ else } M_2 \textbf{ fi}; \; \text{unlock(SEM)}||$$
$$\circ y := 2 \textbf{ coend} \rangle,$$

$$\langle v'; \textbf{cobegin} \otimes \text{lock(SEM)}; x := 1; \; \text{unlock(SEM)}||$$
$$* \textbf{if } \gamma(x) \textbf{ then } M_1 \textbf{ else } M_2 \textbf{ fi}; \; \text{unlock(SEM)}||$$
$$* \textbf{ coend} \rangle. \qquad \square$$

DEFINITION 6.4. *A sequence $\{c_i\}_{0 \leqslant i \leqslant n}$, $n \leqslant \omega$, is a computation of the program M at the valuation v in a data structure $\mathfrak{A}$ in* LIBERAL *semantics iff*

Table 6.1

The instruction	Is replaced by	And the valuation
○ $x := \tau$	*	$v'(x) = \tau_{\mathfrak{A}}(v)$ if Γ' is a non-conflict set undefined otherwise
○ $p := \gamma$	*	$v'(x) = \gamma_{\mathfrak{A}}(v)$ if Γ' is a non-conflict set undefined otherwise
○ if γ then M_1 else M_2 fi	$*M_1$ if $\mathfrak{A}, v \models \gamma$ and Γ' is a non-conflict set $*M_2$ if $\mathfrak{A}, v \models \sim\gamma$ and Γ' is a non-conflict set $*M_1$ or $*M_2$ *otherwise*	$v' = v$
○ while γ do M od	* if $\mathfrak{A}, v \models \sim\gamma$ and Γ' is a non-conflict set $*M$; while γ do M od if $\mathfrak{A}, v \models \gamma$ and Γ' is a non-conflict set * or $*M$; while ... *otherwise*	$v' = v$
○ lock(SEM)	$\otimes$ lock(SEM) *iff* v(SEM) = *closed* * *iff* v(SEM) = *open*	$v' = v$ v'(SEM) = *closed*
○ unlock(SEM)	*choose non-deterministically one instruction marked by* $\otimes$ *and replace it by* *.	v'(SEM) = *open* *iff the set of instructions marked by* $\otimes$ *is not empty.*

(i) $c_0 = \langle v_0; *M \rangle$, where $v_0(x) = v(x)$ for $x \in V_i \cup V_0$ and $v(\text{SEM})$ = open for SEM $\in$ Sem;

(ii) for all i, c_{i+1} is a direct successor of c_i or c_i has no direct successor. $\square$

EXAMPLE 6.4. Let M be the following program:

cobegin
    lock(SEM); $p :=$ **false**; unlock(SEM)||
    $x := x+1$; lock(SEM); **while** p **do** $x := x+1$ **od**;
                                       unlock(SEM)||
    $y := y+1$; lock(SEM); **while** p **do** $y := y+1$ **od**;
                                       unlock(SEM)
coend.

Below we shall present an example of a computation of the program M in LIBERAL semantics in the data structure of natural numbers.

$$\left\langle \frac{\text{SEM } p \; x \; y}{\text{open } 1 \; 0 \; 0} ; \; * M \right\rangle,$$

$$\left\langle \frac{\text{SEM } p \; x \; y}{\text{closed } 1 \; 1 \; 0} ; \text{cobegin } * p := \textbf{false}; \text{unlock(SEM)} \| \right.$$
$$* \text{lock(SEM)}; \textbf{while}... \|$$
$$\circ \, y := y+1; \text{lock(SEM)}; \textbf{while} ... $$
$$\left. \textbf{coend} \right\rangle,$$

$$\left\langle \frac{\text{SEM } p \; x \; y}{\text{closed } 0 \; 1 \; 1} ; \text{cobegin } * \text{unlock(SEM)} \| \right.$$
$$\otimes \text{lock(SEM)}; \textbf{while} ... \|$$
$$\left. * \text{lock(SEM)}; \textbf{while} ... \textbf{coend} \right\rangle,$$

$$\left\langle \frac{\text{SEM } p \; x \; y}{\text{closed } 0 \; 1 \; 1} ; \text{cobegin } * \text{unlock(SEM)} \| \right.$$
$$\otimes \text{lock(SEM)}; \textbf{while} ... \|$$
$$\left. \otimes \text{lock(SEM)}; \textbf{while} ... \textbf{coend} \right\rangle,$$

$$\left\langle \frac{\text{SEM } p \; x \; y}{\text{open } 0 \; 1 \; 1} ; \text{cobegin } * \| * \text{lock(SEM)}; \textbf{while} ... \| \right.$$
$$\left. \otimes \text{lock(SEM)}; \textbf{while} ... \textbf{coend} \right\rangle,$$

$$\left\langle \frac{\text{SEM } p \; x \; y}{\text{closed } 0 \; 1 \; 1} ; \text{cobegin } * \| * \textbf{while} \, p \, \textbf{do} \, x := x+1 \, \textbf{od}; \right.$$
$$\text{unlock(SEM)} \|$$
$$\left. \otimes \text{lock(SEM)}; \textbf{while} ... \textbf{coend} \right\rangle,$$

$$\left\langle \frac{\text{SEM } p \; x \; y}{\text{closed } 0 \; 1 \; 1}; \; \textbf{cobegin} * \; || * \text{unlock(SEM)} || \right.$$

$$\otimes \text{lock(SEM)}; \; \textbf{while } p \; \textbf{do}$$

$$\left. y := y+1 \; \textbf{od} \dots \textbf{coend} \right\rangle,$$

$$\left\langle \frac{\text{SEM } p \; x \; y}{\text{open } 0 \; 1 \; 1}; \; \textbf{cobegin} * \; || * \; || * \text{lock(SEM)}; \; \textbf{while } p \; \textbf{do} \right.$$

$$\left. y := y+1 \; \textbf{od} \dots \textbf{coend} \right\rangle,$$

$$\left\langle \frac{\text{SEM } \; p \; x \; y}{\text{closed } 0 \; 1 \; 1}; \; \textbf{cobegin} * \; || * \; || * \textbf{while } p \; \textbf{do} \; y := y+1 \; \textbf{od}; \right.$$

$$\left. \text{unlock(SEM)} \; \textbf{coend} \right\rangle,$$

$$\left\langle \frac{\text{SEM } p \; x \; y}{\text{open } 0 \; 1 \; 1}; \; * \right\rangle. \qquad\qquad \square$$

A computation of a concurrent program can, as usual, be finite or infinite. However, in LIBERAL semantics, one can observe several other phenomena which could not be observed previously.

One of the most important problems is whether a conflict appears during the execution of the program. Due to LIBERAL semantics a computation in which a conflict has occurred cannot be an object of analysis. We cannot foresee, in general, the behaviour of the whole program. We shall call such computations *conflict* ones. Observe that a conflict computation can be either finite or infinite.

By the definition of the LIBERAL semantics a process can be passivated during the execution of the program. Thus it may happen that in a configuration all processes are passivated, waiting for some semaphores. Obviously such a configuration has no direct successor. We shall call this situation a *deadlock*. Observe that the computation in this case is finite but is not successful. Below we shall present an example of a program with a deadlock computation.

EXAMPLE 6.5. Let I_1, I_2, I_1', I_2' be programs and let SEM1, SEM2 be semaphores. Let us consider the following configuration:

$$\langle v; \; \textbf{cobegin} * \text{lock (SEM1)}; \; I_1 \; \text{lock (SEM2)}; \; I_2||$$

$$* \text{lock (SEM2)}; \; I_1'; \; \text{lock (SEM1)}; \; I_2' \; \textbf{coend} \rangle.$$

If programs I_1 and I_1' do not contain the instructions unlock(SEM1) and unlock(SEM2) and have finite computations then one of the next possible configurations is deadlock

$$\langle v'; \ \textbf{cobegin} \otimes \text{lock(SEM2)}; I_2 \ || \otimes \text{lock(SEM1)}; I_2' \ \textbf{coend}\rangle. \quad \square$$

The next property we should like to mention is *starvation*. This is a property of infinite computations. We shall say that the l-th process of a concurrent program is *starved* if during infinite computation there appears a configuration in which the l-th process is passivated waiting for a semaphore, say SEM, and in the subsequent configurations of this computation the semaphore SEM is opened infinitely many times.

EXAMPLE 6.6. Let x, y be individual variables and let SEM be a semaphore. In the data structure of real numbers with the usual interpretation of functors and relations, we have the following computation of the program M in LIBERAL semantics:

M: **cobegin**
        while $x > 0$ **do** lock(SEM); $x := x+1$;
                                    unlock(SEM) **od** ||
        if $x+y \neq 0$ **then** lock(SEM); $x := 0$; $y := 0$;
                                      unlock(SEM) **fi** ||
        while $y > 0$ **do** lock(SEM); $y := y+1$;
                                    unlock(SEM) **od**
    coend,

$$\left\langle \frac{x \ y \ \text{SEM}}{1 \ 1 \ \text{open}}; * M \right\rangle,$$

$$\left\langle \frac{x \ y \ \text{SEM}}{1 \ 1 \ \text{open}}; \ \textbf{cobegin} * \text{lock(SEM)}; x := x+1; \right.$$
                              unlock(SEM) **while** $x > 0 \ldots$ ||
                        $* \text{lock(SEM)}; x := 0; y := 0;$
                                  unlock(SEM) ||
                        $* \text{lock(SEM)}; \ y := y+1;$
$$\left. \text{unlock(SEM)}; \ \textbf{while} \ y > 0 \ldots \textbf{coend} \right\rangle,$$

. . .

$$\left\langle \frac{x \ y \ \text{SEM}}{2 \ 1 \ \text{closed}}; \ \textbf{cobegin} * \text{unlock(SEM)}; \ \textbf{while} \ x > 0 \ldots \ || \right.$$
$$\left. \otimes \text{lock(SEM)}; \ x := 0; \ldots || \right.$$

$$\otimes \ \text{lock(SEM)}; \ y := y+1;$$

$$\text{unlock(SEM)}; \textbf{while } y > 0 \dots \textbf{coend} \rangle,$$

$$\left\langle \frac{x \ y \ \text{SEM}}{2 \ 1 \ \text{open}} \ ; \textbf{cobegin } * \textbf{ while } x > 0 \textbf{ do } \text{lock(SEM)}; \dots \| \right.$$

$$\otimes \ \text{lock(SEM)}; \ x := 0; \dots \|$$

$$* \ \text{lock(SEM)}; \ y := y+1; \dots \textbf{coend} \Big\rangle,$$

$$\dots$$

$$\left\langle \frac{x \ y \ \text{SEM}}{2 \ 2 \ \text{closed}}; \textbf{cobegin } \otimes \text{lock(SEM)}; \ x := x+1; \dots \| \right.$$

$$\otimes \ \text{lock(SEM)}; \ x := 0; \dots \|$$

$$* \ \text{unlock(SEM)}; \textbf{ while } y > 0 \textbf{ do } \dots$$

$$\textbf{coend} \Big\rangle,$$

$$\left\langle \frac{x \ y \ \text{SEM}}{2 \ 2 \ \text{open}} \ ; \textbf{cobegin } * \text{lock(SEM)}; \ x := x+1; \right.$$

$$\text{unlock(SEM)}; \textbf{ while } \dots \|$$

$$\otimes \ \text{lock(SEM)}; \ x := 0; \dots \|$$

$$* \ \textbf{while } y > 0 \textbf{ do } \dots \textbf{coend} \Big\rangle.$$

In all the next configurations of this computation the second process is awaiting for semaphore SEM while the first and the third process occupy the semaphore in turns alternately.

$$\dots$$

$$\left\langle \frac{x \ y \ \text{SEM}}{i+1 \ i \ \text{closed}} \ ; \textbf{cobegin } * \text{unlock(SEM)}; \textbf{while } x > 0 \textbf{ do} \dots \| \right.$$

$$\otimes \ \text{lock(SEM)}; \ x := 0; \dots \|$$

$$\otimes \ \text{lock(SEM)}; \ y := y+1;$$

$$\text{unlock(SEM)}; \textbf{ while } \dots \textbf{coend} \Big\rangle,$$

$$\left\langle \frac{x \ y \ \text{SEM}}{i+1 \ i \ \text{open}}; \textbf{cobegin } * \textbf{while } x > 0 \textbf{ do} \dots \| \right.$$

$$\otimes \ \text{lock(SEM)}; \ x := 0; \dots \|$$

$$* \ \text{lock(SEM)}; y := y+1; \dots \textbf{coend} \Big\rangle,$$

$$\dots$$

$$\left\langle \frac{x \quad y \quad \text{SEM}}{i+1 \quad i+1 \quad \text{closed}} ; \textbf{cobegin} \otimes \text{lock(SEM)}; \ x := x+1; ... \| \right.$$

$$\otimes \text{lock(SEM)}; \ x := 0; ... \|$$

$$* \text{unlock(SEM)}; \ \textbf{while} \ \ y > 0$$

$$\left. \textbf{do} \ ... \ \textbf{coend} \right\rangle,$$

$$\left\langle \frac{x \quad y \quad \text{SEM}}{i+1 \quad i+1 \quad \text{open}} ; \textbf{cobegin} * \text{lock(SEM)}; x := x+1; ... \| \right.$$

$$\otimes \text{lock(SEM)}; \ x := 0; ... \|$$

$$\left. * \textbf{while} \ \ y > 0 \ \ \textbf{do} \ ... \ \textbf{coend} \right\rangle,$$

$$...$$ □

It is fairly evident that LIBERAL semantics does not assume existence of a central scheduler. We are going to show that under certain assumptions on the form of programs LIBERAL semantics and MAX semantics are equivalent. Hence, MAX semantics does not always require a central synchronizing tool for choosing maximal non-conflict sets.

Let us consider a program K such that for every of its processes and for every atomic instruction At at most one non-local variable of the process occurs in At (i.e. one shared variable). Without loss of generality we can assume that non-local variables do not occur in tests after **if** or after **while**. Let us modify the program K in the following way: for every shared variable x associate a semaphore variable SEM_x, atomic instruction At which contains a shared variable x replace by the three instructions

$$\text{loc(SEM}_x); \quad \text{At}; \quad \text{unlock(SEM}_x).$$

The program obtained in this way will be denoted by K'. Let v be a valuation in A such that values of all semaphore variables are 'open'.

With these assumptions we have the following lemma.

LEMMA 6.1. *Trees of all computations of program K' from the initial valuation v in* MAX *and in* LIBERAL *semantics are equal.*

PROOF. Every MAX computation of K' is equal to a LIBERAL computation of K'. Consider a configuration c of the form

$$\langle v; m_1 a_1 R_1 \| ... \| m_n a_n R_n \rangle,$$

where v is a valuation of variables, $m_1, \ldots, m_n$ are marks, $a_1, \ldots, a_n$ are atomic instructions, $R_1, \ldots, R_n$ are the remaining instructions of processes.

Let $I = \{a_{i_1}, \ldots, a_{i_p}\}$ be the subset of atomic instructions containing all non-passivated instructions.

Every maximal non-conflict subset of I can result, by Definition 6.3 of LIBERAL semantics, and, *vice versa* every set of instructions $J \subset I$ initiated by LIBERAL semantics is a maximal non-conflict set. This is almost self evident. All non-control instructions of I are not conflict, and among the others, i.e. control instructions, one instruction for every semaphore is selected. Thus a maximal non-conflict set will have marks ∘ (under execution).

Observe that every maximal non-conflict set can be defined as a result of the corresponding step in LIBERAL semantics. The remaining details of the proof are straightforward.   □

7. AN ALGORITHMIC THEORY OF REFERENCES

We now proceed to other questions connected with the LOGLAN project; namely, the problems related to the notion of reference and to concatenable declarations of modules.

In Chapter IV we developed algorithmic theories of data structures. In spite of progress made by the application of algorithmic logic, the theories are abstract ones. In § 8 of this chapter the reader will find examples of programming phenomena which cannot be explained on the ground of axiomatic, abstract theories of data structures. In order to understand these phenomena fully, one needs a knowledge of the notion of reference and its properties.

To conclude this chapter, we wish to indicate the rich variety of problems inspired by concatenation rule of module declarations (also called *prefixing*). Introduced for the first time in SIMULA-67, this has found full, unrestricted and efficient implementation in LOGLAN. Problems with the implementation of this concatenation rule are richer than those of the copy rule for elimination of procedure calls (cf. Langmaack, 1979). The numerous applications of prefixing also make it a valuable object of study.

A *reference* is to be understood as an element of a system in which the following operations: reserve a portion of memory cells (frame), release a portion reserved earlier, check whether a frame is reserved, are admissible. These operations lead from one state of memory management to another. Hence, a system of memory management is a two-sorted system with Fr being the sort of frames and St the sort of memory states. On closer examination. we see that the reserve operation splits into two parts: newfr—find a free frame, and insert—an operation which reserves a frame by inserting it into the set of reserved frames. The data structure for memory management is any system with the following signature which satisfies the postulates listed below:

$$\langle \text{Fr} \cup \text{St}; \text{insert}, \text{delete}, \text{newfr}, \textbf{none}, \text{allfree}, \text{member}, = \rangle,$$

where

$$\text{insert}: (\text{Fr} - \textbf{none}) \times \text{St} \rightarrow \text{St}.$$

Given a frame f and a state s, insert (f, s) gives the new state in which frame f is reserved;

$$\text{delete}: \text{Fr} \times \text{St} \rightarrow \text{St}.$$

The value of delete (f, s) is the state s' in which f is freed;

$$\text{newfr}: \text{St} \rightarrow \text{Fr}.$$

newfr(s) brings a frame free in the state s;

> **none** $\in$ Fr—a distinguished frame called *empty frame*;
>
> allfree $\in$ St—a distinguished state of memory in which all frames are free;
>
> member: $\text{Fr} \times \text{St} \rightarrow B_0$.

Relation member(f, s) is satisfied iff frame f is free in state s.

Postulates

P1. For every state $s \in$ St the set of reserved frames is finite.

P2. Operation insert reserves at most one frame f in a given state. Moreover, for every $f' \neq f$ the status of f' in s remains unchanged in $s' = \text{insert}(f, s)$.

P3. Operation delete frees at most one frame f in a given state s. Moreover, for every $f' \neq f$ the status of f' in s remains unchanged in $S' = \text{delete}(f, s)$.

P4. For every state s the value newfr(s) is a frame free in s.

P5. In the state allfree every frame is free.

P6. Frame **none** is not free in any state.

P7. For every frame $f = \textbf{none}$ there exists a state s such that newfr$(s) = f$.

P8. The set of memory frames Fr is denumerable.

P9. The operation insert does not admit frame none as an argument.

The specific axioms of the algorithmic theory of reference ATR contain some postulates, while others can be deduced as theorems of ATR.

ATR1.   **begin** $s' := $ allfree;

        while $s = $ allfree **do**

           $f_i = \text{newfr}(s')$;

           if member(f, s) **then** $s := $ delete(f, d) **fi**;

           $s' := $ insert(f, s')

        od;

     end true

in every state s only a finite number of frames satisfies the member (f, s),

ATR2. $(s' := \text{delete}(f, s))\,(\sim\text{member}\,(f, s')\wedge$

                      $\wedge\,(f' \neq f \Rightarrow (\text{member}(f', s) \equiv \text{member}(f', s'))))$,

ATR3. $(f \neq \textbf{none} \Rightarrow (s' := \text{insert}(f, s))\,(\text{member}(f, s')\wedge$

                      $\wedge\,(f' \neq f \Rightarrow (\text{member}(f', s) \equiv \text{member}(f', s')))))$,

ATR4. $\text{newfr}(s) \neq \textbf{none}$,

ATR5. $\sim\text{member}\,(\text{newfr}(s), s)$

(for every state s, newfr(s) is a free frame in s),

ATR6. $\sim\text{member}(f, \text{allfree})$,

ATR7. $\sim\text{member}\,(\textbf{none}, s)$.

THEOREM 7.1. *Theory* ATR *is consistent.*

PROOF. The following system is a model of ATR:

$$\langle N \cup \text{Fin}\,(N - \{0\}), \text{in, del, nfr, } 0, \varnothing, \text{mb, } = \rangle,$$

where N is the set of natural members, Fin $(N - \{0\})$ is the family of finite subsets of $N - \{0\}$, $\varnothing$ have an obvious meaning and the operations are defined as follows:

$$\text{in}(i, s) = \begin{cases} s \cup \{i\} & \text{if } i \neq 0, \\ \text{undefined} & \text{if } i = 0, \end{cases}$$

$$\text{del}(i, s) = s - \{i\},$$

$$\text{nfr}(s) = \min\,(N - s - \{0\}),$$

$$\text{mb}(i, s) = i \in s.$$

By simple verification we observe that all the axioms ATR1–ATR7 are valid in the above system. $\square$

In the sequel we shall consider an arbitrary model $\mathfrak{M}$ of ATR. We shall study the properties of the set Fr of all frames of the model.

THEOREM 7.2. *For every non-empty frame* $f \neq$ **none** *there exists a state* s *such that* newfr$(s) = f$, *i.e. the formula*

$$(f \neq \textbf{none} \Rightarrow (s := \text{allfree}) \, (\textbf{while} \; \text{newfr}(s) \neq f \; \textbf{do}$$
$$s := \text{insert}(\text{newfr}(s), s) \, \textbf{od}) \, \textbf{true})$$

is a theorem of ATR.

PROOF. Let $f \neq$ **none**. Let $s' = \text{insert}(f, \text{allfree})$. By axiom ATR1 it follows that after a finite iteration of the instruction s' $:= \text{insert}(\text{newfr}(s), s')$, $\text{newfr}(s') = f$. $\qquad \square$

THEOREM 7.3. *The set* Fr *of frames is infinite.*

PROOF. Suppose the contrary, i.e. that $\text{Fr} = \{f_1, \ldots, f_n, \textbf{none}\}$ for some n. Define $s = \text{insert}(f_1, \text{insert}(f_2, \ldots \text{insert}(f_n, \text{allfree}) \ldots))$. By ATR3 and ATR6 it follows that

(1) $(\forall f \neq \textbf{none}) \; \text{member}(f, s)$.

Consider the element newfr(s). By ATR4 we have $\text{newfr}(s) \neq \textbf{none}$ and by ATR5 we obtain $\sim\text{member}(\text{newfr}(s), s)$, which contradicts (1). $\qquad \square$

DEFINITION 7.1.
$$f_1 \leqslant f_2 \stackrel{\text{df}}{\equiv} (f_1 = \textbf{none} \; \vee$$
$$\textbf{begin} \; s := \text{allfree}; \; \text{bool} := \textbf{false}; \; \text{rel} := \textbf{false};$$
$$\textbf{if} f_2 \neq \textbf{none then}$$
$$\textbf{while bool do}$$
$$f := \text{newfr}(s);$$
$$\textbf{if} f = f_1 \textbf{ then } \text{rel} := \text{bool} := \textbf{true else}$$
$$\textbf{if} f = f_2 \textbf{ then } \text{bool} := \textbf{true else}$$
$$s := \text{insert}(f, s) \textbf{ fi}$$
$$\textbf{fi}$$
$$\textbf{od}$$
$$\textbf{fi}$$
$$\textbf{end} \; \text{rel}). \qquad \square$$

LEMMA 7.4. *Relation* $\leqslant$ *is a linear order.* $\qquad \square$

LEMMA 7.5 *The set* Fr *with* $\leqslant$ *is of order type* ω.

PROOF. The first element in Fr is **none** since $(\forall f)$ (**none** $\leqslant f$). We shall define the successor operation f^* in Fr as follows:

$$\mathbf{none}^* \overset{\text{df}}{=} (s := \text{allfree}) \text{ newfr}(s).$$

For $f \neq$ **none** we put

$$f^* \overset{\text{df}}{=} (s := \text{allfree}; \quad \textbf{while} \quad \text{newfr}(s) \neq f \quad \textbf{do}$$
$$s := \text{insert}(\text{newfr}(s), s) \; \textbf{od}; \, s := \text{insert}(f, s)) \text{ newfr}(s).$$

The operation $*$ is defined correctly (cf. Chapter IV, § 2). Consider the set $X = \{f' : f \leqslant f' \wedge f' \neq f\}$. The reader will verify that f^* is the least element in the set X. To complete the proof we should check the following property:

> if **none** $\in Y$ and for every element f of Y, Y contains successor of the element f, then $Y = $ Fr.

Suppose $Y \neq$ Fr, $f_0 \in$ Fr, $f_0 \notin Y$. By Theorem 7.2 we have

$$(\exists i \geqslant 0) \; (s := \text{allfree})$$
$$\left(s := \text{insert}(\text{newfr}(s), \, s)\right)^i \left(\text{newfr}(s) = f_0\right).$$

Since **none** $\in Y$ we obtain $\mathbf{none}^* = (s := \text{allfree}) \text{ newfr}(s) \in Y$. By the definition of a successor it follows that all elements

$$(s := \text{allfree}) \left(s := \text{insert}(\text{newfr}(s), s)\right)^j \text{newfr}(s), \quad \text{for } j < i$$

are in Y. Hence f_0 also belongs to Y, a contradiction. □

COROLLARY. *For every model of* ATR *theory the ordered set* (Fr, $\leqslant$) *is isomorphic with* $(N, \leqslant)$. □

8. REPRESENTATION THEOREM FOR ATR THEORY

In this section we shall justify our choice of axioms by proving that the set of reference can be identified with the set of natural numbers—the addresses of frames.

DEFINITION 8.1. *By a standard model of* ATR *theory we shall understand any model*

$$\langle N \cup \text{Fin}(N - \{0\}), \text{ in, del, nfr, } 0, \emptyset, \text{ mb, } \Rightarrow \rangle$$

as described in the previous section (p. 330) which can differ in the interpretation of an nfr *operation.*   □

DEFINITION 8.2. *Let* $M = \langle$Fr $\cup$ St, insert, delete, newfr, member, **none**, allfree, $=\rangle$ *be a model of* ATR. *By a reduct of* M *we understand the system which results from* M *by omitting operation* newfr, *i.e.*

$$M' = \langle \text{Fr} \cup \text{St}; \text{insert, delete, \textbf{none}, allfree, member,} = \rangle.$$ □

Observe that all reducts of standard models are identical.

DEFINITION 8.3. *We define*

$$s =_{\text{st}} s' \stackrel{\text{df}}{=} (\forall f) \left(\text{member}(f, s) \equiv \text{member}(f, s') \right).$$

We shall say that states s *and* s' *are equal whenever* $s =_{\text{st}} s'$.   □

THEOREM 8.1 (on the Representation of References). *Let* $\mathfrak{M}$ *be a model of* ATR. *Consider the quotient system* $\mathfrak{M}' = \mathfrak{M}/(=, =_{\text{st}})$ *which results from* $\mathfrak{M}$ *by identification of equal states. The reduct of the model* $\mathfrak{M}$ *is isomorphic with the reduct of a standard model.*   □

The question of reducts may seem strange. First of all let us remark that there are other standard models, e.g. the one in which operation newfr is defined as follows

$$\text{newfr}(s) = \max(s) + 1.$$

The standard model defined above and the standard model of the preceeding section are not isomorphic. On the other hand it is hard to argue about the advantages of one model as compared to the other. One can observe a similarity between the theory of references and the theory of dictionaries. Our remarks concerning the effectiveness of the amember operation of dictionaries may be repeated here. Operation newfr is a selector and in general proofs of the existence of this selector are not constructive. All of them have to use the axiom of choice. The theory described above can be used to explain the semantics of those programming languages whose frames are reusable and where at the same time the construction of a language assures safety, e.g. no attempt will be made to access a variable local for a block when the block itself is closed. Moreover for languages like PASCAL, SIMULA, ADA,

LOGLAN one can apply the notion of reference in order to explain why **new** node(e) $\neq$ **new** node(e) (cf. Chapter IV, § 15), or why

$$(a := \textbf{new } \text{node}(e);\ (b := a;\ a.e := \tau)\ (b.e = \tau).$$

The abstract theories of data structures in Chapter IV can now be expanded by introducing references. Suppose we are given a formalized algorithmic theory T which specifies the properties of "abstract" static objects, i.e. elements of a model of T. Making use of the notion of reference, we shall deal with "dynamic" objects to be conceived of as the ordered pairs

$$\langle \text{reference, static object} \rangle.$$

EXAMPLE. For nodes of binary trees (cf. Chapter IV, §15) we have static objects as triples

$$\frac{v \mid l \mid r}{e \mid n_1 \mid n_2}$$

and dynamic objects in pairs $\left\langle \text{ref}, \left\langle \frac{v \mid l \mid r}{e \mid n_1 \mid n_2} \right\rangle \right\rangle.$ □

Let us consider a few simple properties of dynamic objects and states of computation:

1. Every state of a computation contains a finite number of dynamic objects.

2. If every state s of a computation of two dynamic objects have equal references then they are identical.

3. No operation can update a reference in an existing dynamic object.

In LOGLAN the only operations dealing with references are the operation new t which creates a new dynamic object of type t, and the operation kill(x) which deletes the dynamic object x.

For every algorithmic theory T of an abstract data structure we can construct another dynamic theory by "putting together" theories T and ATR (the algorithmic theory of references). An example of this approach can be found in Oktaba (1981).

We shall classify programming languages into two groups; ALGOL and SIMULA belong to the first group, PASCAL, ADA, and LOGLAN to the second. In the case of a language of the first group we observe that objects created during computation exist until there exists a block containing a type declaration for these objects. For these languages the theory of references is adequate, and the treatment of objects is safe.

In languages of the second group there exist instructions disposing an object like kill(x) in LOGLAN. The effect of the kill instruction is to delete the frame associated with an object from the set of occupied frames. Consequently, such a frame can be allocated for another object. This situation is not safe. Let us consider the following example:

```
block
        unit A: class ... ;
        unit B: class ... ;
            variable A1, A2: A, B1: B;
begin
(i1)        A1 := new A;
(i2)        A2 := A1;
(i3)        kill(A1);
(i4)        B1 := new B;
            ...
end.
```

Let us try to interpret this piece of program in ATR theory. Execution of instruction (i1) results in: finding a free frame f in memory (newfr), reservation of this frame for the object **new** A (insert), assigning the frame f to the variable $A1$. The second instruction (i2) assigns the frame f to the variable $A2$. The situation might look like Figure 8.1.

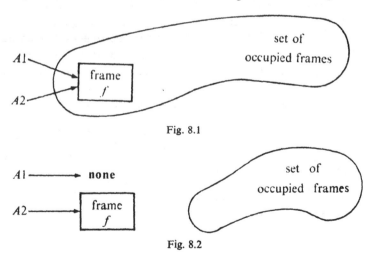

Fig. 8.1

Fig. 8.2

The execution of kill($A1$) would lead to the situation in Figure 8.2. Observe that frame f is no longer reserved but is still accessible via variable $A2$. This could be the conscious decision of the designer of our language but is in contradiction with the assumption that all frames accessible via variables are under the control of the storage management system.

After execution of the subsequent instruction $B1 := \textbf{new } B$ the situation would be shown in Figure 8.3.

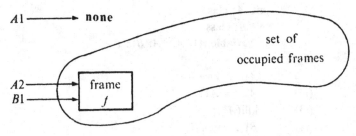

Fig. 8.3

It is now obvious that the proposed solution is not safe. Frame f is accessible via the two different variables $A2$, $B1$, and in different meanings. Since objects of types A and B admit different sets of operations, it is disastrous if one object is interpreted at one point as an object of type A, and at another as an object of type B. We shall not develop this argument, the reader will see all the consequences of such a solution. In this way we have touched on the problem of 'dangling reference'.

Work on LOGLAN has produced another safe solution for the storage management system invented by A. Kreczmar and studied and axiomatized by H. Oktaba. We shall outline it below. The system consists of three sorts: Fr—frames, St—states of reservation and U—univocal references. Variables have references assigned to them.

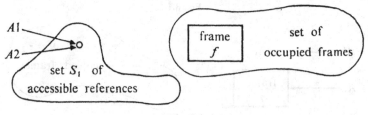

Fig. 8.4

A reference points to a frame. In every state references split into three subsets: S_1—the set of accessible references, S2—the set of used references, S3—the set of fresh references to be used in the future. Let us investigate the four instructions once again.

After the first two instructions the picture is as shown in Figure 8.4. After the kill statement we have the situation in Figure 8.5.

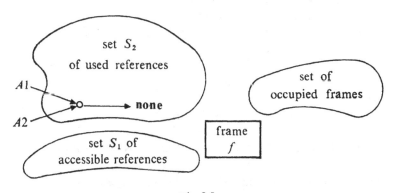

Fig. 8.5

Frame f is not reserved (i.e. it can be used again). The variables $A1$, $A2$ both point to none, a specific frame. It is easy to check that none is the value of $A2$ and to activate handler of exceptional situation, or to program an appropriate test. The execution of $B := \text{new } B$ instruction would lead to the picture in Figure 8.6.

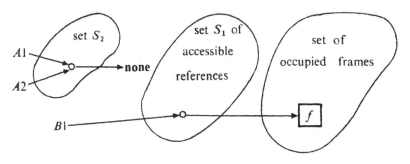

Fig. 8.6

9. SPECIFICATION OF UNIVOCAL REFERENCES

By a *system of univocal references* we shall understand a system of the following signature:

$$\langle UR \cup H; \text{newu}, \text{into}, \text{out}, \text{empty } H, \text{notused}, \text{usable}, \text{used}, = \rangle$$

where UR is a non-empty set called the *set of univocal references*, H is a non-empty set of reference accessibility states disjoint with UR:

newu: $H \to UR$; brings a notused reference,

into: $UR \times H \to H$; converts the status of a notused reference to a usable one,

out: $UR \times H \to H$; converts the status of a usable reference to a used one,

empty $H \in H$; distinguished state, all references are notused,

notused: $UR \times H \to B_0$; notused(u, h) iff the reference u is notused in the state h,

usable: $UR \times H \to B_0$; test if the reference u is usable in the state h,

used: $UR \times H \to B_0$; test if the reference u is waste in the state h.

Moreover the system should satisfy the following specific axioms of the theory of univocal references ATUR.

ATUR1. notused$(u, h) \vee$ usable$(u, h) \vee$ used(u, h).

ATUR2. notused$(\text{newu}(h), h)$.

ATUR3. notused$(u, \text{empty } H)$.

ATUR4. $\sim((\text{notused}(u, h) \wedge \text{usable}(u, h)) \vee (\text{notused}(u, h) \wedge$
$\wedge \text{used}(u, h)) \vee (\text{usable}(u, h) \wedge \text{used}(u, h)))$.

DEFINITION 9.1.

$$u \leqslant u' \overset{\text{df}}{\equiv} \textbf{begin } h := \text{empty } H; \text{bool} := \textbf{false}; \text{rel} := \textbf{false};$$
$$\textbf{while} \sim \text{bool } \textbf{do } u'' := \text{newu}(h);$$
$$\textbf{if } u'' = u \textbf{ then } \text{rel} := \text{bool} := \textbf{true else}$$
$$\textbf{if } u'' = u' \textbf{ then } \text{bool} := \textbf{true else}$$
$$h := \text{into}(u'', h)$$
$$\textbf{fi}$$
$$\textbf{fi}$$
$$\textbf{od}$$
$$\textbf{end } \text{rel}).$$ □

ATUR5. $(\text{usable}(u, h) \vee \text{used}(u, h)) \equiv (u \leqslant \text{newu}(h) \wedge u \neq \text{newu}(h))$.

ATUR6.   notused$(u, h) \Rightarrow (h' := \text{into}(u, h))$ $[\text{usable}(u, h') \wedge u' \neq u$
$$\Rightarrow (\text{notused}(u', h) \equiv \text{notused}(u', h') \wedge$$
$$\wedge \text{usable}(u', h) \equiv \text{usable}(u', h') \wedge$$
$$\wedge \text{used}(u', h) \equiv \text{used}(u', h'))].$$

ATUR7.   usable$(u, h) \Rightarrow (h' := \text{out}(u, h))$ $(\text{used}(u, h') \wedge u' \neq u$
$$\Rightarrow (\text{notused}(u', h) \equiv \text{notused}(u', h') \wedge$$
$$\wedge \text{usable}(u', h) \equiv \text{usable}(u', h') \wedge$$
$$\wedge \text{used}(u', h) \equiv \text{used}(u', h')))).$$

ATUR8.   (**begin** $h' := \text{empty } H; u := \text{newu}(h');$
   while $u \leqslant \text{newu}(h)$ **do**
      if usable(u, h) **then** $h := \text{out}(u, h)$ **fi**;
      $h' := \text{into}(u, h'); u := \text{newu}(h')$
   od
end true).

THEOREM 9.1. *Every two models of the theory* ATUR *which are proper for identity, are isomorphic. Theory* ATUR *is consistent.*   □

A model for ATUR can be found in Oktaba (1981).

10. VIRTUAL MEMORY

A formal specification of a memory management system is presented here as a theory which combines the two latter theories. Access to a memory frame is via a univocal reference. Moreover, references are allocated only once for each $u \in UR$, and in contrast with this, memory frames can be utilized many times over, i.e. one memory frame can be associated with different references. At any moment one reference points to at most one frame.

A *virtual memory system* has three components—a storage management system, a system of univocal references and a memory system, Mem. The latter has a non-empty universe called the set of states of virtual memory. The operations of the Mem subsystem are as follows:

   ref: $UR \times \text{Mem} \rightarrow \text{Fr}$

(Given a univocal reference $u \in UR$ and a memory state $m \in \text{Mem}$ it gives a frame $f \in \text{Fr.}$),

   $\bar{h}: \text{Mem} \rightarrow H$

(For every memory state it gives the state of accessibility.),

$$\bar{s}: \text{Mem} \rightarrow \text{St}$$

(For every memory state it gives the reservation state.),

$$\text{findu}: \text{Mem} \rightarrow UR$$

(In every memory state it gives a notused reference.),

$$\text{reserve}: UR \times \text{Mem} \rightarrow \text{Mem}$$

(For a notused univocal reference u the operation reserve associates with it a free frame $f \in \text{Fr}$. The resulting memory state $m' = \text{reserve}(u, m)$ satisfies three conditions: u is usable, f is reserved and $\text{ref}(u, m') = f$.),

$$\text{kill}: UR \times \text{Mem} \rightarrow \text{Mem}$$

(The frame $f = \text{ref}(u, m)$ is freed, the reference u is used.),

$$\text{freem} \in \text{Mem}$$

(A distinguished memory state in which all references are notused and all frames are free.),

$$\text{inmemory}: UR \times \text{Mem} \rightarrow B_0$$

(Tests whether the given reference u is usable in a given memory state.).

In order to specify the virtual memory system, we shall combine the theories of storage management system and univocal reference system together with the specific axioms of virtual memory.

AVM.1. $\text{inmemory}(u, m) \equiv \text{usable}(u, \bar{h}(m))$,

AVM2. $((\text{ref}(u, m) = \text{ref}(u', m) \wedge \text{ref}(u, m) \neq \textbf{none}) \Rightarrow u = u')$
(Every non-empty frame has exactly one reference.).

AVM3. $\text{ref}(u, m) = \textbf{none} \equiv \sim \text{inmemory}(u, m)$.

AVM4. $\text{find } u(m) = \text{newu}(\bar{h}(m))$.

AVM5. $\text{undef}(u, h(m)) \Rightarrow (m' := \text{reserve}(u, m))[h(m') = \text{into}(u, \bar{h}(m)) \wedge$
$\wedge s(m') = \text{insert}(\text{newfr}(s(m)), s(m)) \wedge \text{ref}(u, m') = \text{newfr}(s(m)) \wedge (u' \neq u$
$\Rightarrow \text{ref}(u', m) = \text{ref}(u', m'))]$

(The operation reserve admits only notused references and consists of associating a free frame $\text{newfr}(s(m))$ to a given reference u and making u and the frame usable in a newly created state m'.).

AVM6. $(m' := \text{kill}(u, m))[(\sim \text{inmemory}(u, m) \Rightarrow m' = m) \vee$
$\vee (\text{inmemory}(u, m) \Rightarrow (\bar{h}(m') = \text{out}(u, \bar{h}(m))) \wedge s(m') = \text{delete}(\text{ref}(u, m),$
$s(m)) \wedge (u' \neq u \Rightarrow \text{ref}(u', m) = \text{ref}(u', m'))]$

(Operation kill changes nothing if it has a used reference as its argument. If it does not, it frees the frame indicated by the reference and makes the reference a used one.).

AVM7. $(\text{usable } (u, \bar{h}(m)) \Rightarrow (\exists f)(\text{member } (f, \bar{s}(m)) \wedge \text{ref}(u, m) = f))\cdot$

AVM8. $(\text{member } (f, \bar{s}(m)) \Rightarrow (\exists u) (\text{usable } (u, \bar{h}(m) \wedge \text{ref}(u, m) = f)).$

(For every memory state operation ref is onto the set of occupied frames).

AVM9. $(\bar{h}(\text{freem}) = \text{empty } H \wedge \bar{s}(\text{freem}) = \text{allfree}).$

AVM10. $(\sim \text{undef}(u, \bar{h}(m)) \Rightarrow (\text{reserve}(u, m) = \textbf{while true do od } m')\cdot$

(Operation reserve is undefined for usable and/or used references.)

The theory of virtual memory is consistent since it has a model. Two of the model's components are standard models for the ATR theory of storage management systems and the ATUR theory of univocal references. The reader can conceive the axioms AVM1, AVM2, AVM6 as definitions of the operations: inmemory, reserve and kill. The system defined in this way will be called a standard model of AVM theory.

Using methods illustrated in earlier sections one can prove that any model of AVM theory is isomorphic to a standard one.

11. CONCATENABLE TYPE DECLARATIONS

The designers of SIMULA-67 have invented prefixing, a new and powerful programming tool which allows one to concatenate type declarations. Concatenation of type declarations plays (or should play) a similar role to that of procedure call, because of its power in defining data structures, program-oriented languages, hierarchies of sets and systems, etc. Prefixing is not widely accepted as a programming tool, because of lack of knowledge concerning its properties, the difficulties involved in proper implementation of concatenation of type declarations, and also because of irrational prejudice. No one doubts the importance of the copy rule for computations with procedures. Similarly, the concatenation rule deserves the attention of researchers and users.

REMARK. It is rare to see a paper describing SIMULA or prefixing in which the author reports all the important properties of concatenation of type declarations. Most authors limit themselves to remarks on encapsulated data types, which may be the least important of the properties of the concatenation rule.   □

This section is intended as an introduction to the concatenation rule. It is informal in character and far from completeness.

We should like to call the reader's attention to the potential applications of prefixing. Here we shall abstract from the dynamics introduced by the storage management system. We shall concentrate on the properties of objects of types which correspond to static semantics. observe that they are full of dynamics.

The central notion is an object. By an *object* we shall understand an ordered pair

$$\langle \text{valuation of variables, sequence of instructions} \rangle.$$

The special object is **none** $= \langle \emptyset, \emptyset \rangle$. Objects may be values of variables. (The careful reader will notice in the light of preceding sections that an object is allocated to a frame and the value of a variable is a univocal reference to the frame, but we shall abstract from these details.) A restriction is posed on valuations, since every variable is declared together with a type name (called its *qualification*) and since the value of the variable has to be an object of appropriate type. There is one object **none** which may be the value of any variable.

The definition of a type has the following structure:

> **unit** T: **class** $(m_1 a_1 : T_1, \ldots, m_n a_n : T_n)$;
> $\qquad at_1, \ldots, at_k$;
> **begin**
> $\qquad I_1$ {prologue instructions} **return**;
> $\qquad I_2$ {instructions 1};
> $\qquad$ **inner**;
> $\qquad I_3$ {instructions 2}
> **end** T,

where $a_1, \ldots, a_n$ are names of formal parameters, $T_1, \ldots, T_n$ are names of types, m_i is information about the mode of transmission of parameters $(i = 1, \ldots, n)$, $at_1, \ldots, at_k$ are declared local attributes and inner and return are special instructions.

Declaration of type T introduces a data structure, or more exactly it extends the existing data structure by a new sort $|T|$ of objects of type T and corresponding operations. The operations are those inherited from the virtual memory system

$$\langle \textbf{new } T, \text{ kill}(x), \ x \textbf{ is } T, \ x \textbf{ in } T \rangle,$$

and the operations declared in the declaration of type T. We associate two operations, read y and update y, with every variable y in the list $at_1, \ldots, at_k$. Let o be an object of type T. The phrases $o.y$ and $o.y := \tau$ are then expressions denoting the operations mentioned above. In a similar way, simple formal parameters can also be conceived of as pairs of operations. Other formal parameters and local attributes also determine operations. This is easy to see if they are functions, but in other cases also (like declaration of types) we can conceive of them as operations.

The operation **new** T creates an object

$$o: \left\langle v: \left.\frac{x_1}{0_1}\right| \ldots \left|\frac{x_l}{0_l}\right.; J_1, \ldots, J_p \right\rangle$$

in the following way:

1° Variables $x_1, \ldots, x_n$ of the object o are formal parameters $a_1, \ldots, a_n$ (we simplify considerations by assuming that all formal parameters are variables).

2° Variables $x_{n+1}, \ldots, x_l$ are local attributes $at_1, \ldots, at_k$ of T.

3° The object o arises from the initial proto-object o',

$$o': \left\langle v: \left.\frac{x_1}{}\right| \ldots \left|\frac{x_l}{}\right.; I \right\rangle$$

where the values of variables $x_1, \ldots, x_l$ are initialized in accordance with the corresponding mode of transmission as values of actual parameters and values of local attributes are initialized in accordance with the general scheme of initialization. The initial values of types are

> Boolean—**false**, integer, read—zero, character—space, all other types—**none**.

Once we have created an proto-object o' it becames a subject, i.e. the prologue instructions are executed until the return instruction is met. In this way the attributes of an object can be initialized in a more specific way.

Objects created by the **new** T operation satisfy the relation **is** between objects and name of type T, i.e. the relation

$$\textbf{new } T \, (act_1, \ldots, act_n) \textbf{ is } T$$

holds.

Another type declaration may be written with T as a prefix

> **unit** T': T **class** $(m_{n+1}a_{nn}: T_{n+1}, ..., m_r a_r: T_r)$;
>
> $at_{k+1}, ..., at_s$;
>
> **begin**
>
> J_1; **return**;
>
> J_2; **inner**;
>
> J_3
>
> **end** T'.

This definition should be viewed as an abbreviation of the full concatenated declaration:

> **unit** T': **class** $(m_1 a_1: T_1, ..., m_n a_n: T_n, \overset{\backslash}{m}_{n+1} a_{n+1}: T_{n+1}, ...$
>
> $..., m_r a_r: T_r)$;
>
> $at_1, ..., at_k, at_{k+1}, ..., at_s$.
>
> **begin**
>
> I_1;
>
> I_2;
>
> J_1; **return**;
>
> J_2; **inner**;
>
> J_3;
>
> I_3;
>
> **end** T'.

12. AN IMPLEMENTATION OF RATIONAL NUMBERS

In this section we shall present an example showing that algebraic operations of structures like product, factorization, etc. have counterparts in programming, and that they can be imitated by means of prefixing.

Product

> **unit** pair: **class** $(L, M$: integer$)$ **begin end** pair.

This declaration introduces the structure

$$\langle |pair|, newpair, L, .M, .L:=, .M:=\rangle,$$

where $|pair|$ denotes the set of objects of the type pair:

newpair: |integer| × |integer| → |pair|,

.L: |pair| → |integer|,

.M: |pair| → |integer|,

.L := : |pair| × |integer| → |pair|,

.M := : |pair| × |integer| → |pair|.

The properties of the structure of pairs are as follows:

.L(**new**pair(a, b)) = a,

.M(**new**pair(a, b)) = b,

.L := (**new**pair(a, b), c) = **new**pair(c, b),

.M := (**new**pair(a, b)c) = **new**pair(a, c),

Subset

The next step in the construction is to define a subset of proper pairs

> **unit** properpair : pair **class**
> **begin if** $M = 0$ **then** ERROR **fi**
> **end** properpair.

The set |properpair| is a subset of |pair| set. ERROR denotes a never-terminating program, e.g. **while true do od**.

Quotient structure

> **unit** rational: properpair **class**
>   **variable** gcd, aux1, aux2 : integer;
> **begin** aux1 := abs(L); aux2 := abs(M);
>   **while** aux1 $\neq$ aux2 **do**
>   **if** aux1 $>$ aux2 **then** aux1 := aux1 $-$ aux2
>   **else** aux2 := aux2 $-$ aux1 **fi**
>   **od**;
>   gcd := aux1;
> L := $L \div$ gcd; M := $M \div$ gcd;
>   **end** rational.

The set |rational| corresponds to irreducible fractions.

Extension

> **unit** RATIONALS: **class**
>   **unit** rational: properpair **class** ... **end** rational;

```
    unit add: function (x, y: rational): rational;
    begin result := new rational (x.L * y.M+y.L * x.M,
    x.M * y.M)
    end add
    unit multiply: function (x, y: rational): rational;
      begin  result := rational(x.L * y.L, x.M * x.M)
      end multiply;
  begin
  end RATIONALS'.
```

In this way we have defined an algebra

$$\langle |rational| \;.add.L \; multiply, ...\rangle$$

which does not satisfy the axioms of the field of rational numbers since the operation.$L :=$ can destroy them.

It is not difficult to forget about the operations $.L, .M, .L :=, .M :=$

```
    unit RATIONALS: class
      hidden rational;
        unit fraction: rational class hidden .L, .M; end fraction;
        unit rational: ...
        unit add: ...
        unit multiply: ...
      begin
      end RATIONALS.
```

The effect of line: **hidden** rational; is that the type rational is invisible outside the **unit** RATIONALS, hence $.L, .M, .L:=, .M; =$ operations are forbidden. One can create objects of type fraction by object expressions like, e.g. **new** fraction (7, 19), the attributes of type fraction are inaccessible to a user. The structure

$$\langle fraction, add, multiply, new \; fraction\rangle$$

corresponds to the field of rationals and the axioms of the field are valid.

BIBLIOGRAPHIC REMARKS

MAX model of concurrent computations was introduced in Salwicki and Müldner (1981b). Axiomatization of the notion of reference was given by Oktaba (1981). Certain results in the semantics of prefixing can be found in Bartol *et al.* (1983) and in Bartol (1981).

The results presented in this chapter form a part of bigger project aimed toward formal specification of LOGLAN programming language, cf. Bartol *et al.* (1983b). The present authors believe that the goal will be reached by the creation of family of algorithmic theories. Each theory is to describe an aspect of language's semantics. Moreover, it is expected that they complement one another and together bring the complete information about behaviour of LOGLAN programs. In our opinion especially the operation of concatenation of program modules (prefixing) deserves more attention. Certain new results in this field are due to Langmaack, cf. Krause *at al.* (1984).

BOOLEAN ALGEBRAS

1. A *Boolean algebra* is an algebra $\langle A, \cup, \cap, - \rangle$ which satisfies the identities:

(l_1) $a \cup b = b \cup a$, $a \cap b = b \cap a$,

(l_2) $a \cup (b \cup c) = (a \cup b) \cup c$, $a \cap (b \cap c) = (a \cap b) \cap c$,

(l_3) $(a \cap b) \cup b = b$, $a \cap (a \cup b) = a$,

(l_4) $a \cap (b \cup c) = (a \cap b) \cup (a \cap c)$, $a \cup (b \cap c) = (a \cup b) \cap (a \cup c)$,

(l_5) $(a \cap -a) \cup b = b$, $(a \cup -a) \cap b = b$,

for every $a, b, c \in A$.

EXAMPLES.

A. The two-element Boolean algebra $B_0 = \langle \{0, 1\}, \cup, \cap, - \rangle$.

B. Field of subsets of a set X $\mathscr{P}(X) = \langle 2^X, \cup, \cap, - \rangle$.

C. The Lindenbaum algebra of a theory T (cf. Chapter III, § 1). ☐

2. Define the relation $a \leqslant b$ putting for every $a, b \in A$

$$a \leqslant b \quad \text{iff} \quad a \cup b = b.$$

Define:

$$0 \overset{\text{df}}{=} a \cap -a,$$
$$1 \overset{\text{df}}{=} a \cup -a,$$
$$a \Rightarrow b \overset{\text{df}}{=} -a \cup b.$$

3. The relation $\leqslant$ is an ordering in A, i.e. for arbitrary $a, b, c \in A$

$$a \leqslant a,$$

if $a \leqslant b$ and $b \leqslant c$, then $a \leqslant c$,

if $a \leqslant b$ and $b \leqslant a$, then $a = b$.

4. $a \cup b$ is the least upper bound of the set $\{a, b\}$.
In fact

$$a \leqslant a \cup b \quad \text{and} \quad b \leqslant a \cup b \quad \text{by}(l_3).$$

If $a \leqslant c$ and $b \leqslant c$ then by definition $a \cup c = c$ and $b \cup c = c$. Making use of (l_1) and (l_2) we have $(a \cup b) \cup c = (a \cup b) \cup (c \cup c) = (a \cup c) \cup (b \cup c) = c \cup c = c$, hence $a \cup b \leqslant c$.

Similarly, $a \cap b$ is the greatest lower bound of the set $\{a, b\}$.

5. A nonempty set V of elements of a Boolean algebra A is said to be a *filter* in A provided that for every element $a, b \in A$ the following two conditions are satisfied:

> If $a, b \in V$, then $a \cap b \in V$.
> If $a \in V$ and $a \leqslant b$, then $b \in V$.

6. Let A_0 be a non-empty subset of A. The set of all elements $a \in A$, such that $a \geqslant a_1 \cap \dots \cap a_n$ for some elements $a_1, \dots, a_n \in A_0$, is a filter. Moreover, this set is the least filter containing A_0.

7. A filter is said to be *proper* if it is a proper subset of Boolean algebra A. It is easy to observe that a filter is proper if and only if it does not contain the zero element $\mathbf{0}$.

8. A subset A_0 of A is said to have the *finite intersection property* if for every elements $a_1, \dots, a_n \in A$

$$a_1 \cap \dots \cap a_n \neq \mathbf{0}.$$

9. Every subset A_0 which possesses the finite intersection property is contained in a proper filter.

10. A filter V is said to be *prime* provided it is a proper filter and the condition $a \cup b \in V$ implies that either $a \in V$ or $b \in V$.

11. A prime filter is *maximal*, i.e. it is not any proper subset of a proper filter.

12. By a chain of filters we mean a non-empty family of filters linearly ordered by the relation of inclusion.

13. The union of any chain of proper filters is a proper filter.

14. Every proper filter can be extended to a prime filter. Consider the family of all proper filters. By 13 every chain of filters has an upper bound. By the Kuratowski–Zorn Lemma (cf. Rasiowa and Sikorski, 1968) there exists a maximal element in the family which is a prime filter.

15. Let A_t be an infinite subset of A:

$$A_t = \{a_{t,s}\}_{s\in S}.$$

If the least upper bound of the set A_t exists then we shall denote it by

$$\text{l.u.b.}_{s\in S} (a_{t,s}).$$

Similarly, we shall use the notation

$$\text{g.l.b.}_{s\in S} (a_{t,s})$$

for the greatest lower bound if it exists.

16. Let $T\cup U$ be a set of indices. By Q we shall denote the set of infinite operations described below:

$$(Q)\qquad \begin{aligned} a_t &= \text{l.u.b.}_{s\in S} (a_{t,s}), \quad t\in T, \\ b_u &= \text{g.l.b.}_{r\in R} (b_{u,r}), \quad u\in U. \end{aligned}$$

17. A filter V is said to be a Q-filter provided it is a prime filter such that for every $t\in T$, $u\in U$:

If $a_t \in V$, then there exists $s_0 \in S$ such that $a_{t,s_0} \in V$.

If $b_u \notin V$, then there exists $r_0 \in R$ such that $b_{u,r_0} \notin V$.

18. THE RASIOWA–SIKORSKI LEMMA (Rasiowa and Sikorski, 1968). *If the set Q is denumerable then every non-zero element $a_0 \in A$ is contained in a Q-filter.*

PROOF. We shall construct a subset C of Boolean algebra such that it possesses finite intersection property and contains a_0. By 9 and 14 the set C will be contained in a Q-filter.

The construction of C will assure us that for every $t\in T$ if $a_t \in V$, then there exists $s_0 \in S$ such that $a_{t,s_0} \in V$. Without loss of generality we can assume the set U is empty. The set T is denumerable so without loss of generality we can assume $T = 1, 2, \ldots$ Consider the sequence

$$a_0, a_1, a_2, \ldots$$

The set C is defined by induction. on $t \in \{0\}\cup T$.

$a_0 \in C$. (We recall that $a_0 \neq 0$).

Let c be an g.l.b. of all elements already included in C, $c \neq 0$. Consider $a_t = \text{l.u.b.}_{s\in S} (a_{t,s})$, $t\in T$. We shall prove that there exists $s_0 \in S$

such that

$$c\cap(a_t \Rightarrow a_{t.s_0}) \neq 0.$$

Suppose, on the contrary, $c \cap (-a_t \cup a_{t,s}) = 0$ for all $s \in S$, then $c \cap -a_t = 0$ and $c \cap a_{t,s} = 0$ for all $s \in S$, i.e. $c \leqslant a_t$ and l.u.b. $\underset{s \in S}{(c \cap a_{t,s})} = 0$, and consequently $c \cap a_t = 0$, $c \leqslant -a_t$.

Hence

$$c \leqslant a_t \cap -a_t$$

i.e. $c = 0$, which contradicts our assumption. In this way we have proved that for every $t \in T$ there exists $s_t \in S$ such that

$$a_0 \cap (a_1 \Rightarrow a_{1,s_1}) \cap \ldots \cap (a_t \Rightarrow a_{t,s_t}) \neq 0.$$

Hence the set $C = \{a_0, (a_1 \Rightarrow a_{1,s_1}), (a_2 \Rightarrow a_{2,s_2}), \ldots\}$ has the finite intersection property. By 9 and 14 it can be extended to a prime filter V. By the definition of C, V is a Q-filter.

19. The Representation Theorem. *Every Boolean algebra is isomorphic to a field of subsets of a set.*

THE PROOF OF LEMMA 2.2 FROM CHAPTER III

Let f be a function which to every formula and every program of an algorithmic language L assigns an ordinal number in the following way:

$f(\alpha) = 1$ for every propositional variable or elementary formula α,

$f(s) = 1$ for every assignment instruction s.

If α, β are arbitrary formulas and K, M—arbitrary programs, then

$$f(\sim\alpha) = f(\alpha)+1,$$
$$f(\alpha\vee\beta) = f(\alpha\wedge\beta) = f(\alpha \Rightarrow \beta) = \max\left(f(\alpha), f(\beta)\right)+1,$$
$$f(M\alpha) = f(\alpha)+f(K),$$
$$f(\text{if } \gamma \text{ then } K \text{ else } M \text{ fi}) = \max\left(f(\gamma), f(K), f(M)\right)\cdot 3+1,$$
$$f(\text{begin } K; M \text{ end}) = \max\left(f(M), f(K)\right)\cdot 2+1,$$
$$f(\text{while } \gamma \text{ do } K \text{ od}) = \omega\cdot\max\left(f(\gamma), f(K)\right)+1,$$
$$f(\bigcup K\alpha) = f(\bigcap K\alpha) = \omega\cdot\max\left(f(\alpha), f(K)\right)+1,$$
$$f\left((\forall x)\alpha(x)\right) = f\left((\exists x)\alpha(x)\right) = f(\alpha)+2.$$

Observe that for every classical formula α of the language L, $f(\alpha) < \omega$.

Let us put $\bar{f}(\alpha) = f(\alpha)$ for every classical formula of the language L (i.e., for every formula in which programs do not appear) and $\bar{f}(\alpha) = \omega+f(\alpha)$ for any other formula.

LEMMA. *For every algorithmic formula α, β*

(∗)   if $\alpha \prec \beta$ then $\bar{f}(\alpha) < \bar{f}(\beta)$.

PROOF. We shall prove that property (∗) holds for any pair of formulas from the set Z defined in Chapter III, § 2.

1. Consider a simple formula of the form $s\varrho(\tau_1, ..., \tau_n)$, where ϱ is an n-argument predicate, $\tau_1, ..., \tau_n$ are terms and s is an assignment instruction:

$$\bar{f}(s\varrho(\tau_1, ..., \tau_n)) = \omega+f(\varrho(\tau_1, ..., \tau_n))+f(s) = \omega+2.$$

But $f(\varrho(\eta_1, ..., \eta_n)) = 1$ for arbitrary terms $\eta_1, ..., \eta_n$ and therefore

$$\bar{f}\big(\varrho(\bar{s\tau}_1, ..., \bar{s\tau}_n)\big) < \bar{f}\big(s\varrho(\tau_1, ..., \tau_n)\big).$$

Let α, β, γ denote algorithmic formulas and s, K, M denote programs.

2. By the definition of the function f we have

$$\bar{f}\big(s(\alpha \vee \beta)\big) = \omega + f(\alpha \vee \beta) + f(s) = \omega + \max\big(f(\alpha), f(\beta)\big) + 2$$

and

$$\bar{f}(s\alpha) = \omega + f(\alpha) + 1, \quad \bar{f}(s\beta) = \omega + f(\beta) + 1.$$

By the properties of the ordering relation $\leqslant$ in the set of ordinal numbers $f(F)$ we have

$$\bar{f}(s\alpha) < \bar{f}\big(s(\alpha \vee \beta)\big), \quad \bar{f}(s\beta) < \bar{f}\big(s(\alpha \vee \beta)\big).$$

3. Consider the pair $(s\alpha, s{\sim}\alpha)$. It is obvious that (∗) holds, since

$$\bar{f}(s{\sim}\alpha) = \omega + f({\sim}\alpha) + f(s) = \omega + f(\alpha) + 2$$
$$> \omega + f(\alpha) + 1 = \bar{f}(s\alpha).$$

4. Let us denote $\max\big(f(K), f(M)\big)$ by a. By the definition of the function f we have;

$$f(s\ \mathbf{begin}\ K;\ M\ \mathbf{end}\ \alpha)$$
$$= f(\alpha) + f(\mathbf{begin}\ K;\ M\ \mathbf{end}) + 1 = f(\alpha) + a \cdot 2 + 2$$
$$f\big(s(K(M\alpha))\big) = f(K(M\alpha)) + 1 = f(M\alpha) + f(K) + 1$$
$$= f(\alpha) + f(M) + f(K) + 1.$$

As a consequence of $f(M) + f(K) + 1 < a \cdot 2 + 2$ we have

$$\bar{f}\big(s(K(M\alpha))\big) < \bar{f}\big(s(\mathbf{begin}\ K;\ M\ \mathbf{end}\ \alpha)\big).$$

5. By definition

$$f(s\ \mathbf{if}\ \gamma\ \mathbf{then}\ K\ \mathbf{else}\ M\ \mathbf{fi}\ \beta)$$
$$= f(\beta) + \max\big(f(\gamma), f(K), f(M)\big) \cdot 3 + 2,$$
$$f\big(s(\gamma \wedge K\beta)\big) = \max\big(f(\gamma), f(\beta) + f(K)\big) + 2,$$
$$f\big(s({\sim}\gamma \wedge M\beta)\big) = \max\big(f(\gamma) + 1, f(\beta) + f(M)\big) + 2.$$

If at least one of the maxima that appear above is equal to its first argument, then $f(\gamma), f(K), f(M)$ are finite and obviously (∗) holds. If

$$\max\big(f(\gamma), f(K), f(M)\big) = f(M),$$
$$\max\big(f(\gamma) + 1, f(\beta) + f(M)\big) = f(\beta) + f(M),$$
$$\max\big(f(\gamma), f(\beta) + f(K)\big) = f(\beta) + f(K),$$

then

$$f(M) \cdot 3 + 2 > f(M) + 2 \quad \text{and} \quad f(K) + 2 < f(K) \cdot 3 + 2.$$

It follows that

$$\bar{f}(s(\gamma \wedge K\beta)) < \bar{f}(s(\text{if } \gamma \text{ then } K \text{ else } M \text{ fi } \beta))$$

and

$$\bar{f}(s(\sim\gamma \wedge M\beta)) < \bar{f}(s(\text{if } \gamma \text{ then } K \text{ else } M \text{ fi } \beta)).$$

An analogous proof of the remaining cases is omitted.

6. Let us denote $\max(f(\gamma), f(K))$ by a. By the definition of the function f we have:

$$f(s(\text{while } \gamma \text{ do } K \text{ od } \beta)) = f(\beta) + \omega \cdot a + 2$$

and, for every natural number i,

$$f(s(\text{if } \gamma \text{ then } K \text{ fi})^i(\beta \wedge \sim\gamma)) = (a \cdot 3 + 1) \cdot i + 1 < \omega \cdot a + 2.$$

Consider two cases:

(a) $a < \omega$. In this case it is obvious that $(a \cdot 3 + 1) \cdot i + 1 < \omega \cdot a + 2$.

(b) $a \geq \omega$. By the definition of the function f, a is less than ω^ω. So, there exists $n < \omega$ and $b_i < \omega$ for $j = 1, 2, ..., n$ such that $a = \omega^n \cdot b_n + \omega^{n-1} \cdot b_{n-1} + ... + b_0$.

Since i is a finite ordinal number, we have

$$(a \cdot 3 + 1) \cdot i + 1$$
$$= \omega^n \cdot (b_n \cdot i) + \omega^{n-1} \cdot (b_{n-1} \cdot i) + ... + (b_0 \cdot i + 1).$$

As a consequence

$$(a \cdot 3 + 1) \cdot i + 1 \leq \omega^{n+1} + 1 = \omega \cdot \omega^n + 1.$$

However, $\omega^n \leq a$ and thus $\omega \cdot \omega^n + 1 \leq \omega \cdot a + 1$ and finally

$$(a \cdot 3 + 1) \cdot i + 1 < \omega \cdot a + 2.$$

It follows from (a) and (b) that for every natural number i,

$$\bar{f}(s(\text{if } \gamma \text{ then } K \text{ fi})^i(\sim\gamma \wedge \beta)) < \bar{f}(s(\text{while } \gamma \text{ do } K \text{ od } \beta)).$$

7. Consider the pair $(s \sim \alpha(\tau), s(\exists x)\alpha(x))$:

$$f(s((\exists x)\alpha(x))) = f(s\alpha) + 2 = f(\alpha) + 3,$$
$$f(s \sim \alpha(\tau)) = f(\sim\alpha(\tau)) + 1 = f(\alpha) + 2.$$

Thus, $f(\alpha) + 2 < f(\alpha) + 3$. $\square$

LEMMA. *For every set of formulas Z there exists a formula   which is a minimal element of that set with respect to the relation $\prec$.*

PROOF. Let us consider the set of ordinal numbers $\bar{f}(Z) = \{\bar{f}(\alpha): \alpha \in Z\}$, and let a_0 be the first element of $\bar{f}(Z)$.

Every formula $\alpha \in Z$ such that $\bar{f}(\alpha) = a_0$ is a minimal element of Z. In fact, if $\bar{f}(\alpha) = a_0$ and for some $\beta \in Z$, $\beta \prec \alpha$, then by the property (∗)

$$\bar{f}(\beta) < \bar{f}(\alpha).$$

Hence $\bar{f}(\beta) < a_0$ and therefore a_0 is not the first element of $f(Z)$, contrary to the assumption.   □

An analogous reasoning can be repeated for algorithmic formulas with non-deterministic programs (cf. Chapter VI).

Let us extend the relation $\prec$ as follows:

$$\lozenge K\beta \prec \lozenge \textbf{ either } K \textbf{ or } M \textbf{ ro } \beta, \quad \square K\beta \prec \square \textbf{either } K \textbf{ or } M \textbf{ ro } \beta,$$

$$\lozenge M\beta \prec \lozenge \textbf{ either } K \textbf{ or } M \textbf{ ro } \beta, \quad \square M\beta \prec \square \textbf{either } K \textbf{ or } M \textbf{ ro } \beta,$$

$$(\gamma \wedge \lozenge K\alpha) \prec \lozenge \textbf{ if } \gamma \textbf{ then } K \textbf{ else } M \textbf{ fi } \alpha,$$

$$(\sim\gamma \wedge \lozenge M\alpha) \prec \lozenge \textbf{ if } \gamma \textbf{ then } K \textbf{ else } M \textbf{ fi } \alpha,$$

$$(\gamma \wedge \square K\alpha) \prec \square \textbf{if } \gamma \textbf{ then } K \textbf{ else } M \textbf{ fi } \alpha,$$

$$(\sim\gamma \wedge \square K\alpha) \prec \square \textbf{if } \gamma \textbf{ then } K \textbf{ else } M \textbf{ fi } \alpha,$$

$$\lozenge \ (\textbf{if } \gamma \textbf{ then } M \textbf{ fi})^i(\alpha \wedge \sim \gamma) \prec \lozenge \textbf{ while } \gamma \textbf{ do } M \textbf{ od } \alpha,$$
$$\text{for all } i \in N,$$

$$\square (\textbf{if } \gamma \textbf{ then } M \textbf{ fi})^i(\alpha \wedge \sim \gamma) \prec \square \textbf{while } \gamma \textbf{ do } M \textbf{ od } \alpha,$$
$$\text{for all } i \in N,$$

$$\lozenge \textbf{ begin } K; \ M \textbf{ end } \alpha \prec \lozenge K(\lozenge M\alpha),$$

$$\square \textbf{ begin } K; \ M \textbf{ end } \alpha \prec \square K(\square M\alpha),$$

$$\lozenge K^i\alpha \prec \bigvee K\alpha, \quad \square K^i\alpha \prec \bigsqcup K\alpha,$$

$$\lozenge K^i\alpha \prec \bigwedge K\alpha, \quad \square K^i\alpha \prec \sqcap K\alpha$$

for an arbitrary natural number i.

Let us put

$$f(\textbf{either } K \textbf{ or } M \textbf{ ro}) = \max\big(f(K), f(M)\big) \cdot 2 + 1,$$
$$f(\lozenge M\alpha) = f(\square M\alpha) = f(\alpha) + f(M).$$

It can be proved now that, for arbitrary formulas α, β of non-deterministic algorithmic language,

(∗∗)   if $\alpha \prec \beta$ then $\bar{f}(\alpha) \prec \bar{f}(\beta)$.

As a consequence of this fact we have the following result.

LEMMA. *For every set Z of non-deterministic algorithmic formulas there exists an element α such that for every $\beta \in Z$, $\sim\beta \prec \alpha$. i.e. α is a minimal element of Z.*   □

BIBLIOGRAPHY

ABBREVIATIONS

ACM   Association for Computing Machinery
Bull PAS Bulletin de l'Académie Polonaise des Sciences
CACM   Communications of ACM
FOCS   IEEE Symp. on Foundations of Computer Science
IPL   Information Processing Letters
JACM   Journal of ACM
JCSS   Journal of Computer and System Science
LNCS   Lecture Notes in Computer Science
POPL   Symp. on Principles of Programming Languages
STOC   Symp. on Theory of Computing
TCS   Theoretical Computer Science
TOPLAS ACM Transactions on Programming Languages and Systems

Aho A., Hopcroft J., Ullman J. (1974), *The Design and Analysis of Computer Algo-rithms*, Addison-Wesley, Reading, Massachusetts.
Andreka H., Németi I. (1981), A Characterization of Floyd Provable Programs, Proc. Mathematical Logic in Computer Science, Salgótarián 1978, in: *Colloquia Mathematica Societatis János Bolyai* **26**, North-Holland.
Andreka H., Németi I., Sain I. (1979), Completeness Problem in Verification of Pro-grams and Program Schemes, in: *Proc. MFCS'79* (J. Becvar ed.), LNCS 74, Springer Verlag, Berlin, 208–218.
Andreka H., Németi I., Sain I. (1979b), Henkin-Type Semantics for Program Schemes to Turn Negative Results to Positive, in: *Proc. FCT'79* (L. Budach ed.), Akademie Verlag, Berlin, Band 2,18–24.
Andreka H. (1983), Sharpening the Characterization of the Power of Floyd Method, in: *Proc. Logics of Programs and Their Applications, Poznań 1980* (A. Salwicki ed.), LNCS 148, Springer Verlag, Berlin, 1–26.
Apt K.R. (1979), Ten Years of Hoare's Logic: A Survey—Part 1, *TOPLAS* 3, 431–483.
Apt K. R., Olderog E.-R. (1982), *Proof Rules Dealing with Fairness in Logics of Programs* (D. Kozen ed.), LNCS 131, Springer Verlag, Berlin, 1–8.
de Bakker J. W. (1976), Semantics and Termination of Non-deterministic Recursive Programs, in: *Automata Languages and Programming, Edinburgh*, 435–477.

de Bakker J. W. (1977), A Sound and Complete Proof System for Partial Program Correctness, in: *Proc. MFCS'79 Olomouc* (J. Becvar ed.), LNCS 74, 1–12.

de Bakker J. W. (1979), A Sound and Complete Proof System for Partial Program Correctness, in: *Proc. MFCS'79* (J. Becvar ed.), LNCS 74, Springer Verlag, Berlin, 1–12.

de Bakker J. W. (1980), *Mathematical Theory of Program Correctness*, Prentice Hall, Englewood Cliffs.

Banachowski L. (1975), Modelar Approach to the Logical Theory of Programs, in: *Proc. MFCS'74*, LNCS 28, Springer Verlag, Berlin.

Banachowski L. (1975b), An Axiomatic Approach to the Theory of Data Structures, *Bull. PAS* 23, 315–323.

Banachowski L. (1977), Investigations of Properties of Programs by Means of the Extended Algorithmic Logic, *Fundamenta Informaticae* 1, 93–119, 167–193.

Banachowski L. (1983), On Proving Program Correctness by Means of Stepwise Refinement Method, in: *Proc. Logics of Programs and Their Applications, Poznań 1980* (A. Salwicki ed.), LNCS 148, Springer Verlag, Berlin, 27–45.

Banachowski L., Kreczmar A., Mirkowska G., Rasiowa H., Salwicki A. (1977), An introduction to Algorithmic Logic, Mathematical Investigations in the Theory of Programs, in: *Math. Foundations of Computer Science* (A. Mazurkiewicz and Z. Pawlak eds.), Banach Center Publications, PWN, Warsaw, 7–99.

Bartol W. M. (1981), *Application of Static Structure of Type Declarations and the System of Dynamic Configurations in a Definition of Semantics of a Universal Programming Language* (in Polish), Doct. Diss., Dept. Math. Inform., University of Warsaw.

Bartol W. M., Kreczmar A., Litwiniuk A. I., Oktaba H. (1983), Semantics and Implementation of Prefixing at Many Levels, in: *Proc. Logics of Programs and Their Applications, Poznań 1980* (A. Salwicki ed.), LNCS 148, Springer Verlag, Berlin, 45–80.

Bartol W. M. *et al.* (1983b), *Raport of LOGLAN Programming Language*, PWN, Warsaw.

Barzdin J. M. (1979), The Problem of Reachability and Verification of Programs, in: *Proc. MFCS'79* (J. Becvar ed.), LNCS 74, Springer Verlag, Berlin, 13–26.

Bergstra J., Tiuryn J., Tucker J. (1982), Floyd's Principle Correctness Theories and Program Equivalence, *TCS* 17, 113–149.

Bergstra J., Tiuryn J. (1981), Implicit Definability of Algebraic Structures by Means of Program Properties, *Fundamenta Informaticae* 4, 661–674.

Bergstra, J., Tiuryn J. (1981b), Algorithmic Degrees of Algebraic Structures, *Fundamenta Informaticae* 4, 851–863.

Bergstra J., Tucker J. V. (1982), The Refinement of Specifications and the Stability of Hoare's Logic, in: *Logics of programs 1981* (D. Kozen ed.), LNCS 131, Springer Verlag, Berlin, 24–36.

Bergstra J., Tucker J. V., (1984) Hoare's Logic for Programming Languages with Two Data Types, *TCS* 28, 215–222.

Berman F. (1979), A Completeness Technique for D-Axiomatizable Semantics, in: *Proc. 11th ACM STOC*, 160–166.

Berman P., Halpern J., Tiuryn J. (1982), On the Power of Non-determinism in Dynamic Logic, *Proc. ICALP'82* (M. Nielsen, E. M. Schmidt eds.), LNCS 140, Springer Verlag, Berlin, 48–61.

Birkhoff G., Lipson J. (1970), Heterogeneous Algebras, *Journal of Combinatorial Theory* 8, 115–133.

Blikle A., Mazurkiewicz A. (1972), An Algebraic Approach to the Theory of Programs, Algorithms and Recursiveness, in: *Proc. MFCS'72*, Reports of the Computer Center of the Polish Academy of Sciences, Warsaw.

Blikle A. (1977), An Analysis of Programs by Algebraic Means, in: *Math. Foundations of Computer Science* (Z. Pawlak, A. Mazurkiewicz eds.), Banach Center Publications, vol. 2, PWN, Warsaw, 167–213.

Burkhard H. D. (1981), Ordered Firing in Petri Nets, *Elektron. Informationsverarbeitung und Kybernetik* 17, 71–86.

Burkhard H. D. (1981b), Two pumping lemmata for Petri nets, *Elektron. Informationsverarbeitung und Kybernetik* 17, 349–362.

Burkhard H. D. (1983), On Priorities of Parallelism: Petri Nets under the Maximum Firing Strategy, in: *Proc. Logics of Programs and Their Applications, Poznań 1980* (A. Salwicki ed.), LNCS 148, Springer Verlag, Berlin, 86–98.

Burkhard H. D. (1984), *An Investigation of Controls for Concurrent Systems by Abstract Control Languages*, LNCS 176, Springer Verlag, 223–231

Burstall R. M. (1969), Proving Properties of Programs by Structural Induction, *Computing* 12, 41–48.

Cartwright R., McCarthy J. (1979), First Order Programming Logic, in: *Proc 6th ACM POPL, San Antonio*, 68–80.

Cartwright R. (1982), Toward a Logical Theory of Program Data, in: *Proc. Logics of Programs 1981* (D. Kozen ed.), LNCS 131, Springer Verlag, Berlin, 37–51.

Chandra A., Halpern J., Meyer A., Parikh R. (1981), Equations Between Regular Terms and Application to Process Logic, in: *Proc. ACM STOC 1981*, 384–390.

Chlebus B. (1982), Completeness Proofs for Some Logics of Programs, *Zeitschrift für Math. Logic* 28, 49–62.

Chlebus B. (1982b), On Decidability of Propositional Algorithmic Logic, *Zeitschrift für Math. Logic* 28, 247–261.

Chlebus B. (1983), On Four Logics of Programs and the Complexity of Their Satisfiability Problems: Extended Abstract, in: *Proc. Logics of Programs and Their Applications, Poznań 1980* (A. Salwicki ed.), LNCS 148, Springer Verlag, Berlin, 98–109.

Church A. (1936), An Unsolvable Problem of Elementary Number Theory, *Amer. J. Math.* 58, 345–363.

Clarke E. M. (1979), Programming Language Constructs for Which It Is Impossible to Obtain Good Hoare-Like Axioms, *JACM* 26, 129–147.

Constable R. L. (1977), A *Constructive Programming Logic IFIP'77*, North Holland, Amsterdam, 733–738.

Constable R. L. (1977b), On the Theory of Programming Logics, in: *Proc. 9th ACM STOC*, 269–285.

Constable R. L., O'Donnell M. J. (1978), *A Programming Logic*, Wintkrop, Cambridge, Massachussets.

Constable R. L., Zlatin D. R. (1982), The Type Theory of PL/CV3, in: *Proc. Logics of Programs 1981* (D. Kozen ed.), LNCS 131, Springer Verlag, Berlin, 72–93.

Cook S. A. (1978), Soundness and Completeness of an Axiom System for Program Verification, *SIAM J. Comput.* 7, 70–90.

Cousineau G., Enjalbert P. (1979), Program Equivalence and Provability, in: *Proc. MFCS'79* (J. Becvar ed.), LNCS 74, Springer Verlag, Berlin, 237–245.

Dahl O.-J., Hoare C. A. R. (1972), Hierarchical Program Structures, in: O.-J. Dahl, E. W. Dijkstra, C. A. R. Hoare, *Structured Programming*, Academic Press, 197–220.

Dańko W. (1974), Not Programmable Function Defined by a Procedure, *Bull. PAS* 22, 587–594.

Dańko W. (1978), Algorithmic Properties of Programs with Tables, *Fundamenta Informaticae* 1, 379–398.

Dańko W. (1979), Definability in Algorithmic Logic, *Fundamenta Informaticae* 2, 271–287.

Dańko W. (1980), A Criterion of Undecidability of Algorithmic Theories, in: *Proc. MFCS'80* (P. Dembiński ed.), LNCS 88, Springer Verlag, Berlin, 205–216.

Dańko W. (1983), Interpretability of Algorithmic Theories, *Fundamenta Informaticae* 6, 217–233.

Dańko W. (1983b), Algorithmic Properties of Finitely Generated Structures, in: *Proc. Logics of Programs and Their Applications, Poznań 1980* (A. Salwicki ed.), LNCS 148, Springer Verlag, Berlin, 118–131.

Dijkstra E. W. (1975), On Guarded Commands, Non-determinacy and Formal Derivation of Programs, *CACM* 18, 453–457.

Dijkstra E. W. (1976), *Discipline of Programming*, Prentice Hall, Englewood Cliffs.

van Emde Boas P., Janssen T. M. (1977), The Expressive Power of Intentional Logic in the Semantics of Programming Languages, in: *Proc MFCS'77* (J. Gruska ed.), LNCS 53, Springer Verlag, Berlin, 303–312.

van Emde Boas P., Janssen T. (1978), *Intensional Logic and Programming*, Amsterdam, preprint No. ZW 98/78.

Engeler E. (1967), Algorithmic Properties of Structures, *Math. Systems Theory* 1, 183–195.

Engeler E. (1968), Remarks on the Theory of Geometrical Constructions, in: *Syntax and Semantics of Infinitary Languages*, Lecture Notes on Mathematics 72, Springer Verlag, Berlin, 64–76.

Engeler E. (1971), Structure and Meaning of Elementary Programs, in: *Proc. Symp. Semantics of Algorithmic Languages*, Lecture Notes in Mathematics 188, Springer Verlag, Berlin, 89–101.

Engeler E. (1973), On the Solvability of Algorithmic Problems, in: *Logic Colloquium'73*, (H. E. Rose and J. C. Shepherdson eds.), Studies in Logic 80, North-Holland, 231–251.

Engeler E. (1975), Algorithmic Logic, in: *Mathematical Centre Tracts* (J. de Bakker ed.), Amsterdam, 57–85.

Enjalbert P. (1981), *Contribution à l'étude de la logique algorithmique: systèmes de deduction pour les arbres et les schemas de programmes*, doct. diss., Université Paris VII.

Enjalbert P. (1983), Algebraic Semantics and Program Logics: Algorithmic Logic for Program Trees, in: *Proc. Logics of Programs and Their Applications, Poznań 1980* (A. Salwicki ed.), LNCS 148, Springer Verlag, Berlin, 132–147.

Enjalbert P., Michel M. (1984), *Many-Sorted Temporal Logic for Multiprocesses Systems*, LNCS 176, Springer Verlag, 273–281.

Fischer M. J., Ladner R. E. (1979), Propositional Dynamic Logic of Regular Programs, *JCSS* **18**, 194–211.

Floyd R. W. (1967), Assigning Meanings to Programs, in: *Proc. Symp. Appl. Math. AMS* 19, *Mathematical Aspects of Computer Science* (J. T. Schärtz ed.), 19–32.

Fraenkel A., Bar-Hillel Y. (1958), *Foundations of Set Theory*, North–Holland, Amsterdam.

Glushkov V. M. (1965), Automata theory and formal transformation of microprograms (in Russian), *Kibernetika* **1**, 1–10.

Glushkov V. M., Tseytlin G. E., Yoshchenko E. L. (1978), *Algebra Languages, Programming* (in Russian), 2nd edition, Naukova Dumka, Kiev.

Goguen J. A., Thatcher J. W., Wagner E. G. (1977), An Initial Algebra Approach to the Specification, Correctness and Implementation of Abstract Data Types, *IBM Res. RC* 6487.

Goldblatt R. (1982), *Axiomatising the Logic of Computer Programming*, LNCS 130, Springer Verlag, Berlin.

Grabowski M. (1972), The Set of All Tautologies of Zero-Order Algorithmic Logic is Decidable, *Bull. PAS* **20**, 575–582.

Grabowski M., Kreczmar A. (1978), Dynamic Theories of Real and Complex Numbers, in: *Proc. MFCS'78* (J. Winkowski ed.), LNCS 64, Springer Verlag, Berlin, 239–249.

Grabowski M. (1981), Full Weak Second-Order Logic versus Algorithmic Logic, Proc. Mathematical Logic in Computer Science, Salgótarján 1978, in: *Colloquia Mathematica Societatis János Bolyai* **26**, North-Holland, Amsterdam, 471–483.

Grabowski M. (1983), Some Model Theoretical Properties of Logic for Programs with Random Control, in: *Proc. Logics of Programs and Their Applications, Poznań 1980* (A. Salwicki ed.), LNCS 148, Springer Verlag, Berlin, 148–155.

Greibach S. (1975), *Theory of Program Structures, Schemes, Semantics, Verification*, LNCS 36, Springer Verlag, Berlin.

Greif I., Meyer A. (1980), Specifying Programming Language Semantics, in: *7th Proc. ACM POPL 1980*, 180–189.

Guttag J. (1977), Abstract Data Types and the Development of Data Structures, *CACM* **20**, 396–404.

Góraj A., Mirkowska G., Paluszkiewicz A. (1970), On the Notion of Description of Program, *Bull. PAS* **18**, 499–506.

Habasinski Z.: (1984), Process Logic: Two Decidability Results, in: *Proc. MFCS'84* (M. Chityl ed.), LNCS 176, Springer Verlag, 282–290.

Harel D. (1978), *First Order Dynamic Logic*, LNCS 68, Springer Verlag, Berlin.

Harel D. (1978b), Arithmetical Completeness in Logics of Programs, in: *Automata, Languages and Programming, Udine 1978* (G. Ausiello and C. Böhm eds.), LNCS 62, Springer Verlag, Berlin, 286–289.

Harel D. (1979), Recursion in Logics of Programs, in: *Proc 6th ACM POPL, San Antonio*, 81–92.

Harel D.: (1980), On Folk Theorems, *CACM* **23**.

Harel D. (1982), *Dynamic Logic*, manuscript

Harel D., Kozen D., Parikh R. (1980b), Process Logic: Expresiveness, Decidability, Completeness, in: *Proc. FOCS 1980*, 129–142.

Harel D., Meyer, A. R., Pratt V. R. (1977), Computability and Completeness in Logics of Programs, in: *Proc. 9th ACM STOC*, 261–268.

Harel D., Pnueli A., Stavi J. (1977b), A Complete Axiomatic System for Proving Deductions About Recursive Programs, in: *Proc. 9th ACM STOC*, 249–260.

Harel D., Pratt V. (1978), Non-determinism in Logics of Programs, in: *Proc. 5th ACM POPL, Tucson Ariz.*, 203–213.

Hajek P. (1981), Making Dynamic Logic First-Order, in: *Proc. MFCS'81* (J. Gruska, M. Chytil eds.), LNCS 118, Springer Verlag, Berlin, 287–295.

Hennessy M. C. B., Plotkin G. D. (1980), A Term Model for CCS, in: *Proc. MFCS'80* (P. Dembiński ed.), LNCS 88, Springer Verlag, Berlin, 261–274.

Hermes H. (1965), *Enumerability, Decidability, Computability*, Academic Press, New York.

Hoare C. A. (1969), An Axiomatic Basis for Computer Programming, *CACM* **12**, 576–583.

Hoare C. A. R. (1972), Proof of Correctness of Data Representation, *Acta Informatica* **1**, 271–281.

Hoare C. A. R., Wirth N. (1973), An Axiomatic Definition of the Programming Language PASCAL, *Acta Informatica* **2**, 335–355.

Hoare C. A. R. (1978), Communicating Sequential Processes, *CACM* **21**, 666–677.

Igerashi S. (1968), An Axiomatic Approach to the Equivalence Problems of Algorithms with Applications, *Rep. Comp. Centre of University of Tokyo* **1**.

Karp R. A. (1984), Proving Failure–Free Properties of Concurrence Systems Using Temporal Logic, *TOPLAS* **6**, 239–253.

Kawai H. (1983), A Formal System for Parallel Programs in Discrete Time and Space, in: *Proc. Logics of Programs and Their Applications, Poznań 1980* (A. Salwicki ed.), LNCS 148, Springer Verlag, 155–165.

Kfoury D. (1972), Comparing Algebraic Structures up to Algorithmic Equivalence, in: *Proc. ICALP*, North Holland, Amsterdam, 253–264.

Kfoury A. J., Park D. M. (1975), On the Termination of Program Schemas, *Information and Control* **29**, 243–251.

Kluzniak F., Szpakowicz S. (1985), *Prolog for Programmers*, Academic Press, Orlando.

Knuth D. E. (1968), *The Art of Computer Programming*, vols 1–3, Addison-Wesley, 1968, 1969, 1973.

Knuth D. E. (1974), Structured Programming with 'go to' Statements, *Computing Surveys* **6**, 261–301.

Kotov V. E. (1978), An Algebra for Parallelism Based on Petri Nets, in: *Proc. MFCS'78* (J. Winkowski ed.), LNCS 64, Springer Verlag, 39–56.

Kozen D. (1980), A Representation Theorem for Models of *-Free PDL, in: *Proc. 7th ICALP* (J. de Bakker, J. van Leeuwen eds.), LNCS 85, Springer Verlag, Berlin, 351–362.

Kozen D. (1981), On the Duality of Dynamic Algebras and Kripke Models in Logics of Programs (E. Engeler ed.), LNCS 125, Springer Verlag, Berlin, 1–11.

Kozen D., Parikh R. (1981b), An Elementary Completeness Proof for PDL, *TCS* 14, 113–118.

Kozen D. (1982), On Induction Versus—Continuity in Logics of Programs. in: *Proc. Logics of Programs 1981* (D. Kozen ed.), LNCS 131, Springer Verlag, Berlin, 167–176.

Krause M., Kreczmar A., Langmaack H., Salwicki A. (1984), *Specification and Implementation Problems of Programming Languages Proper for Hierarchical Data Types*, Raport no. 8410, Institut für Informatik Christian Albrecht Universität Kiel.

Kreczmar A. (1972), Degree of Recursive Unsolvability of Algorithmic Logic, *Bull. PAS* 20, 615–617.

Kreczmar A. (1974), Effectivity Problems of Algorithmic Logic, in: *ICALP'74* (J. Loeckx ed.), LNCS 14, Springer Verlag, Berlin, 584–600.

Kreczmar A. (1977), Effectivity Problems of Algorithmic Logic, *Fundamenta Informaticae* 1, 19–32.

Kreczmar A. (1977b), Programmability in fields, *Fundamenta Informaticae* 1, 195–230.

Kreczmar A., Müldner T. (1983), Coroutines and Processes in Block Structured Languages, in: *Proc. 6 GI Dortmund, Jan. 1983*, LNCS 145, Springer Verlag, Berlin, 231–243.

Kröger F. (1976), Logical Rules for Natural Reasoning about Programs, in: *ICALP'76* (S. Michaelson, R. Milner eds.), Edinburgh, 87–98.

Kröger F. (1977), A Logic of Algorithmic Reasoning, *Acta Informatica* 8, 243–266.

Kröger F. (1978), A Uniform Logical Basis for the Description, Specification and Verification of Programs, in: *Formal Description of Programming Concepts* (E. J. Neuhold ed.), North Holland, Amsterdam, 441–459.,

Kuratowski K., Mostowski A. (1967), *Set Theory*, North Holland, Amsterdam, PWN, Warsaw.

Lamport L. (1980), "Sometimes" is sometimes "not never", in: *Proc. 7th ACM POPL, Las Vegas*, 174–185.

Lamport L. (1984), Using Time instead of Timeout for Foult Tolerant Distributed Systems, *TOPLAS* 6, 254–280.

Lamport L., Scheider F. (1984), The "Hoare Logic" of CSP and All That, *TOPLAS* 6, 281–295.

Langmaack H. (1979); *On Termination Problems for Finitely Interpreted ALGOL-like Programs*, Rep. 7904, Institut für Informatik und Praktische Mathematik, Christian Albrechts Universität Kiel, Sept. 1979.

Langmaack H. (1982), On Termination Problems for Finitely Interpreted ALGOL--like Programs, *Acta Informatica* 18, 79–108.

Lipton R. J. (1977), A Necessary and Sufficient Condition for the Existence of Hoare Logics, in: *Proc. 18th FOCS'77*.

Liskov B. H., Zilles S. N. (1975), Specification Techniques for Data Abstractions. *IEEE Trans. Software Engrg.*

Liskov B. H., Zilles S. N. (1979), Programming with Abstract Data Types, in: *Proc. ACM SIGPLAN Symp. on Very High Level Languages, SIGPLAN Notices* 4, 50–59.

Luckham D. C., Park D. M., Paterson M. S. (1970), On formalized computer programs, *JCSS* 4, 220–249.

Machtey M., Young P. (1978), *An Introduction to the General Theory of Algorithms*, North Holland, New York.

Malcev A. I. (1965), *Algorithms and Recursive Functions* (in Russian), Nauka, Moscow,

Malcev A. I. (1970), *Algebraic Systems* (in Russian), Nauka, Moscow.

Manna Z. (1969), The Correctness of Programs, *JCSS* 3.

Manna Z. (1974), *Mathematical Theory of Computation*, McGraw-Hill, New York.

Manna Z., Pnuelli A. (1979), The Modal Logic of Programs, in: *Automata, Language and Programming*, LNCS 71, Springer Verlag, Berlin, 385–405.

Markov A. (1954), *Theory of Algorithms* (in Russian), Proc. Steklov Math. Inst., Moscow.

Mazur S. (1963), Computable Analysis, *Dissertationes Math.* 33.

Mazurkiewicz A. (1975), Parallel Recursive Program Schemes, in: *Proc. MFCS'75* (J. Becvar ed.) LNCS 32, Springer Verlag, Berlin, 75–87.

McCarthy J. (1963), A Basis for Mathematical Theory of Computation, in: *Computer Programming and Formal Systems*, North–Holland, Amsterdam.

Meyer A. R., Winklmann K. (1979), On the Expressive Power of Dynamic Logic, in: *Proc. 11th ACM STOC, Atlanta.*

Meyer A., Halpern J. (1980), Axiomatic Definitions of Programming Languages: a Theoretical Assessment, in: *Proc. 7th ACM POPL, Las Vegas 1980*, 203–212.

Meyer A. R., Parikh R. (1980b), Definability in Dynamic Logic, in: *Proc. 12th ACM STOC, Los Angeles 1980*, 1–7.

Meyer A., Streett R. S., Mirkowska G. (1981), The Deducibility Problem in Propositional Dynamic Logic, in: *Logics of Programs, Zürich 1979* (E. Engeler ed.), LNCS 125, Springer Verlag, Berlin, 12–23.

Meyer A. R., Tiuryn J. (1982), A Note on Equivalences Among Logics of Programs, in: *Proc. Logics of Programs, Yorktown Heights 1981* (D. Kozen ed.), LNCS 131, Springer Verlag, 282–299.

Milner R. (1980), *A Calculus of Communication Systems*, LNCS 92, Springer Verlag, Berlin.

Mirkowska G. (1971), On Formalized Systems of Algorithmic Logic, *Bull. PAS* 19, 421–428.

Mirkowska G., Salwicki A. (1976), A Complete Axiomatic Characterization of Algorithmic Properties of Block-Structured Programs with Procedures, in: *Proc. MFCS'76* (A. Mazurkiewicz ed.), LNCS 45, Springer Verlag, 602–606.

Mirkowska G. (1977), Algorithmic Logic and Its Applications in the Theory of Programs, *Fundamenta Informaticae* 1, 1–17, 147–165.

Mirkowska G. (1980), Algorithmic Logic with Non-deterministic Programs, *Fundamenta Informaticae* 3, 45–64.

Mirkowska G. (1980b), Model Existence Theorem for Algorithmic Logic with Non-deterministic Programs, *Fundamenta Informaticae* 3, 157–170.

Mirkowska G. (1980c), Complete Axiomatization of Algorithmic Properties of Program Schemes with Bounded Non-deterministic Interpretations, in: *Proc. 12th STOC, Los Angeles 1980*, 14–21.

Mirkowska G. (1981), PAL-Propositional Algorithmic Logic, in: *Logics of Programs, Zürich 1979* (E. Engeler ed.), LNCS 125, Springer Verlag, Berlin, 12–22, *Fundamenta Informaticae* 4, 675–757.

Mirkowska G. (1982), The Representation Theorem for Algorithmic Algebras, in: *Proc. Logics of Programs, Yorktown Heights 1981* (D. Kozen ed.), LNCS 131, Springer Verlag, Berlin, 300–310.

Mirkowska G. (1983), On the Algorithmic Theory of Arithmetic, in: *Proc. Logics of Programs and Their Applications, Poznań 1980* (A. Salwicki ed.), LNCS 148, Springer Verlag, Berlin, 166–185.

Mostowski A. (1948), *Mathematical Logic* (in Polish), Mathematical Monographs Series, no 18, Warszawa—Wrocław.

Müldner T., Salwicki A. (1978), Computational Processes Generated by Programs with Recursive Procedures and Block Structures, *Fundamenta Informaticae* 1, 305–323.

Müldner T. (1981), On the Synchronizing Tools for Parallel Programs, *Fundamenta Informaticae* 4, 95–134.

Müldner T. (1981b), On Semantics of Parallel Programs, *Fundamenta Informaticae* 4, 35–82.

Naur P. (1966), Proof of Algorithms by General Snapshots, *BIT* 6, 310–316.

Nemeti I. (1982), Non-standard Dynamic Logic, in: *Proc. Logics of Programs, Yorktown Heights 1981* (D. Kozen ed.), LNCS 131, Springer Verlag, Berlin, 311–348.

Nemeti I. (1983), Non-standard Runs of Floyd-Provable Programs, in: *Proc. Logics of Programs and Their Applications, LOGLAN 77, Poznań 1980* (A. Salwicki ed.), LNCS 148, Springer Verlag, Berlin, 186–204.

Nishimura H. (1979), Sequential Method in Propositional Dynamic Logic, *Acta Informatica* 12, 377–400.

Nishimura H. (1980), Descriptively Complete Process Logic, *Acta Informatica* 14, 359–369.

O'Donnell M. J. (1982), A Critique of the Foundations of Hoare-Style Programming Logics, in: *Proc. Logics of Programs 1981* (D. Kozen ed.), LNCS 131, Springer Verlag, Berlin, 349–374.

Oktaba H. (1981), *Formalization of the Notion of Reference and Its Applications in Theory of Data Structures* (in Polish), Doct. Diss., Univ. of Warsaw.

Orłowska E. (1983), Program Logic with Quantifiable Propositional Variables, in: *Proc. Logics of Programs and Their Applications, Poznań 1980* (A. Salwicki ed.), LNCS 148, Springer Verlag, Berlin, 205–212.

Owicki S., Gries D. (1976), Verifying Properties of Parallel Programs: An Axiomatic Approach, *CACM* 19, No 5, 279–285.

Parikh R. (1978), A Completeness Result for PDL, in: *Proc. MFCS'78* (J. Winkowski ed.), LNCS 64, Springer Verlag, Berlin, 403–416.

Parikh R. (1980), Propositional Logics of Programs: System Models and Complexity, *7th ACM POPL, Las Vegas*, 186–192.

Parikh R. (1981), Propositional Dynamic Logics of Programs: A Survey, in: *Logics of Programs, Zürich 1979* (E. Engeler ed.), LNCS 125, Springer Verlag, Berlin, 102–144.

Park D. (1969), Fixed Point Induction and Proofs of Program Properties, *Machine Intelligence Workshop* **5**, 59–78.

S. Passy, T. Tinchev (1985), PDL with Data Constants, *IPL* **20**, 35–42.

Perkowska E. (1972), On Algorithmic *m*-Valued Logics, *Bull. PAS* **20**, 717–719.

Petermann U. (1983), On Algorithmic Logic with Partial Operations, in: *Proc. Logics of Programs and Their Applications, Poznań 1980* (A. Salwicki ed.), LNCS 148, Springer Verlag, Berlin, 213–223.

Pnueli A. (1977), The Temporal Logic of Programs, in: *Proc. 18th FOCS'71*, 46–57.

Pnueli A. (1979), Temporal Semantics of Concurrent Programs, in: *Semantics of Concurrent Computation* (G. Kahn ed.), LNCS 70, Springer Verlag, Berlin, 1–20.

Poythress V. S. (1973), Partial Morphisms on Partial Algebras, *Algebra Universalis* **3**, 182–202.

Pratt V. R. (1976), Semantical Considerations on Floyd–Hoare Logic, in: *Proc. 17th FOCS'76*, 109–121.

Pratt V. R. (1978), A Practical Decision Method for Propositional Dynamic Logic, in: *Proc. 10th ACM STOC*, 326–337.

Pratt V. R. (1979), *Dynamic Algebras: Examples, Constructions, Applications*, Raport MIT/LCS/TM.

Pratt V. R. (1979b), Process Logic, in: *Proc. 6th ACM POPL, San Antonio*, 93–100.

Radev S. (1983), Infinitary Propositional Modal Logic and Program Language, in: *Proc. Logics of Programs and Their Applications, Poznań 1980* (A. Salwicki ed.), LNCS 148, Springer Verlag, 253–258.

Rasiowa H. (1972), On Logical Structure of Programs, *Bull. PAS* **20**, 319–324.

Rasiowa H. (1975), ω^+-Valued Algorithmic Logic as a Tool to Investigate Procedures, *Proc. MFCS'74* (A. Blikle ed.), LNCS 28, Springer Verlag, Berlin.

Rasiowa H. (1975b), Completeness Theorem for Extended Algorithmic Logic, in: *Proc. 5th Intern. Congress of Logic, Methodology and Philosophy of Science*, III, D. Reidel, Dordrecht, 13–15.

Rasiowa H. (1975c), Many Valued Algorithmic Logic, in: *Proc. ASL Symp. Kiel 1974*, Lecture Notes in Mathematics 499, Springer Verlag, Berlin, 543–565.

Rasiowa H. (1977), Algorithmic Logic—Notes From Seminar in Simon Fraser University 1975, *Reports of the Computer Center of the Polish Academy of Sciences*, no 281, Warsaw.

Rasiowa H. (1979), Algorithmic Logic, Multiple-Valued Extensions, *Studia Logica* **38**, 317–335.

Rasiowa H. (1979b), Logic of Complex Algorithms, in: *Proc. FCT'79* (L. Budach ed.), Akademie Verlag, Berlin, 371–380.

Rasiowa H., Sikorski R. (1968), *Mathematics of Metamathematics*, PWN, Warsaw.

Reif J. H., Peterson G. L. (1980), A Dynamic Logic of Multiprocessing with Incomplete Information, in: *7th ACM POPL, Las Vegas*, 193–202.

Reif H. J. (1980b), Logics for Probabilistic Programming, in: *Proc. 12th STOC*, Los Angeles, 8–13.

Reiterman J., Trnkova V. (1980), Dynamic Algebras which are not Kripke Structures, in: *Proc. MFCS'80* (P. Dembiński ed.), LNCS 88, Springer Verlag, Berlin, 528–538.

Rice H. G. (1954), Recursive Real Numbers, *Proc. Amer. Math. Soc.* **5**, 784–791.

Rogers H., Jr. (1967), *Theory of Recursive Functions and Effective Computability*, McGraw-Hill, New York.

Salwicki A. (1970), Formalized Algorithmic Languages. *Bull. PAS* **18**, 227–232.

Salwicki A. (1975), Procedures, Formal Computations and Models, in: *Proc. MFCS'74* (A. Blikle ed.), LNCS 28, Springer Verlag, Berlin, 464–484.

Salwicki A. (1977), Applied Algorithmic Logic, in: *Proc. MFCS'77*, (J. Gruska ed.), LNCS 53, Springer Verlag, Berlin, 122–134.

Salwicki A. (1977b); An Algorithmic Approach to Set Theory, in: *Proc. FCT'77* (M. Karpiński ed.), LNCS 56, Springer Verlag, Berlin, 499–510.

Salwicki A. (1977c), Algorithmic Logic, a Tool for Investigation of Programs, in: *Logic, Foundations of Mathematics and Computability Theory, Part One of the Proceedings of the Fifth International Congress of Logic, Methodology and Philosophy of Science, London, Ontario, 1975* (R. E. Butts, J. Hintikka eds.), D. Reidel Publ., Dordrecht, 281–295.

Salwicki A. (1980), On Algorithmic Theory of Stacks, *Fundamenta Informaticae* 3, 311–332.

Salwicki A. (1981), On the Algorithmic Theory of Dictionaries, in: *Logics of Programs, Zürich 1979* (E. Engeler ed.), LNCS 125, Springer Verlag, Berlin, 145–168.

Salwicki A., Mülldner T. (1981b), On the Algorithmic Properties of Concurrent Programs, in: *Logics of Programs, Zürich 1979* (E. Engeler ed.), LNCS 125, Springer Verlag, Berlin, 169–197.

Salwicki A. (1982), Algorithmic Theories of Data Structures, in: *Proc. ICALP'82 Aarhus* (M. Nielsen, E. Schmidt eds.), LNCS 140, Springer Verlag, Berlin, 458–472.

Salwicki A. (1982b), Critical Remarks on MAX Model of Concurrency, in: *Proc. Logics of Programs, Yorktown Heights 1981* (D. Kozen ed.), LNCS 131, Springer Verlag, Berlin, 397–405.

Scott D. (1970), *Outline of a Mathematical Theory of Computation*, Oxford Monographs PRG-2, Oxford University Press.

Scott O. (1976), Data Types as Lattices, *SIAM J. Comput.* **5**, 522–587.

Scott D. (1982), Domains For Denotational Semantics, in: *Proc. ICALP'82, Aarhus* (M. Nielsen, E. Schmidt eds.), LNCS 140, Springer Verlag, Berlin, 577–613.

Scott D., Strachey C. (1971), Towards a Mathematical Semantics for Computer Languages, *Technical Monograph PRG 6*, Oxford University.

Segerberg K. (1982), A Completeness Theorem in the Modal Logic of Programs, in: *Universal Algebra and Applications* (T. Traczyk ed.), PWN, Warszawa, 31–46.

Shepherdson J. C., Sturgis H. E. (1963), Computability of Recursive Functions, *JACM* **10**, 217–255.

Shoenfield J. R. (1967), *Mathematical Logic*, Addison-Wesley, Reading, Massachusetts.

Skowron A. (1983), Concurrent Programs, in: *Proc. Logics of Programs and Their Applications, Poznań 1980* (A. Salwicki ed.), LNCS 148, Springer Verlag, 258–270.

Skowron A., Radev S., Vakarelov D. (1980), Propositional Computational Logic, *Reports of the Institute of Computer Science of the Polish Academy of Sciences*, no. 411, Warsaw, 64–66.

Spitzen J., Wegbreit B. (1975), The Verification and Synthesis of Data Structures, *Acta Informatica* **4**.

Szczerba L. W. (1977), Interpretability of Elementary Theories, in: *Logic, Foundations of Mathematics and Computability Theory, Part One of the Proceedings of the Fifth International Congress of Logic, Methodology and Philosophy of Science, London, Ontario 1975* (R. E. Butts, J. Hintikka eds.), D. Reidel, Dordrecht.

Thiele H. (1966), *Wissenschaftstheoretische Unitersuchungen in algorithmischen Sprachen*, VEB Deutscher Verlag der Wissenschaften, Berlin.

Tiuryn J. (1981), Unbounded Program Memory Adds to Expressive Power of First-Order Dynamic Logic, in: *Proc. 22nd FOCS'81, Nashville*, 335–339.

Tiuryn J, (1981b), Logic of Effective Definitions, *Fundamenta Informaticae* **4**, 629–660.

Tiuryn J. (1981c), A Survey of the Logic of Effective Definitions, in: *Logics of Programs 1979* (E. Engeler ed.), LNCS 125, Springer Verlag Berlin, 198–245.

Trakhtenbrot B. A. (1979), On Relaxation Rules in Algorithmic Logic, in: *Proc. MFCS'79* (J. Becvar ed.), LNCS 74, Springer Verlag, Berlin, 453–462.

Urzyczyn P. (1981), Algorithmically Triviality of Abstract Structures, *Fundamenta Informaticae* **4**, 819–849.

Urzyczyn P. (1982), On the Unwinding of Flow-Charts with Stacks, *Fundamenta Informaticae* **4**, 119–126.

Vakarelov D. (1982), *Reduction of Dynamic Logic to Modal Logic*, manuscript.

Vakarelov D. (1983), Filtration Theorem for Dynamic Algebras with Tests and Inverse Operator, in: *Proc. Logics of Programs and Their Applications, Poznań 1980* (A. Salwicki ed.), LNCS 148, Springer Verlag, Berlin, 314–324.

Valiev M. K. (1979), On Axiomatization of Deterministic Propositional Dynamic Logic, in: *Proc. MFCS'79* (J. Becvar ed.), LNCS 74, Springer Verlag, Berlin, 482–491.

Valiev M. K. (1980), Decision Complexity of Variants of Propositional Dynamic Logic, in: *Proc. MFCS'80* (P. Dembiński ed.), LNCS 88, Springer Verlag, Berlin, 656–664.

Valiev M. K. (1983), On Axiomatization of Process Logic, in: *Proc. Logics of Programs and Their Applications, Poznań 1980* (A. Salwicki ed.), LNCS 148, Springer Verlag, Berlin, 304–313.

Vaught R. L. (1973), Some Aspects of the Theory of Models, *Amer. Math. Monthly* **80**, 3–37.

Wand M. (1978), A New Incompleteness Result for Hoare's Systems, *JACM* **25**, 168–175.

Wegbreit B. (1976), Verifying Program Performance, *JACM* **23**, 691–700.

Winklmann K. (1977), *Equivalence of DL and DL+ for Regular Programs without Array Assignments but with DL-Formulas in Tests*, Manuscript, Lab. for Comp. Sci. MIT, Dec. 1977.

Winkowski J. (1977), A Natural Method of Proving Properties of Programs, *Fundamenta Informaticae* 1, 33–49.

Winkowski J. (1979), An Algebraic Approach to Concurrence, in: *Proc. MFCS'79* (J. Becvar ed.), LNCS 74, Springer Verlag, Berlin 523–532.

Wirsing M., Broy M. (1980), *Abstract Data Types as Lattices of Finitely Generated Models*, LNCS 88, Springer Verlag, 673–685.

Wirth N. (1971), Program Development by Stepwise Refinement, *CACM* 14, 221–227.

Yanov Y. I. (1959), The Logical Schemes in Algorithms, *Problems of Cybernetics* 1, Pergamon Press, New York, 82–140.

Yeh R. (1977), *Current Trends in Programming Methodology*, v. 1, 2, Prentice Hall, Englewood Cliffs.

INDEX